Metamaterial-Based Electromagnetic Radiations and Applications

Zhaoyun Duan

Metamaterial-Based Electromagnetic Radiations and Applications

Science Press
Beijing

Springer

Zhaoyun Duan
School of Electronic Science
and Engineering
University of Electronic Science
and Technology of China
Chengdu, Sichuan, China

ISBN 978-981-97-8107-2 ISBN 978-981-97-8108-9 (eBook)
https://doi.org/10.1007/978-981-97-8108-9

Jointly published with Science Press
The print edition is not for sale in China (Mainland). Customers from China (Mainland) please order the print book from: Science Press.

This Springer imprint is published by the registered company Springer Nature Singapore Pte Ltd.
The registered company address is: 152 Beach Road, #21-01/04 Gateway East, Singapore 189721, Singapore

Foreword by Edl Schamiloglu

Aristotle (384 BC–322 BC) ascribed a difference between natural and artificial things. He attributed the difference to motion and change. Natural things have a source of motion or change within them. Artificial things don't have any source of change in them, so they need an external cause. In this monograph authored by Prof. Zhaoyun Duan, the changes in artificial materials caused by high-power electromagnetic waves are explored to identify new physical phenomena and new classes of vacuum electron devices (VEDs).

In 1967, Veselago[1] theoretically investigated the hypothetical exotic electromagnetic properties of an assumed homogeneous isotropic electromagnetic material in which the real parts of the complex permittivity (ε) and permeability (μ) are both negative. Examples of exotic electromagnetic properties include negative refractive index, the reversed Doppler effect, and reversed Cherenkov radiation (RCR). Veselago referred to this type of material as a left-handed medium (LHM) which has subwavelength features that are engineered to have unique properties not typically found in nature. It took more than 30 years for such materials to receive renewed attention. Pendry et al.[2,3] theoretically analyzed the negative of a regular array of conducting wires and the negative of a split-ring resonator (SRR) array; they also anticipated the possibility of realizing an LHM in practice. Smith et al.[4] followed the work of Pendry et al. and constructed and tested the first LHM in the microwave part of the electromagnetic spectrum.

---

[1] V.G. Veselago, "The electrodynamics of substances with simultaneously negative values of ε and μ," Sov. Phys. Uspekhi, vol. 47, pp. 509–514, 1968, 10.1070/PU1968v010n04ABEH003699.

[2] J.B. Pendry, A.J. Holden, D.J. Robbins, and W.J. Stewart, "Low frequency plasmons in thin-wire structures," J. Phys., Condens. Matter, vol. 10, pp. 4785–4809, 1998, 10.1088/0953-8984/28/48/481002.

[3] J.B. Pendry, A.J. Holden, D.J. Robbins, and W.J. Stewart, "Magnetism from conductors and enhanced nonlinear phenomena," IEEE Trans. Microw. Theory Techn., vol. 47, pp. 2075–2084, 1999, 10.1109/22.798002.

[4] D.R. Smith, W.J. Padilla, D.C. Vier, S.C. Nemat-Nasser, and S. Schultz, "Composite medium with simultaneously negative permeability and permittivity," Phys. Rev. Lett., vol. 84, pp. 4184–4187, 2000, 10.1103/PhysRevLett.84.4184.

Professor Duan and his group have been actively exploring the use of metamaterials in VEDs since 2008. He and his group were the first to directly observe RCR, where the Cherenkov radiation emitted predominantly in the opposite direction to the movement of a single sheet electron beam bunch in the experiment in a square waveguide loaded with a layer of complementary electric SRRs (CeSRRs). In addition, he and his group have been exploring new classes of VEDs exploiting the RCR effect.

In this monograph, Prof. Duan describes the historical development of metamaterials as they are used in the microwave and millimeter-wave range of the electromagnetic spectrum. He then reviews how the theoretical concepts of Veselago were put into practice by Pendry, Smith, and others. He then focuses on theoretical research into RCR, followed by its experimental observations. He then describes a new class of VEDs based on the RCR effect. Finally, he concludes by discussing theoretical, simulation, and experimental studies of coherent and enhanced transition radiation inspired by metamaterials.

This monograph by Prof. Duan, a leader in the field, is a tremendous resource to researchers, both young and established alike.

October 2022

Edl Schamiloglu
IEEE Fellow
American Physical Society Fellow
Distinguished Professor of Electrical and Computer Engineering
Associate Dean for Research and Innovation
School of Engineering
University of New Mexico
Albuquerque, New Mexico, USA

Foreword by Min Chen

To ordain conscience for heaven and earth, to secure life and fortune for the people, to perpetuate the lost teachings of past sages, and to establish peace for future generations, these aspirations were the ideally envisioned by the Chinese thinker Zai Zhang during the Northern Song Dynasty in China. For many centuries, this ideal has remained the goal that scholars have aspired to in their hearts.

Before the 2022 Thanksgiving Day, I received a monograph manuscript entitled *Metamaterial-Based Electromagnetic Radiations and Applications* written by Zhaoyun. I carefully read the manuscript with joy in my heart. After reading, I was reminded of a comprehensive scientific work from the Ming Dynasty in China, named T'ien-Kung K'ai-Wu: Chinese Technology in the Seventeenth Century. This book extensively documented various technologies of ancient China before the mid-Ming period, describing more than 130 production techniques and tools, along with detailed explanations of the operation methods of production techniques, the shapes of production tools, and the manufacturing process. What particularly impressed me was the prevalent bathing silkworm technique in Jiangsu and Zhejiang areas.

The bathing silkworm technique was used by silk farmers living in Jiangsu and Zhejiang areas in ancient times to breed silkworms, resulting in silkworms that were both highly productive in silk and less in consuming mulberry leaves. Silk farmers used natural materials throughout the process. In an era when scientific and technological advancements were yet to flourish on a large scale, natural materials were abundant but often lacked the properties required for technological innovation. To overcome these limitations, successive generations of scientists and engineers designed artificial materials and conducted various experiments to develop new properties which are suitable for technological advancement, making natural materials more widely applicable.

In this regard, this monograph shares a similar spirit with the book entitled T'ien-Kung K'ai-Wu: Chinese Technology in the Seventeenth Century. It provides a comprehensive account of development, classification, unique properties, realization, and characterization of metamaterials. The theoretical and experimental research on reversed Cherenkov radiation (RCR), RCR oscillators (RCROs), and amplifiers, as well as coherent enhanced transition radiation and the related devices, were entirely developed by Zhaoyun and his research group.

Zhaoyun, like the silk farmers of ancient times, used natural materials to develop various metamaterials by designing subwavelength artificial unit cells and structuring their spatial configurations. This led to the demonstration of numerous novel and extraordinary physical properties related to electricity, magnetism, and more, creating many novel applications. This is the essence and purpose of this monograph.

After reading this monograph, I am proud of Zhaoyun's achievements and cannot help but recall his years of research and outstanding contributions in the field of metamaterials and electromagnetic radiation. I first met Zhaoyun fifteen years ago. It was the Thanksgiving Day in 2007 when I met Zhaoyun in the house of my colleague, Prof. J. A. Kong, who had made outstanding contributions in the field of electromagnetic wave theory, microwave remote sensing, electromagnetic wave scattering, and so on.

At that time, Prof. Kong was studying RCR in left-handed materials (LHMs), and my research was focusing on particle physics. Hence, Prof. Kong and I had regular exchanges and collaborations. In the early 1970s, I designed a double-arm vertical spectrometer, with six large Cherenkov detectors that played a crucial role in detecting the weak signals of positive and negative electrons from the background of hadrons with billions of times intensities. The working principle of this detector was just based on Cherenkov radiation. It was this instrument that directly led to the discovery of the J particle in 1974. This work received the Nobel Prize in Physics in 1976.

In 2002, Prof. Kong and I predicted that high-speed charged particles moving in LHMs could generate RCR, where the momentum direction was opposite to the energy flow. It is a highly unusual characteristic. In 2006, my Ph.D. student, Jie Lu designed an LHM suitable for generating RCR. In 2007, our research was ongoing. After a brief conversation with Zhaoyun, I knew that he, funded by the China Scholarship Council, just came to MIT in August to do postdoctoral research. His research project was RCR based on LHMs. During a brief conversation, Zhaoyun's clear research ideas, keen problem-solving abilities, and outstanding methods for addressing problems left a profound impression on me.

My classroom was on the third floor of Building 26, adjacent to Zhaoyun's office. After the Thanksgiving Day, we had more contact. Closer contact made me admire Zhaoyun's dedication to his remarkable research abilities even more. Not long after, Zhaoyun's research made significant progress. Relevant papers were published in journals such as *Journal of Applied Physics* and *Journal of Physics D: Applied Physics*. Subsequently, Zhaoyun conducted systematic research on the excitation of RCR by using charged particles in anisotropic metamaterial-filled waveguides, achieving encouraging results. His papers were published in Optics Express.

In my close contact with Zhaoyun, I was deeply moved by his unwavering spirit on the path of research. He had already achieved outstanding results, yet he showed no complacency. He continuously improved LHMs suitable for exciting RCR and conducted a series of cutting-edge experiments.

In 2012, he proposed a metamaterial high-frequency structure and developed a terahertz sheet beam metamaterial radiation source. Zhaoyun's research group showed that compared with traditional dielectric material terahertz radiation sources, this new radiation source could increase the power by 100–1000 times. The paper was published in *Physics of Plasmas*. Following that, Zhaoyun began researching how to achieve all-metal LHMs which are suitable for high-vacuum environments. After a series of in-depth studies and continuous experiments, he finally proposed a new type of all-metal LHM suitable for high-vacuum environments in 2013.

In 2015, Zhaoyun's research made a significant breakthrough. In July, we were invited to provide an academic review of the research results from the group of Prof. Federico Capasso at Harvard University. This paper was published in *Nature Nanotechnology*.

Prior to this, Zhaoyun had designed a new energy coupler suitable for LHMs and began to do high-energy charged particle experiments in the latter half of 2015 to demonstrate the unique left-handed properties of this metamaterial. They observed for the first time that rapidly moving electrons in LHMs produced Cherenkov radiation opposite to their motion direction. When the energy of the electrons was lower than the critical threshold for RCR, the radiation disappeared as expected. In RCR, the frequency of the new electromagnetic radiation can also be changed by changing the kinetic energy of charged particles. Thus, the final puzzle of the three novel electromagnetic properties in LHMs predicted by Veselago was experimentally solved. This original work was published in *Nature Communications* in March 2017.

As systematically described in this monograph, RCR is used to develop a high-power microwave radiation source, with the original achievements published in *Applied Physics Letters*, *IEEE Electron Device Letters*, *IEEE Transactions on Electron Devices*, and so on.

These new high-power microwave radiation sources mentioned in the monograph can be widely used in large scientific facilities, radar systems, biomedical imaging, microwave heating, microwave sterilization and disinfection, industrial irradiation, and other fields.

This academic monograph reports the achievements of Zhaoyun's research on metamaterials and their applications, marking a milestone in this field. This monograph is self-contained and provides valuable information and research directions for advanced researchers in the field of metamaterials. I hope that metamaterials will benefit humanity more widely in the future. Talented individuals emerge in every generation, and I also hope that, together with readers, we can work tirelessly from generation to generation to fulfill the millennium-old aspirations of Zai Zhang.

Thanksgiving 2022

Min Chen
The Contributor of Nobel Physics Prize
Emeritus Professor of Physics
Department of Physics
MIT, Cambridge
Massachusetts, USA

Preface

Metamaterials are a type of subwavelength artificial structures with novel physical properties that cannot be achieved or are difficult to be obtained in natural materials. In 1999, the term metamaterial was first coined by R. M. Walser, with the prefix meta-meaning beyond in Greek. Walser did not use the prefix super. Perhaps his intention was to signify that metamaterials were not conventional natural materials but rather a new class of subwavelength unit cells. After that, many different terms such as left-handed materials (LHMs), double-negative material, negative refractive index material, artificial electromagnetic material, metamaterial, and metasurface appeared. With the advance of research, the narrow concept of LHMs has expanded to the broad concept of metamaterials. In this monograph, the term metamaterials are employed. However, in some places in the monograph, other terms are used for following the original expression.

Since the experimental verification of metamaterials in 2001, researchers have incorporated the concept of microwave metamaterials into their respective research areas such as physics, electronics, optics, acoustics, materials science, mechanics, and thermodynamics. This has led to a cross-disciplinary and sparked research enthusiasm in metamaterials, resulting in fruitful research achievements. The affinity between metamaterials and vacuum electronics is from their novel electromagnetic properties such as reversed Cherenkov radiation (RCR) and enhanced transition radiation mechanisms. The introduction of metamaterial has brought new development opportunities to vacuum electronics. Metamaterial vacuum electron devices (VEDs) have significant scientific values and application prospects in the future.

This monograph focuses on the fundamental theory, implementation, and characterization of metamaterials. It reports RCR, enhanced transition radiation, and their applications in VEDs systematically, including my research results over more than a decade. It is suitable for senior undergraduate and graduate students in electronic engineering and physics, as well as the related scientists and engineers. Zhaoyun Duan is responsible for the content arrangement, writing, and proofreading of the entire monograph. Zhifang Lyu and Chuanchao Wang assisted in Chap. 1; Jibo Dong and Zhifang Lyu assisted in Chap. 2; Mingcheng Han assisted in Chap. 3; Zhifang Lyu assisted in Chap. 4; Zhifang Lyu and Chuanchao Wang assisted in Chap. 5; and Xuanming Zhang assisted in Chap. 6. Jibo Dong, Xuyuan Chen, and Ning Li carefully proofread the monograph. I thank them for their help. In addition, my students, Yanshuai Wang, Xianfeng Tang, Fei Wang, Shifeng Li, Xin Wang, Shengkun Jiang, Xuanming Zhang, Zhifang Lyu, Chuanchao Wang, Jucheng Lu, Xutong Mao, Chen Guo, Junpu Ling, Xin Guo, Yan Nie, Xiang Huang, Xinwu Ma, Sen Yang, Xirui Zhan, Xiaoyi Li, and Hengyu Luo, contributed to some of the contents of this monograph. I would like to take this chance to express my gratitude to them.

The successful publication of this monograph would not have been possible without the generous assistance of various experts. I would like to thank Academician Shenggang Liu and the older scientists who pioneered vacuum electronics at the University of Electronic Science and Technology of China (UESTC). I am grateful for the careful guidance and strong support from Profs. Yubin Gong, Wenxiang Wang, Bingzhong Wang, Zhiqin Zhao, Yong Fan, and Bin Li at UESTC. I appreciate the guidance and help from Profs. J. A. Kong, Min Chen, Richard J. Temkin, Nicholas Xuanlai Fang, Drs. B.-I. Wu and Michael A. Shapiro during my stay at MIT. I also thank Profs. K. R. Chu from the Taiwan University, Professor Manfred Thumm from the KIT in Germany, Dr. Steven Gold from the U.S. Navy Laboratory, Prof. Edl Schamiloglu from the University of New Mexico, Prof. John P. Verboncoeur from the Michigan State University, Prof. John H. Booske from the University of Wisconsin, Dr. Monica Blank from CPI in USA, Prof. Xiaodong Chen from the Queen Mary University of London in UK, Profs. Levi Schachter and Yakov Krasik from the Technion in Israel, Prof. B. N. Basu from IIT-BHU in India, Academician Tie Jun Cui from the Southeast University, Dr. Jinjun Feng from the Beijing Vacuum Electronics Research Institute, Prof. Hongsheng Chen from the Zhejiang University, and Prof. Runhua Fan from the Shanghai Maritime University.

In particular, I would also like to thank Prof. Edl Schamiloglu and Prof. Min Chen for writing the forewords during their busy time.

Due to the space limitation, I cannot express my gratitude individually to all the leaders, teachers, colleagues, and students who ever helped me.

I apologize for any omissions or errors despite our multiple proofreadings. Let us work hard to promote the research on metamaterials and their applications and achieve great breakthroughs in the future.

Finally, I would like to dedicate this monograph to my parents who untimely passed away. May they be well in heaven. I would also like to dedicate it to my beloved wife and cute son, who have made me feel the warmth of family and have given me the courage and motivation to keep me moving forward.

Written in October 2022
Updated in June 2024

Zhaoyun Duan
Chinese Institute of Electronics Fellow
IET Fellow
Professor of Electronics
Deputy Dean of the School of
Electronic Science and Engineering
University of Electronic Science and
Technology of China (UESTC)
Chengdu, Sichuan, China

Contents

1 Introduction 1
- 1.1 Development History of Metamaterials 2
- 1.2 Classification of Metamaterials 8
  - 1.2.1 Negative Permittivity Material 9
  - 1.2.2 Negative Permeability Material 11
  - 1.2.3 Left-Handed Material 12
  - 1.2.4 Near-Zero Refractive Index Material 13
  - 1.2.5 High Refractive Index Material 14
- 1.3 Unique Properties of Metamaterial 16
  - 1.3.1 Negative Refraction Phenomenon 16
  - 1.3.2 Reversed Doppler Effect 18
  - 1.3.3 Reversed Cherenkov Radiation 19
  - 1.3.4 Anomalous Radiation Pressure Characteristics 20
- 1.4 Prospective Applications of Metamaterials 21
- References 27

2 Implementation and Characterization of Metamaterials 33
- 2.1 Implementation of Metamaterials 33
  - 2.1.1 Classical Combination Method and Its Extension 34
  - 2.1.2 Natural Material Composite Method 38
  - 2.1.3 Transmission Line Method 39
- 2.2 Effective Medium Theory 42
- 2.3 Methods of Retrieving Effective Electromagnetic Parameters 45
  - 2.3.1 Basic Principle for Retrieving S-Parameters 46
  - 2.3.2 S-Parameters Retrieval in Free Space 50
- 2.4 Overview of Metasurfaces 55
  - 2.4.1 Basic Concept and Development 55
  - 2.4.2 Implementation and Characterization of Metasurfaces 58
  - 2.4.3 Application Prospectives of Metasurfaces 59
- References 60

3 Basic Theory of Reversed Cherenkov Radiation 65
3.1 Reversed Cherenkov Radiation in Unbounded Anisotropic Double Negative Metamaterial 66
3.1.1 Single-Particle Model 67
3.1.2 Numerical Calculations 70
3.2 Reversed Cherenkov Radiation in Semi-Unbounded Double Negative Metamaterial 74
3.2.1 Theoretical Analysis of Single-Particle Model 74
3.2.2 Numerical Calculations of Single-Particle Model 78
3.2.3 Theoretical Analysis of Multi-Particles Model 81
3.2.4 Numerical Calculations of Multi-Particles Model 87
3.3 Reversed Cherenkov Radiation in a Circular Waveguide Filled with Double Negative Metamaterial 90
3.3.1 Theoretical Analysis of Single-Particle Model 90
3.3.2 Numerical Calculations of Single-Electron Model 93
3.3.3 Theoretical Analysis of Pencil Beam Model 98
3.3.4 Numerical Calculations of Pencil Beam Model 100
References 102

4 Experimental Studies on Reversed Cherenkov Radiation 105
4.1 Metamaterial Slow-Wave Structure 107
4.1.1 Effective Electromagnetic Parameters of Metamaterials 107
4.1.2 Electromagnetic Properties of Metamaterial Slow-Wave Structure 110
4.2 Formation and Transportation of Sheet Beam 113
4.2.1 Sheet Beam Electron Gun 113
4.2.2 Transportation of Sheet Beam in Metamaterial Slow-Wave Structure 114
4.3 Simulation Analysis of Reversed Cherenkov Radiation 119
4.3.1 Beam-Wave Interaction 119
4.3.2 Mode Analysis in Metamaterials 122
4.3.3 Analysis and Discussion of Simulated Results 123
4.4 Experimental Verification of Reversed Cherenkov Radiation 125
4.4.1 Experimental Setup and Platform 125
4.4.2 Transmission Characteristics Analysis of Output Couplers 127
4.4.3 Design of the Solenoid Magnetic Focusing Configuration 131
4.4.4 Experimental Measurement and Analysis 137
References 142

5 Oscillator and Amplifier Based on Reversed Cherenkov Radiation ... 145
5.1 Novel Metamaterial Slow-Wave Structures ... 146
5.1.1 Circuit Theory ... 146
5.1.2 High-Frequency Characteristics ... 150
5.1.3 Transmission Characteristics ... 155
5.1.4 Cold Test Experiment ... 158
5.2 Electron-Optical System ... 163
5.2.1 Grid-Controlled Electron Gun ... 163
5.2.2 Uniform Magnetic Focusing Configuration ... 166
5.3 Reversed Cherenkov Radiation Oscillator ... 169
5.3.1 Beam-Wave Interaction Analysis ... 170
5.3.2 Hot Test Experiment ... 172
5.4 Reversed Cherenkov Radiation Amplifier ... 176
References ... 180

6 Vacuum Electron Devices Based on Transition Radiation ... 183
6.1 Resonant Cavity Loaded with Metamaterial Unit Cells ... 185
6.1.1 Electromagnetic Parameters of Resonant Cavity ... 185
6.1.2 Electromagnetic Properties of Resonant Cavity Loaded with Single-Bridge CeSRRs ... 189
6.1.3 Experimental Study of Resonant Cavities Loaded with Single-Bridge CeSRRs ... 191
6.2 Transition Radiation with High Coherence and Strong Radiation ... 194
6.2.1 Theoretical Analysis of Transition Radiation ... 194
6.2.2 Experimental Verification of Transition Radiation ... 200
6.3 P-Band Metamaterial-Inspired Klystron ... 207
6.3.1 High-Frequency Characteristics of Metamaterial-Inspired Klystron ... 208
6.3.2 Beam-Wave Interaction of Metamaterial-Inspired Klystron ... 210
6.3.3 Thermal Analysis of Metamaterial-Inspired Klystron ... 213
6.4 Metamaterial-Inspired Extended Interaction Oscillator ... 215
6.4.1 High-Frequency Characteristics of Metamaterial-Inspired Extended Interaction Oscillator ... 215
6.4.2 Beam-Wave Interaction of Metamaterial-Inspired Extended Interaction Oscillator ... 216
6.5 Metamaterial-Inspired Extended Interaction Amplifier ... 219
6.5.1 High-Frequency Characteristics of Metamaterial-Inspired Extended Interaction Amplifier ... 219
6.5.2 Experimental Investigation of Metamaterial-Inspired Extended Interaction Amplifier ... 221

6.5.3 Beam-Wave Interaction of Metamaterial-Inspired Extended Interaction Amplifier 224
References 227

Epilogue 231

Index 237

Symbols

$\boldsymbol{A}$	Magnetic vector potential
a	Broad side of rectangular waveguide cross-section
$\boldsymbol{B}$	Magnetic induction intensity
b	Narrow side of rectangular waveguide cross-section
$\boldsymbol{C}$	Capacitance
C_{all}	Perimeter of inner wall of slot-line
C_s'	Capacitance of the slot-line of the double-bridge CeSRR considering the metal bridge
C_c	Capacitance of cavity
C_g	Capacitance of the metal bridge of the double-bridge CeSRR
C_i	Additional capacitance of grooves on both sides of the metal bridge of the double-bridge CeSRR
C_s	Capacitance of the slot-line of the double-bridge CeSRR
C_s''	Capacitance of the slot-line of the double-bridge CeSRR considering the metal bridge and grooves on both sides of the metal bridge
C_{pul}	Capacitance per unit length of the double-bridge CeSRR
c_0	Speed of light in vacuum
$\boldsymbol{D}$	Electric displacement
D	Duty cycle of power supply
d	Thickness of the effective medium
e	Electron charge
$\boldsymbol{E}$	Electric field intensity
$\boldsymbol{E}_0$	Initial electric field intensity
$\boldsymbol{E}_1$	Electric field on the incident wave side of the effective medium
$\boldsymbol{E}_2$	Electric field on the outgoing wave side of the effective medium
E_{z0}	Amplitude of electric field intensity of the zeroth-order spatial harmonic of the fundamental mode
$\boldsymbol{F}$	Electromagnetic field incident on the incident wave side of the effective medium
F	Filling factor

$\boldsymbol{F}'$	Electromagnetic field on the outgoing wave side of the effective medium
f	Frequency
f_s	Resonant frequency of the CeSRR
f_{mnp}	Resonant frequency of a rectangular resonant cavity
f_{TEmnp}	Resonant frequency of the TE mode in a circular resonant cavity
f_{TMmnp}	Resonant frequency of the TM mode in a circular resonant cavity
f_0	Resonant frequency of empty resonant cavity
G	Equivalent conductance of resonant cavity
G_e/G_0	Normalized electron beam loading conductance
$\boldsymbol{H}$	Magnetic field intensity
$\boldsymbol{H}_0$	Initial magnetic field intensity
$\boldsymbol{H}_1$	Magnetic field on the incident wave side of the effective medium
$\boldsymbol{H}_2$	Magnetic field on the outgoing wave side of the effective medium
h	Height of cavity
$\boldsymbol{I}$	Current
I_h	Heater current of cathode
$\boldsymbol{J}$	Current density
$J_1(2.405)$	First-order Bessel function of first kind with respect to its argument 2.405
$K(k)$	Complete elliptic integral of the first kind
k_{eff}	Extinction coefficient
k_{mnp}	Amplitude of wave number of the resonant cavity
K_{rough}	Correction factor
$K_{rough\text{-}HB}$	Correction factor with the H-B model
$K_{rough\text{-}G}$	Correction factor with the Groiss model
K_c	Interaction impedance
K'	Coupling factor with CeSRR loading
K	Coupling factor between CeSRRs
$\boldsymbol{k}$	Wave vector
l	Length of rectangular waveguide
L_c	Inductance of a cavity
L_c'	Modified cavity inductance with CeSRR loading
L_{pul}	Inductance per unit length
L_s'	Inductance of slot-line
L	Inductance
l_1	Length of transit gap
L_3	Series inductance with the coupling between slot-lines
L_g	Inductance of metal bridge connecting inner and outer loops of a CeSRR unit cell
l_s	Effective electrical length of slot-line
L_s	Inductance of slot-line in a CeSRR unit cell
M	Coupling coefficient
M^2R/Q	Effective characteristic impedance
$m_{e\text{-}eff}$	Effective electron mass

m_e	Electron mass
n	Refractive index
n_i	Imaginary part of refractive index
n_r	Real part of refractive index
n_{eff}^0	Effective refractive index corresponding to principal branch of logarithmic function
n_e	Electron density
$n_{e\text{-}eff}$	Effective electron density
N_{eff}	Complex effective refractive index
n_{eff}	Effective refractive index
$\boldsymbol{P}$	Electric polarization intensity
P	Principal value
P_e	Power loss on external circuit load coupled to resonant cavity
P_r	Prandtl number
P_s	Average power loss of resonant cavity during one high-frequency period
P_w	Total power flow
p	Period of SRRs
Q	Quality factor
Q_0	Intrinsic quality factor
Q_e	External quality factor
Q_L	Loaded quality factor
q	Particle charge
q_T	Total charge
$\boldsymbol{r}$	Position vector
R/Q	Characteristic impedance
r_{av}	Average radius of slot-line in a double-bridge CeSRR
$\boldsymbol{S}$	Poynting vector
$<S>$	Time-averaged Poynting vector
$\boldsymbol{T}$	Transmission matrix
t	Time
t_{kyr}	Warm-up time of cathode
U	Voltage
U_c	Cathode voltage
U_{gp}	Electrode voltage for the control grid
U_h	Heater voltage of cathode
U_m	Voltage amplitude at two ports of equivalent circuit
u	Flow speed of water
$\langle \boldsymbol{v}_0 \rangle$	Average velocity
$\boldsymbol{v}$	Velocity
$\boldsymbol{v}_g$	Group velocity
$\boldsymbol{v}_p$	Phase velocity
v_0	Electron beam speed
W	Total average energy storage in a resonant cavity
W_s	Energy loss in one high-frequency period of resonant cavity

$\boldsymbol{x}$	Displacement vector
$\boldsymbol{x}_0$	Initial displacement vector
X_n	Normalized data
X	Original data
X_{max}	Maximum value of X
Z_{eff}	Effective wave impedance
z_w	Relative wave impedance
α	Heat transfer coefficient
β	Phase constant
β_e	Propagation coefficient of electron beam
β_0	Phase constant of the zeroth-order spatial harmonic of the fundamental mode
Γ	Dissipation coefficient
$\delta(x)$	Dirac function
δ	Skin depth
γ	Damping coefficient
$\gamma_{m\rho}$	Collision frequency associated with magnetic loss in the ρ direction in DNMs
$\gamma_{m\theta}$	Collision frequency associated with magnetic loss in the θ direction in DNMs
γ_{mz}	Collision frequency associated with magnetic loss in the z direction in DNMs
$\gamma_{e\rho}$	Collision frequency associated with electric loss in the ρ direction in DNMs
$\gamma_{e\theta}$	Collision frequency associated with electric loss in the θ direction in DNMs
γ_{ez}	Collision frequency associated with electric loss in the z direction in DNMs
ε	Permittivity
ε_{eff}	Effective permittivity
ε_i	Imaginary part of permittivity
ε_0	Permittivity in vacuum
ε_r	Real part of permittivity
θ	Angle
λ	Wavelength in free space
λ_t	Thermal conductivity
μ	Permeability
μ_{eff}	Effective permeability
μ_i	Imaginary part of permeability
μ_{mn}	nth root of mth-order Bessel function of the first kind
μ'_{mn}	Derivative of the nth root of mth-order Bessel function of the first kind
μ_0	Permeability in vacuum
μ_r	Real part of permeability
ρ	Resistance per unit length
ρ_0	Minimum average distance from field source

σ	Conductivity
σ_s	Conductivity of metal with the smooth surface
σ_{eff}	Effective conductivity
τ	Group delay
φ	Scalar potential
ω	Angular frequency of the incident wave
ω_{mnp}	Angular frequency of the resonant cavity
ω_{m0}	Cutoff angular frequency of a specific TM mode in a hollow waveguide
ω_{e0}	Cutoff angular frequency of a specific TE mode in a hollow waveguide
ω_p	Effective plasma angular frequency
$\omega_{p\rho}$	Angular plasma frequency in the ρ direction
$\omega_{p\theta}$	Angular plasma frequency in the θ direction
ω_{pz}	Angular plasma frequency in the z direction
ℓ	Period of horizontally arranged SRRs
Δ	Surface roughness
Δf	Offset of Resonant frequency

Chapter 1
Introduction

Since the establishment of Maxwell's equations in 1865 [1], the theory of electromagnetic fields has undergone approximately 160 years of development. Building upon the work of H. C. Oersted, A. M. Ampère, M. Faraday, and others, J. C. Maxwell proposed the hypothesis of displacement current, and created the electromagnetic theory. This theory not only predicted the existence of electromagnetic wave but also illustrated the intrinsic unity of electricity, magnetism, and light, marking a major synthesis in physics. In 1888, H. R. Hertz confirmed the existence of electromagnetic wave through spark gap oscillations in experiment, explicitly pointing out that light is also an electromagnetic phenomenon in 1889. After that, the electromagnetic wave theory had developed rapidly, and various devices and applications emerged.

The development and application of the electromagnetic field theory gave rise to vacuum electron devices (VEDs). In 1904, J. A. Fleming invented the first vacuum diode for the wireless radio detection. In 1906, L. D. Forest created the vacuum triode, which could amplify the electromagnetic wave, becoming one of the fundamental and crucial electron devices in radio technology and one of the greatest inventions of the twentieth century. Subsequently, vacuum tetrode, vacuum pentode, and related devices emerged, and used in wireless transmitters, which significantly promoted the early development of the communication system and inaugurated the era of the wireless communication. Due to the electrostatic controlling principle underlying vacuum diodes and triodes, their operational frequency was greatly limited. To overcome this limitation, microwave VEDs based on the dynamic controlling principle were gradually developed. In 1921, A. W. Hull invented the magnetron and observed microwave oscillations. In 1935, A. L. Samuel first manufactured a multi-cavity magnetron. In 1937, R. Varian and S. Varian developed the first klystron, achieving the microwave amplification. In 1939, J. Randall and H. Boot produced a fully practical multi-cavity magnetron, which was equipped in Britain's first-generation microwave early-warning radar during World War II and made a significant contribution to the severe damage inflicted on the German Air Force. After the war, researchers recognized the advantages of VEDs and vigorously promoted

Z. Duan, *Metamaterial-Based Electromagnetic Radiations and Applications*,
https://doi.org/10.1007/978-981-97-8108-9_1

their development. Simultaneously, electron devices based on semiconductor technology emerged. In 1947, W. Brattain, J. Bardeen, and W. Shockley invented the point-contact germanium transistor. In 1958, J. S. Kilby created the first integrated circuit (IC). In 1959, R. Noyce developed the first commercial IC, which marked the transition of the semiconductor industry from the invention era to the commerce era and the beginning of the semiconductor industry. VEDs experienced their prime development period in the first half of the twentieth century, playing an indispensable role in the invention of radio, television, long-distance telephone, computer, and more. However, with the rapid development of semiconductor devices, especially large-scale ICs, the application of VEDs faced significant challenges in terms of low-frequency and small power. Consequently, they had to evolve towards high-frequency, high-power, wide bandwidth, and high-efficiency [2, 3]. Today, VEDs still exhibit distinct advantages in civilian applications such as microwave heating, high-fidelity audio, medical accelerators, large scale scientific devices, as well as military applications including space communications, high-resolution radars, electronic countermeasures, and high-power microwave devices.

The development of the electromagnetic field theory and its numerical computation has strongly promoted the development of VEDs. As a new type of artificial electromagnetic materials, metamaterials possess characteristics that are challenging or impossible for natural materials to achieve. They have a great degree of freedom in unit cell design and array arrangement, resulting in some novel electromagnetic properties [4]. The introduction of metamaterials has brought new development opportunities to VEDs. This chapter primarily introduces the history of development, classification, novel electromagnetic properties, and applications of metamaterials.

1.1 Development History of Metamaterials

In the field of electromagnetics, Maxwell's equations and their boundary conditions determine the characteristics of electromagnetic fields. Since the response of a material to the electromagnetic field depends on the permittivity and permeability, which are commonly used for material classification. The real part of permittivity ε_r is taken as the abscissa, and the real part of permeability μ_r is taken as the ordinate, establishing a Cartesian coordinate system, as shown in Fig. 1.1a. For materials in the first quadrant (such as natural materials), electromagnetic wave can propagate in these media. Since the cross product of electric field intensity and magnetic field intensity $\boldsymbol{E} \times \boldsymbol{H}$ satisfies the right-hand rule with the wave vector $\boldsymbol{k}$ and the direction of Poynting vector $\boldsymbol{S}$, V. G. Veselago referred to these materials as right-handed materials (RHMs) [4], as shown in Fig. 1.1b. For materials in the second quadrant (such as metals and plasmas in optics), the existing evanescent wave cannot penetrate these media, as the real part of permittivity is less than zero and the real part of permeability is greater than zero, making them termed as negative permittivity materials. For materials in the third quadrant, theoretically, the electromagnetic wave can propagate, but such materials have not yet been discovered in the natural materials.

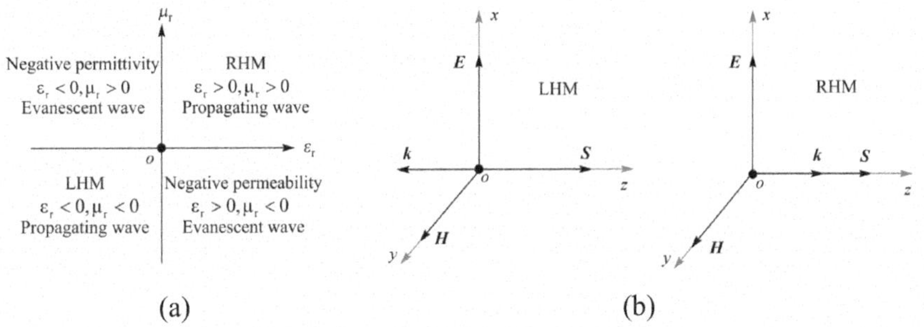

Fig. 1.1 Schematic of the material classification and the propagation of the electromagnetic wave. **a** The material classification using its permittivity and permeability. **b** The relationships among these vectors when electromagnetic wave propagates in LHMs and RHMs

This is because $\boldsymbol{E} \times \boldsymbol{H}$ and $\boldsymbol{S}$ still satisfy the right-handed rule, while $\boldsymbol{E} \times \boldsymbol{H}$ and $\boldsymbol{k}$ satisfy the left-handed rule. Therefore, V. G. Veselago referred to these as left-handed materials (LHMs), which were also called as double-negative materials or negative refractive index materials later [4], as shown in Fig. 1.1b. For materials in the fourth quadrant (such as ferromagnet and ferrimagnet that are operating near their resonant frequencies), since the real part of permeability is less than zero and the real part of permittivity is greater than zero, the electromagnetic wave cannot propagate within them and these materials are called as negative permeability materials.

When a plane wave propagates in a linear, isotropic, homogeneous, and lossy medium, its two curl equations are:

$$\nabla \times \boldsymbol{E} = -\frac{\partial \boldsymbol{B}}{\partial t}, \quad \nabla \times \boldsymbol{H} = \frac{\partial \boldsymbol{D}}{\partial t}, \tag{1.1}$$

where the constitutive relation of the medium is:

$$\boldsymbol{D} = \varepsilon \boldsymbol{E}, \quad \boldsymbol{B} = \mu \boldsymbol{H}. \tag{1.2}$$

The time-harmonic electromagnetic field can be expressed as:

$$\boldsymbol{E} = \boldsymbol{E}_0 \mathrm{e}^{\mathrm{j}(\boldsymbol{k}\cdot\boldsymbol{r}-\omega t)}, \quad \boldsymbol{H} = \boldsymbol{H}_0 \mathrm{e}^{\mathrm{j}(\boldsymbol{k}\cdot\boldsymbol{r}-\omega t)}, \tag{1.3}$$

where $\boldsymbol{r}$ is the position vector, $\boldsymbol{E}$ is the electric field intensity, $\boldsymbol{D}$ is the electric displacement vector, $\boldsymbol{H}$ is the magnetic field intensity, $\boldsymbol{B}$ is the magnetic induction intensity, $\boldsymbol{E}_0$ and $\boldsymbol{H}_0$ are the initial electric field intensity and initial magnetic field intensity, respectively, ω is the angular frequency of the incident wave, and t is time. Substituting (1.2) and (1.3) into (1.1), we can obtain:

$$\boldsymbol{k} \times \boldsymbol{E}_0 = \omega\mu \boldsymbol{H}_0, \quad \boldsymbol{k} \times \boldsymbol{H}_0 = -\omega\varepsilon \boldsymbol{E}_0. \tag{1.4}$$

According to (1.4), when the real parts of the permittivity ε and the permeability μ of the medium are both positive, $\boldsymbol{E}$, $\boldsymbol{H}$, and $\boldsymbol{k}$ satisfy the right-hand rule, as shown in Fig. 1.1b. As mentioned earlier, this type of medium is called RHM, and the majority of materials in natural world belongs to RHMs. When the real parts of the permittivity and permeability are both negative, $\boldsymbol{E}$, $\boldsymbol{H}$, and $\boldsymbol{k}$ satisfy the left-hand rule, as depicted in Fig. 1.1b. Therefore, it is referred to as LHM. When the electromagnetic wave propagates in LHMs, the direction of $\boldsymbol{S}$ is opposite to the direction of $\boldsymbol{k}$, indicating that the energy propagating direction of the electromagnetic wave in LHMs is opposite to the phase velocity direction.

Metamaterials are artificial subwavelength structures that have novel electromagnetic properties, which are either unattainable or challenging to achieve with conventional natural materials. These properties include negative refraction, reversed Cherenkov radiation (RCR), reversed Doppler effect, anomalous radiation pressure characteristics [5, 6]. In 1999,[1] R. M. Walser coined the term metamaterial [7]. When the constituent materials of a metamaterial are determined, its electromagnetic properties are primarily determined by the dimensions of the subwavelength unit cells, shapes, arrangement, and other characteristics. LHMs are considered a crucial category within metamaterials. After LHMs were proposed in the microwave field, they rapidly expanded into other fields, including optics, acoustics, materials science, mechanics, thermodynamics, and so on.

The history of metamaterials can be traced back to the early twentieth century. H. Lamb and H. C. Pocklington [8, 9] theoretically pointed out the possibility of waves with opposite directions between phase and group velocities by their studies on mechanical waves in mechanical systems. A. Schuster briefly mentioned H. Lamb's work in his book in 1904 and speculated the impact of media with the backward-wave property on the optical refraction [10], which indicated an early awareness of the phenomenon of negative refraction. L. I. Mandelshtam described the phenomenon of negative refraction in his lectures [11]. G. D. Malyuzhinets studied Sommerfeld radiation conditions in backward wave media, and pointed out that the phase velocity of the wave points from infinity to the source [12]. D. V. Sivukhin published an article discussing the energy flow of the electromagnetic wave in dispersive media [13], which was the first consideration of the fact that when both permittivity and permeability of the medium are simultaneously negative, the direction of the energy flux density of the electromagnetic wave is opposite to the direction of the wave vector. He analyzed this phenomenon, and pointed out the lack of understanding in the scientific community at that time about media with both negative permittivity and permeability, indicating the uncertainty about their existence in nature. In 1959, V. E. Pafomov discovered RCR in materials with negative permittivity and negative permeability [14]. In 1967,[2] V. G. Veselago systematically studied the electromagnetic properties of the LHM in principle [4], and predicted some novel electromagnetic properties

[1] ① R. W. Ziolkowski explained the historical process of the term "Metamaterial," which was first introduced by R. M. Walser in an informal occasion in 1999 (see reference [7]).

[2] ② V. G. Veselago first theoretically proposed the concept of left-handed materials in 1967 (in Russian). The English translation can be found in reference [4].

such as negative refraction, RCR, reversed Doppler effect, and anomalous radiation pressure characteristics.

Since LHMs were not found in natural materials, the challenge was to develop artificial methods to realize them. Artificial materials have a long history of research [7]. For example, in 1962, W. Rotman proposed a wire grid medium to simulate the wave propagation characteristics of in plasmas [15]. In 1981, W. N. Hardy and L. A. Whitehead proposed a metal split-ring resonator (SRR) and studied its magnetic response characteristics [16]. In 1995, D. F. Sievenpiper and his colleagues proposed a metal lattice structure and studied its plasma characteristics [17]. These pioneering studies laid the theoretical foundation for achieving LHMs. In 1996 and 1999, J. B. Pendry and others inspired by the work of W. N. Hardy and D. F. Sievenpiper and based on the research of photonic crystals [18], proposed the use of periodic metal wire arrays [19] and SRR arrays [20] to construct LHMs with negative real parts of the effective permittivity and effective permeability, as shown in Fig. 1.2. These theoretical works made the artificial realization of LHMs possible. In 2000, D. R. Smith et al., creatively combined metal wires and SRRs to realize LHMs based on the theoretical work of J. B. Pendry et al. and experimentally verified their transmission of the electromagnetic wave [21, 22]. In 2001, R. A. Shelby et al. used LHMs as shown in Fig. 1.3a and the experimental platform shown in Fig. 1.3b experimentally verifying the phenomenon of the negative refraction in the microwave frequency band [23]. In 2003, the reversed Doppler effect of LHMs was experimentally confirmed [24]. In 2017, the RCR in LHMs was experimentally verified using real charged particles for the first time [25].

For conventional materials, their permittivity and permeability are determined by the response of the material's atoms to applied electromagnetic fields. At scales much larger than the atomic size and interatomic distance, the electromagnetic information about the material can be described by permittivity and permeability. Metamaterials take this idea by considering subwavelength unit structures to artificial atoms [26]. These subwavelength unit structures are arranged in a certain pattern to construct a metamaterial, as shown in Fig. 1.4. Based on the effective medium theory (EMT)

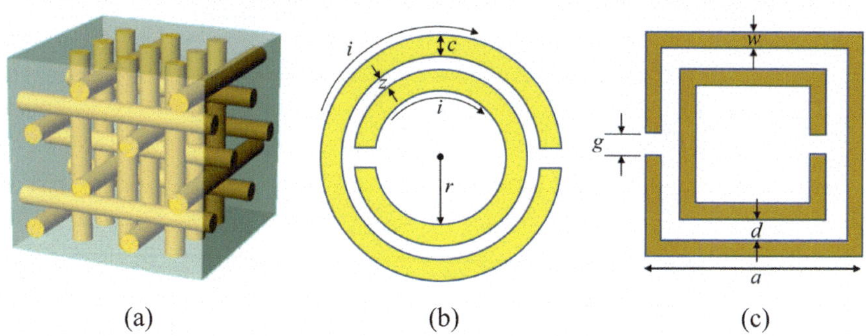

Fig. 1.2 Implementation of single-negative material. **a** Metal wire array. **b** Circular SRR. **c** Rectangular SRR

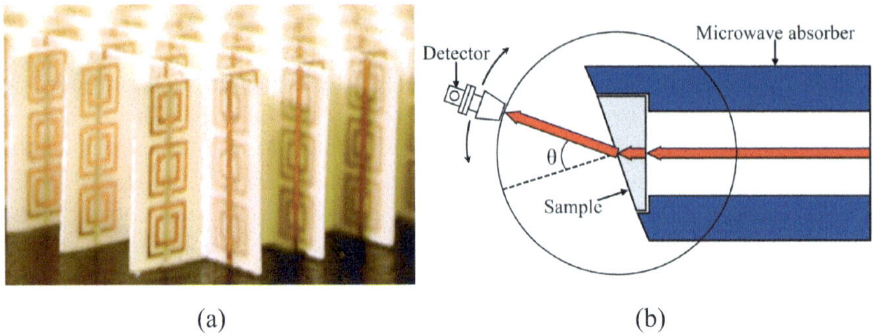

Fig. 1.3 LHM sample and verification experiment of the negative refraction. **a** LHM composed of metal wires and metal SRRs. **b** The experimental verification of the negative refraction

(see Sect. 2.2 for details), when the operating wavelength is much larger than the subwavelength unit cell and the spacing between the two subwavelength unit cells, the electromagnetic properties can be described by effective permittivity ε_{eff} and effective permeability μ_{eff}. In other words, it implies that the size of the subwavelength unit cells must be much smaller than the wavelength of the electromagnetic wave so that ε_{eff} and μ_{eff} can characterize the effective electromagnetic properties of metamaterial.

Due to the novel electromagnetic properties of metamaterial, research on the theory, experiments, and applications has attracted the widespread attention from scientists and engineers [27–34]. In 2003, the realization of LHMs was recognized by *Science* as one of the top ten breakthroughs of the year. In 2006, the invisibility cloak

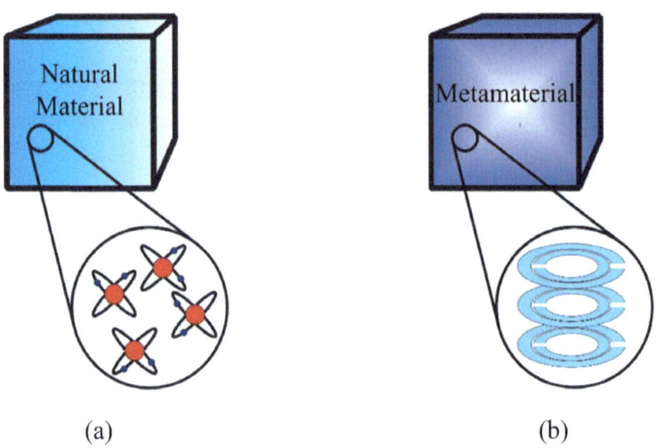

Fig. 1.4 Schematic diagram of natural material and metamaterial. **a** Atoms of natural material. **b** Subwavelength unit structure of metamaterial

made from metamaterial was listed as one of the top ten breakthroughs of the year by *Science*. In 2010, metamaterial was also listed as one of the ten most remarkable developments in *Science* over the past decade. In 2012, the U.S. Department of Defense classified metamaterial as one of the six disruptive foundational research areas and a field strongly advancing growth. In 2016, the metamaterial lens was selected as one of the top ten breakthroughs by *Science* again. Metamaterial has been extensively studied around the world and is being developed rapidly.

As a new interdisciplinary field, the emergence of metamaterial naturally raised numerous skepticisms and sparked intense debates within the academic community. Discussions mainly focused on whether the constructed materials were true LHMs and the theoretical and experimental verification of the novel electromagnetic properties of LHMs, such as negative refraction, perfect imaging, and invisibility. The negative refraction phenomenon is an example. Through the analysis of the dispersion of LHMs, P. M. Valanju and colleagues theoretically concluded that the wavefront was refracted in the negative direction at the interface between RHMs and LHMs. According to the principle of causality and that group velocity couldn't exceed light velocity, the energy of the wave could only be refracted in the positive direction. The difference between the direction of the wavefront and the direction of enphaseergy refraction would lead to the strong dispersion with the non-uniformity. They believed that the early assumption that the refracted of the wave represented the refraction of energy was incorrect. Simultaneously, they doubted R. A. Shelby and colleagues' observation of the negative refraction phenomenon in microwave LHMs, suggesting that it was merely a near-field diffraction of the material itself [35]. In 2002, N. Garcia and colleagues argued that the negative refraction experimentally observed by R. A. Shelby and others in microwave LHMs was due to the high loss characteristics of LHMs. They pointed out that when loss dominates, the real part of the negative refractive index must be determined by the imaginary part, and the wave is non-uniform. It makes no sense to differentiate LHMs and RHMs with the high loss [36]. In the same year, they also stated that the theoretical analysis of the perfect imaging with planar metamaterial was incomplete. The theoretical analysis indicated that perfect imaging could only exist in finite thickness and lossless planar metamaterial [37].

During the skepticism process, the related research was gradually deepened. In 2002, J. A. Kong's research group theoretically pointed out that P. M. Valanju and colleagues mistakenly considered the direction of the interference fringes of the wave as the direction of energy propagation in their research. They conducted theoretical calculations using both discrete and continuous signals, and concluded that the negative refraction would occur at the interface between the uniform, isotropic LHMs and RHMs [38]. In the same year, J. B. Pendry and D. R. Smith responded to P. M. Valanju and colleagues' skepticism about the microwave negative refraction experiment. They started from the original definition of the group velocity in negative refractive index materials and concluded that the group velocity is refracted in the negative direction [39]. In 2003, R. W. Ziolkowski used time-domain simulations to reveal that the propagation characteristics of the electromagnetic wave in LHMs comply with the law of causality. When electromagnetic waves are incident on the interface between RHM and LHM, they experience a long delay before reconstructing new wavefronts, and refract in the direction of negative angles [40, 41].

Building on the response to the theoretical skepticism, the phenomenon of the negative refraction in metamaterial was subsequently confirmed by different research groups in the microwave frequency band. In 2003, C. G. Parazzoli and others conducted experiments using metal SRRs and metal wires and observed the negative refraction in the frequency range of 12.6–13.2 GHz [42]. In the same year, I. L. Chuang's group also conducted experiments, and the results showed that LHMs using metal wires and metal SRRs in the microwave frequency band followed the Snell's law and exhibited a negative refractive index [43]. In 2004, J. Huangfu et al. experimentally confirmed the negative refraction phenomenon using a metamaterial composed of Ω-like metallic patterns [44]. In 2008, X. Zhang's group combined arrays of 60 nm diameter silver wires and the aluminum oxide to create three-dimensional metamaterial, providing the experimental evidence for the negative refraction in the optical frequency band [45].

It should be mentioned that in 2009, B. A. Munk specifically wrote a monograph, presenting his views on LHMs, particularly providing in-depth discussions on the phenomenon of the negative refraction [46]. These perspectives contribute to a deeper understanding of metamaterial, helping readers to critically evaluate their merits and limitations.

1.2 Classification of Metamaterials

As mentioned in Sect. 1.1, there are two parameters that characterize the electromagnetic properties of a medium: One is the medium's permittivity, which reflects the medium's response to an applied electric field, and the other is the permeability, which reflects the medium's response to an applied magnetic field. Considering the electromagnetic loss, the permittivity ε and permeability μ of an isotropic medium can be expressed in the complex forms:

$$\begin{cases} \varepsilon = \varepsilon_{\mathrm{r}} + \mathrm{j}\varepsilon_{\mathrm{i}} \\ \mu = \mu_{\mathrm{r}} + \mathrm{j}\mu_{\mathrm{i}} \end{cases}, \tag{1.5}$$

where, ε_{r} and μ_{r} are the real parts of the permittivity and permeability, ε_{i} and μ_{i} are the imaginary parts of the permittivity and permeability, respectively. ε_{r} characterizes the polarization degree of the dielectric medium under an applied electric field. The response of the dielectric medium to an applied electric field depends on the frequency of the applied electric field. When the frequency is high, polarization is not instantaneous but delayed, resulting in a phase difference with the applied electric field. μ_{r} characterizes the magnetization degree of the magnetic medium under an applied magnetic field. Similar to the polarization, the magnetization also depends on the frequency of the applied magnetic field and involves a phase difference. The real parts ε_r and μ_r of the permittivity and permeability respectively determine the phase change of the electromagnetic wave during the electromagnetic propagation in the medium, while the imaginary parts ε_i and μ_i represent the electrical loss and the magnetic loss, respectively, determining the amplitude change of the electromagnetic wave during the electromagnetic propagation in the medium [5]. Note that the real and imaginary parts of the medium's permittivity are determined by the Kramers–Kronig relation [47].

As shown in Fig. 1.1a, in the second and fourth quadrants, one of the real parts of the permittivity or permeability is negative. According to the electromagnetic wave theory, the electromagnetic wave cannot propagate in such media and appear as evanescent wave, whose amplitude rapidly decays exponentially. As mentioned in Sect. 1.1, materials in the second quadrant are referred to as negative permittivity materials, and those in the fourth quadrant are referred to as negative permeability materials. Based on the above classification, the following sections provide a brief introduction to negative permittivity materials, negative permeability materials, LHMs, near-zero refractive index materials, and high refractive index materials.

1.2.1 Negative Permittivity Material

Negative permittivity material refers to a class of metamaterial whose real part of the permittivity is only negative within a specific frequency range. As mentioned in Sect. 1.1, specific electromagnetic structures can be artificially designed to achieve a negative real part of the effective permittivity. In the early stages, metal wires were commonly used to design artificial subwavelength unit cells. J. B. Pendry et al. started from a simple and effective physical model which is the free electron gas model [48] to discuss the electromagnetic properties of metal wires. In this model, the valence electrons of the metal are similar to an ideal gas. They proposed a structure with the metal wire array, as shown in Fig. 1.2a. The electron density in the metal is denoted as n, the electron mass is m_e, the electron charge is e, and the equation of the motion for electrons in an electric field $\boldsymbol{E}$ is given by [49]:

$$m_e \frac{\partial^2 \boldsymbol{x}}{\partial t^2} + m_e \gamma \frac{\partial \boldsymbol{x}}{\partial t} = -e\boldsymbol{E}. \tag{1.6}$$

Here, $\boldsymbol{x}$ is the displacement vector of electrons, γ is the damping coefficient. When an alternating electric field $\boldsymbol{E}(\omega, t) = \boldsymbol{E}_0 \mathrm{e}^{-\mathrm{j}\omega t}$ is applied, where $\boldsymbol{E}_0$ is the initial electric field intensity, the solution to (1.6) is:

$$\boldsymbol{x} = \boldsymbol{x}_0 \mathrm{e}^{-\mathrm{j}\omega t} = \frac{\mathrm{e}}{\mathrm{m_e}(\omega^2 + \mathrm{j}\gamma\omega)} \boldsymbol{E}(\omega, \mathrm{t}). \tag{1.7}$$

In this context, where $\boldsymbol{x}_0$ represents the initial displacement vector, and considering that the electric dipole moment of a single electron is $-e\boldsymbol{x}$, the vector sum of the electric dipole moments per unit volume, namely the electric polarization intensity, is given by:

$$\boldsymbol{P} = -ne\boldsymbol{x} = -\frac{ne^2}{m_e(\omega^2 + j\gamma\omega)}\boldsymbol{E}(\omega, t). \tag{1.8}$$

The electric displacement vector $\boldsymbol{D}$ is defined as:

$$\begin{aligned} \boldsymbol{D} &= \varepsilon_0\boldsymbol{E} + \boldsymbol{P} = \varepsilon(\omega)\varepsilon_0\boldsymbol{E} \\ \varepsilon(\omega) &= 1 - \frac{\omega_p^2}{\omega^2 + j\gamma\omega}. \end{aligned} \tag{1.9}$$

In this context, where ω_p represents the plasma angular frequency, $\omega_p^2 = ne^2/\varepsilon_0 m_e$. $\varepsilon(\omega)$ is the complex permittivity and is given by:

$$\begin{aligned} \varepsilon(\omega) &= \varepsilon_r(\omega) + j\varepsilon_i(\omega) \\ \varepsilon_r(\omega) &= 1 - \frac{\omega_p^2}{\omega^2 + \gamma^2} \\ \varepsilon_i(\omega) &= \frac{\omega_p^2\gamma}{\omega(\omega^2 + \gamma^2)}. \end{aligned} \tag{1.10}$$

The above equation reveals that when the angular frequency ω of the incident wave is less than the plasma angular frequency ω_p, the real part of the effective permittivity ε_r manifested by the metal wire array is negative. If it is possible to control the plasma angular frequency ω_p of the metal wire array, such that $\omega < \omega_p$, it can achieve the negative effective permittivity in a specific frequency range.

The metal wire array can not only reduce the effective electron density but also significantly increase the effective electron mass, thereby greatly reducing the plasma angular frequency [19]. The following is a brief analysis of the effective plasma frequency of the metal wire array.

When the radius of the metal wire is r, the period is p, and the free electron density inside the metal is n, the effective electron density of the metal is given by:

$$n_{\text{eff}} = n\frac{\pi r^2}{p^2}. \tag{1.11}$$

From (1.11), it can be seen that the effective electron density is only related to the wire radius and the period. Since the radius r of the metal wire is much smaller than p, thus significantly reduces n_{eff}.

Due to the inductance effect of the metal wire itself, the magnetic field induced by the induced current generated inside the wire is given by:

$$H(R) = \frac{I}{2\pi R} = \frac{\pi r^2 n\upsilon e}{2\pi R}, \tag{1.12}$$

where I is the induced current inside the wire, υ is the average electron velocity, and R is the distance from the field point to the center of the metal wire. Thus, the expression for the induced magnetic field can be written as:

$$\boldsymbol{H}(R) = \frac{1}{\mu_0}\nabla \times A(R)\hat{z}, \tag{1.13}$$

$$A(R) = \frac{\mu_0 \pi r^2 n\upsilon e}{2\pi}\ln\left(\frac{p}{R}\right). \tag{1.14}$$

The additional momentum generated by electrons moving in the magnetic field is given by eA. Assuming that all electrons flow on the surface of the conductor, the total additional momentum produced by the metal wire per unit length is:

$$n\pi r^2 eA(r) = \frac{\mu_0(\pi r^2 n)^2 e^2 \boldsymbol{\upsilon}}{2\pi}\ln\left(\frac{p}{r}\right) = nm_{\text{eff}}\pi r^2 \boldsymbol{\upsilon}. \tag{1.15}$$

The effective electron mass can be obtained from the previous expression as:

$$m_{\text{eff}} = \frac{\mu_0 n\pi r^2 e^2}{2\pi}\ln\left(\frac{p}{r}\right). \tag{1.16}$$

Thus, the effective plasma frequency for the metal wire array can be obtained as:

$$\omega_{\text{p}} = \sqrt{\frac{n_{\text{eff}} e^2}{\varepsilon_0 m_{\text{eff}}}} = \frac{c_0}{p}\sqrt{\frac{2\pi}{\ln\left(\frac{p}{r}\right)}}. \tag{1.17}$$

Here, c_0 represents the velocity of light in vacuum. From (1.17), it can be observed that by appropriately designing the radius and period of the metal wires, it is possible to obtain $\omega < \omega_{\text{p}}$, thereby leading to the metal wire array with the negative effective permittivity in a specific frequency range.

1.2.2 Negative Permeability Material

Negative permeability material refers to a class of metamaterial where the real part of the effective permeability is only negative within a specific frequency range. In 1999, J. B. Pendry et al. proposed a metal SRR structure, which could achieve a negative real part of the effective permeability, as shown in Fig. 1.2b. After a complex derivation, the expression for the effective permeability of the array of metal SRRs was obtained [20, 22]:

$$\mu_{\text{eff}} = 1 - \frac{\pi r^2 / p^2}{1 - 3\ell / \pi^2 \mu_0 \omega^2 C r^3 + \mathrm{j}(2\ell\rho / \omega r \mu_0)}. \tag{1.18}$$

Here, r is the structural parameter of the SRR, p is the period of vertically arranged SRRs, ℓ is the period of horizontally arranged SRRs, ρ is the resistance per unit length measured around the circumference of the ring, and C is the capacitance related to the gap between the rings, which is expressed as:

$$C = \frac{\varepsilon_0}{\pi} \ln \frac{2c}{z} = \frac{1}{\pi \mu_0 c_0^2} \ln \frac{2c}{z}, \tag{1.19}$$

where c and z are structural parameters of the SRR.

$$F = \frac{\pi r^2}{p^2}, \tag{1.20}$$

$$\omega_0^2 = \frac{3\ell c_0^2}{\pi r^3 \ln(2c/z)} = \frac{3\ell}{\pi^2 \mu_0 C r^3}, \tag{1.21}$$

$$\Gamma = \frac{2\ell\rho}{r\mu_0}, \tag{1.22}$$

where F is called the filling factor, ω_0 is the resonance angular frequency, and Γ is the dissipation coefficient. According to (1.20)–(1.22), (1.18) can be further simplified as:

$$\mu_{\text{eff}} = 1 - \frac{F\omega^2}{\omega^2 - \omega_0^2 + \mathrm{j}\omega\Gamma}. \tag{1.23}$$

The metallic SRR structure can achieve the tunable magnetic response, which aroused researchers' interest in the artificial magnetic resonance response materials. Subsequently, various modified metal SRR structures have been proposed, such as the rectangular SRR, as shown in Fig. 1.2c [50].

1.2.3 Left-Handed Material

LHM, called as negative index material or double-nagative material, pertains to a category of metamaterial characterized by both negative real parts of the effective permittivity and the effective permeability within a specific frequency range. The following analysis approaches this concept from an optical standpoint. When an electromagnetic wave propagates from the medium A to the medium B, refraction occurs at the interface between the two media. The refractive index of the medium B relative to the medium A is defined as the ratio of the sine of the angle of the incidence to the sine of the angle of the refraction. The refractive index of any medium relative to the vacuum is named as the absolute refractive index. For dispersive media, the absolute refractive index n can be further expressed as the ratio of c_0 to the phase velocity υ_p of the electromagnetic wave in that specific medium [5]:

$$n = \frac{c_0}{\upsilon_p}. \tag{1.24}$$

If considering the electromagnetic loss of the medium, the absolute refractive index can be expressed as a complex number form:

$$n = n_r + jn_i = \pm\sqrt{(\varepsilon_r\mu_r - \varepsilon_i\mu_i) + j(\varepsilon_r\mu_i + \varepsilon_i\mu_r)}, \tag{1.25}$$

where n_r is the real part of the refractive index, and n_i is the imaginary part. For a normal lossy medium, the imaginary part of the refractive index is greater than zero. Therefore, to achieve a material with a negative refractive index, it must satisfy the condition $n_r < 0$, as indicated [51]:

$$\varepsilon_r\mu_i + \varepsilon_i\mu_r < 0. \tag{1.26}$$

From the above analysis, it can be seen that it is not necessary for both ε_r and μ_r to be negative to obtain a negative refractive index. It is sufficient for either one of them to be negative and satisfy (1.26). Clearly, as shown in Fig. 1.3a, the LHM constructed from metal wires and SRRs satisfies these conditions [20]. The figure of merit (FOM), which is defined as FOM $= -n_r/n_i$, is used to characterize the quality of negative refractive index materials. The larger the FOM, the better the quality of the material [52].

1.2.4 Near-Zero Refractive Index Material

Near-zero refractive index material refers to a class of metamaterial where the effective absolute refractive index is approximately zero. When either or both the real parts of the permittivity and permeability approach zero, the effective refractive index can be approximated to zero. Therefore, near-zero refractive index materials include epsilon-near-zero metamaterial (ENZM), mu-near-zero metamaterial (MNZM), and mu-and-epsilon-near-zero metamaterial (MENZM).

ENZMs have been a research focus within the near-zero refractive index materials category, exhibiting unconventional properties such as the electromagnetic compression, supercoupling characteristics, angle filtering, and waveform transformation [31, 53, 54]. In 2002, S. Enoch et al. utilized the metamaterial to achieve near-zero permittivity in the microwave frequency range and experimentally demonstrated its ability to control the direction of electromagnetic wave propagation through antenna experiments [31]. In 2008, R. P. Liu et al. achieved near-zero permittivity in the microwave

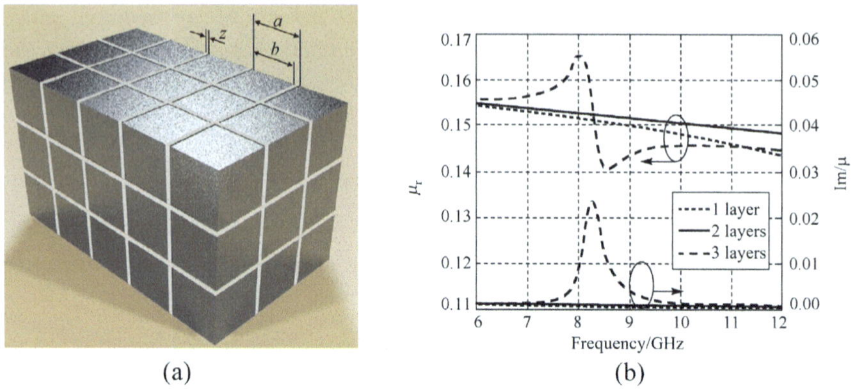

Fig. 1.5 Near-zero effective permeability metamaterial. **a** Metallic cubic metamaterial. **b** The relationship between effective permeability and frequency

frequency range by designing rectangular metal SRRs and experimentally confirmed its electromagnetic tunneling effect [53]. In 2013, R. Maas et al. achieved near-zero permittivity materials in the visible light frequency range through non-resonant silver-nitride-silicon multilayer film structures [54].

MNZMs can achieve impedance matching with air, exhibiting characteristics of the impedance matching and continuous propagation constants at the boundary with air. In 2013, P. A. Belov et al. proposed a metallic cubic metamaterial with a real part of the effective permeability around 0.15 in the microwave frequency range, as shown in Fig. 1.5 [55].

In 2007, M. Silveirinha et al. achieved MENZM by embedding dielectric or open perfect electric conductor rings into ENZM [56], as shown in Fig. 1.6. MENZM achieves both permittivity and permeability near zero around normalized frequencies, exhibiting isotropic characteristics. It can be used as a high-directional coherent radiation source.

1.2.5 High Refractive Index Material

High refractive index material refers to a class of metamaterial that can achieve a high effective refractive index in a specific frequency range. Apart from materials like ferroelectric ceramics such as barium strontium titanate and silicon carbide in the microwave frequency range, and lead sulfide in the far-infrared range [5], there are not many natural materials with ultra-high refractive indices. The design concept of metamaterial provides a possibility to create artificially engineered materials with the high refractive index. By constructing special artificial unit cells to enhance the electric or magnetic response, it is possible to achieve a high effective permittivity or effective permeability, thereby realizing a high effective refractive index.

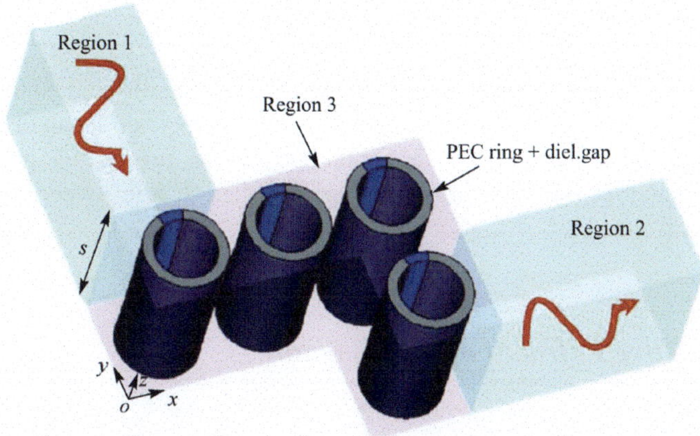

Fig. 1.6 Metamaterial with simultaneously near-zero effective permittivity and effective permeability

In 2011, M. Choi et al. utilized I-shaped thin metal unit cells embedded in a dielectric substrate, producing large-area, low-loss, flexible, and bendable high refractive index materials in the terahertz frequency range, as shown in Fig. 1.7 [57]. By effectively suppressing the magnetic response of the system while enhancing the electric response, they achieved an effective refractive index as high as 33.2, whereas the refractive index of conventional materials typically ranges from 1 to 4.

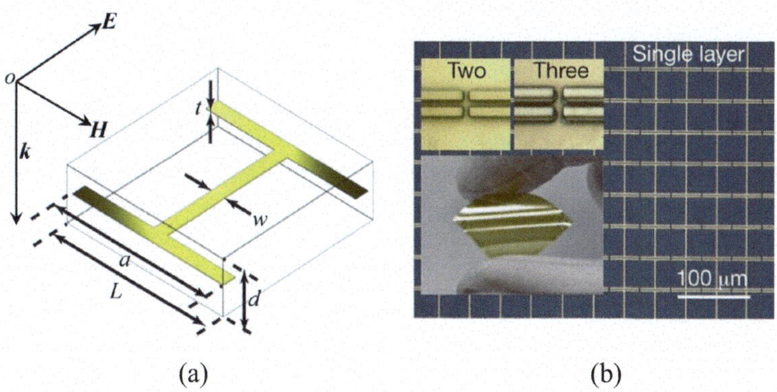

Fig. 1.7 High effective refractive index material. **a** I-shaped metal unit cell. **b** Sample

1.3 Unique Properties of Metamaterial

V. G. Veselago assumed the existence of LHMs and theoretically studied the interaction between electromagnetic wave and LHMs [4]. He predicted novel electromagnetic properties including negative refraction phenomena, reversed Doppler effect, RCR, and anomalous radiation pressure characteristics. A brief overview of these properties is given below.

1.3.1 Negative Refraction Phenomenon

When a light beam is incident from the RHM to the interface with the LHM, the refracted light is on the same side as the incident light concerning the normal direction. This is called as the negative refraction phenomenon [20, 22]. Although this phenomenon differs from the conventional refraction, it does not violate the basic law of refraction. According to (1.26), when both ε_r and μ_r are negative, the law of refraction still holds. It's just that the real part of the refractive index n is negative.

The negative refraction phenomenon in LHMs was experimentally confirmed by R. A. Shelby et al. in 2001 [23]. They conducted experiments using the LHM shown in Fig. 1.3a, and the experimental platform is illustrated in Fig. 1.3b. To achieve the oblique incidence of the incident wave, they fabricated a wedge shape LHM, where the incident wave is perpendicular to the right-angle plane of the LHM. No refraction occurs on this interface, and only a small amount of reflection exists. The reflected wave can be absorbed by surrounding absorbing materials. The transmitted wave enters the inclined surface of the wedge shape LHM, resulting in the oblique incidence on this interface. The dashed line represents the normal to the interface of the wedge. A detection device is set on the plane of the refracted wave, stepping on both sides of the normal to detect the propagation direction of the refracted wave. If the refracted wave and the incident wave are both on the same side of the normal, it indicates that the negative refraction phenomenon occurs in the LHM. Otherwise, it is a traditional refraction phenomenon (here referred to as a positive refraction phenomenon). They used the RHM (polytetrafluoroethylene (PTFE)) to conduct the experiments for comparison. The experimental results showed that in the LHM, when the refracted wave is on the same side as the normal, the angle between the refracted wave and the normal is negative. In the RHM, the angle between the refracted wave and the normal is positive, indicating a positive refraction phenomenon. This experiment confirmed the existence of the negative refraction phenomenon in LHMs.

Due to the presence of the negative refraction phenomenon, imaging systems do not need to be fabricated as curved surfaces. Imaging functionality can be achieved using flat-type LHMs to superimpose the phase of the electromagnetic wave propagation, enabling the high-quality imaging [4]. The information of the point light source exists not only in the transmitted wave but also in the evanescent wave. However, the evanescent wave can only propagate along the interface for about one wavelength in depth and propagate a wavelength-order distance along the interface. When the evanescent wave leaves the interface, the amplitude in conventional prism materials exponentially decays, and its information cannot be transmitted to the image point. Therefore, images formed by conventional prism materials lack information from the evanescent wave. Metamaterial prisms can not only converge transmitted wave to achieve imaging but also converge evanescent wave onto the image plane. Moreover, image points can maintain the same phase as the source point, theoretically achieving perfect imaging, as shown in Fig. 1.8. Figure 1.8a illustrates a schematic diagram of the propagation wave emitted by a light source through metamaterial with an effective refractive index $n = -1$. It can be seen from the figure that at the interface between the air and the metamaterial, negative refraction of electromagnetic waves occurs, and the incident angle and the refracted angle are equal, both being θ. Figure 1.8b illustrates a schematic diagram of the evanescent wave emitted by the light source through metamaterial with $n = -1$. The evanescent wave also undergoes a negative refraction at the interface between air and the metamaterial, eventually converging onto the image point. At this point, theoretically, the image point contains all the information emitted by the light source. However, due to the inevitable presence of the electromagnetic loss, the ideal metamaterial prism cannot be truly realized. The negative refraction phenomenon has important applications in overcoming the diffraction limit imaging, electromagnetic invisibility, spectrometer, particle detector, and biochemical detection [21].

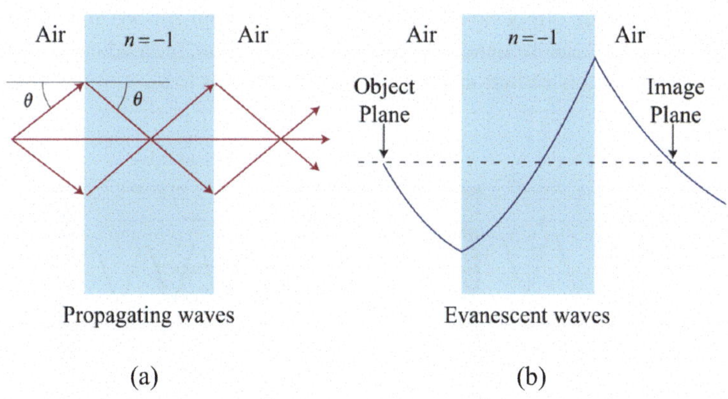

Fig. 1.8 Optical path of propagating and evanescent waves through a planar metamaterial prism. **a** The propagating waves through a flat plate of metamaterial prism. **b** The evanescent waves through a flat plate of metamaterial prism

1.3.2 Reversed Doppler Effect

When a moving source radiates the electromagnetic wave with the angular frequency ω, the angular frequency received by a stationary receiver is denoted as ω_1, where $\omega \neq \omega_1$. In the case of the RHM, if the source approaches the receiver, ω_1 is greater than ω. When the source moves away from the receiver, ω_1 is less than ω. This is the well-known Doppler effect in physics, as shown in Fig. 1.9a. In the case of the LHM, when the source and the receiver are moving toward each other, the received angular frequency decreases ($\omega > \omega_1$), whereas when they are moving away from each other, the received angular frequency increases ($\omega < \omega_1$). This phenomenon is known as the reversed Doppler effect [4, 24], as shown in Fig. 1.9b.

In 2003, N. Seddon and T. Bearpark conducted the first experimental verification of the reversed Doppler effect in LHMs [24]. In their experiment, a special transmission line was employed to achieve dispersion characteristics similar to LHMs. This transmission line consisted of a series of serial inductances L, transmission line-to-ground capacitances C, and transmission line cross capacitances C^*. By designing an initial frequency pulse signal into the transmission line, the reflections from different capacitance and inductance components in the circuit were observed. The periodically arranged capacitors, inductors, and other components in the model effectively simulated the reflections from a receding boundary. The experimental results demonstrated that, at a fixed observation point, the frequency of the reflected wave from the receding boundary increases, which is contrary to the conventional Doppler effect. The reversed Doppler effect has significant potential applications in radar detection, laser vibration measurement, astrophysical exploration, biomedical monitoring, multi-frequency tunable radiation sources, and other fields [58].

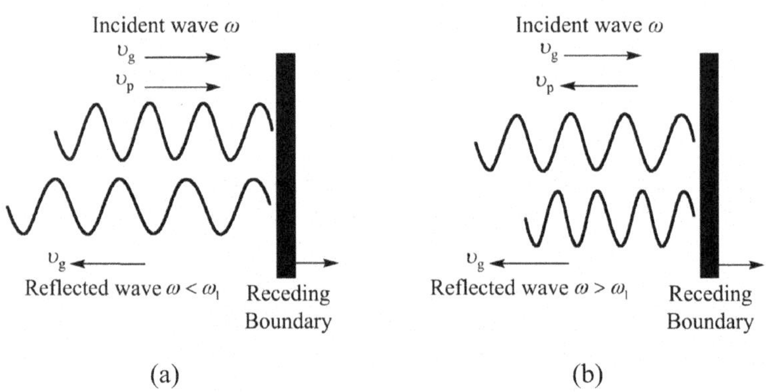

Fig. 1.9 Doppler effect illustration. **a** Doppler effect. **b** Reversed Doppler effect

1.3.3 Reversed Cherenkov Radiation

When a charged particle moves through a medium, it induces currents in the surrounding medium, creating a series of sub-wave sources along its path, emitting sub-wave signals. If the velocity of the charged particle exceeds the speed of light in the surrounding medium, these sub-wave sources interfere with each other, forming radiated electromagnetic wave, which is known as Cherenkov radiation (CR). In RHMs, the interference forms a wavefront or equal-phase surface that resembles a cone, and the energy of the electromagnetic wave is radiated along the normal direction to this cone surface, creating a forward cone angle, as shown in Fig. 1.10a. This type of the electromagnetic radiation was discovered by P. A. Cherenkov in 1934, and it was named after him. In 1937, I. M. Frank and I. Y. Tamm successfully explained the cause of CR using the classical electromagnetic theory. The three physicists won the Nobel Prize in Physics in 1958. In LHMs, the direction of the energy propagation is opposite to that of the phase velocity. Consequently, the electromagnetic wave will be radiated in the opposite direction of the charged particle's motion, forming a backward cone angle, as shown in Fig. 1.10b. This phenomenon is called as the RCR [59, 60].

The research process for experimentally confirming the RCR in LHMs has involved both indirect and direct experimental methods. Due to the significant challenges associated with directly using real charged particles to verify the RCR, early studies employed electromagnetic radiation as a substitute for charged particles. As early as 2002, G. V. Eleftheriades's group observed a similar RCR in an LHM using an indirect experimental method [61]. However, the authenticity of LHMs was questioned, and the equivalency of the results remained uncertain due to the indirect nature of the experiment. In the case of direct experimental methods, the use of real charged particles necessitates the design of entirely new LHMs suitable for a vacuum environment, significantly increasing the experimental complexity. In 2017, Duan's group proposed a novel all-metal LHM, and experimentally observed the RCR using

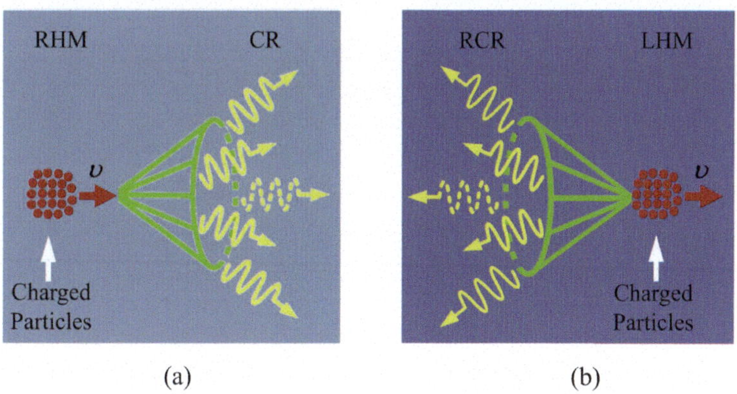

Fig. 1.10 CR and RCR. **a** CR in RHM. **b** RCR in LHM

real free electrons (see Chap. 4 for details) for the first time. RCR holds significant applications such as in novel high-power and miniaturized VEDs, particle detection, accelerators, and materials science [25].

1.3.4 Anomalous Radiation Pressure Characteristics

In conventional materials, the radiation pressure refers to the force exerted on an object when it is interacted by incident light. Since photons lack the rest masses but possess the momentums, their pressure characteristics are described in terms of the momentum. The monochromatic wave can be approximated as a beam of photons, and their impact on the illuminated object results in pressure, which is commonly referred to as the radiation pressure. V. G. Veselago theoretically predicted that LHMs would exhibit radiation pressure characteristics, which are distinct from those of conventional materials [4]. When a beam of light propagates through an LHM and illuminates an ideal reflector, the light undergoes refraction at the interface of the two media without reflection. The ideal reflector effectively provides twice the momentum of the incident light, pointing toward the side of the radiation source. The effect is manifested as the tension exerted by the incident light on the ideal reflector, as illustrated in Fig. 1.11.

In 2016, C. T. Chan's group furthered the theoretical research based on the work of V. G. Veselago. They discovered that a beam of light propagating through LHMs could exert the pressure or tension on a reflector, depending on the parameters of the microcrystal lattice structure composing the LHM. In the case of a triangular lattice arrangement, the force at the boundary manifests as tension, while for a square lattice, the force at the boundary manifests as the pressure [62]. The phenomenon of the anomalous radiation pressure in LHMs has many potential applications in areas such as nanoparticle manipulation, optical sensors, and optical detectors.

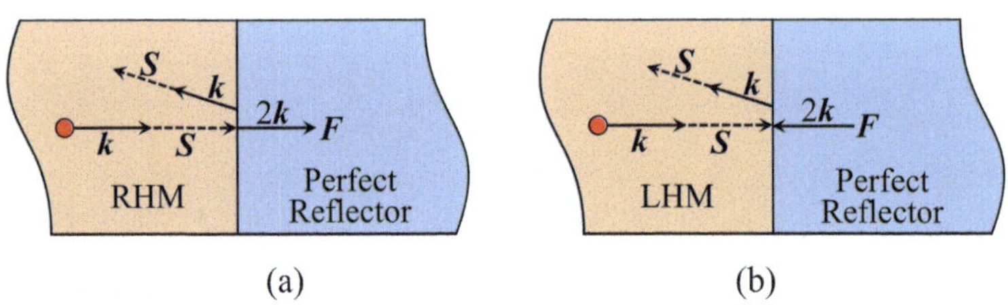

Fig. 1.11 The principle of radiation pressure. **a** Radiation pressure in RHMs. **b** Anomalous radiation pressure in LHMs

1.4 Prospective Applications of Metamaterials

After more than 20 years of research and exploration since the metamaterial was proposed, relevant theories have matured and the research scope has been expanding. The design concept of metamaterial has extended from the microwave domain to other fields such as optics, acoustics, materials science, mechanics, and thermodynamics, with a wide impact. Due to the unique properties, metamaterial can offer significant prospects for various applications.

In the field of imaging, in 2007, X. Zhang's group proposed an optical metamaterial lens with a sub-diffraction limit resolution of 130 nm, which consists of a curved periodic stack of Ag and Al_2O_3 deposited on a half-cylindrical cavity fabricated on a quartz substrate [63]. The group of S. Zhu and T. Li utilized the optical phase dispersion of metamaterial to develop a fully medium, wide-field, highly integrated microscopic imaging system. This system was integrated into the Complementary Metal Oxide Semiconductor (CMOS), realizing a centimeter-scale microscopic imaging system [64].

In the cloaking field, D. R. Smith and his group conducted cloaking experiments in 2009 using thousands of I-shaped metamaterial unit structures. The results indicated the successful cloak within the frequency from 13 to 16 GHz [65]. In 2020, H. Chen's group proposed a metasurface cloaking device using a deep learning technique. Experimental results demonstrated that the metasurface cloaking device could achieve the electromagnetic response in just 15 ms when the applied electromagnetic environment changes [66].

In passive devices such as antennas and absorbers, Y. Hao's group implemented a metamaterial substrate microstrip antenna based on complementary SRRs [67]. Z. Chen's group proposed an antenna based on a metasurface. By implementing the metasurface on both sides of the medium, they reduced the antenna thickness and loss, while maintaining the same performance [68]. S. J. Lim and his colleagues introduced a hexagonal metamaterial structure, demonstrating an excellent absorption performance in the frequency from 6.79 to 14.96 GHz [69]. L. Deng's group used an SRR to realize a wide-angle metamaterial absorber. Experimental results showed the absorption rate exceeding 90% within a bandwidth of 1.6 GHz [70].

In addition to the applications mentioned above, metamaterial also plays a crucial role in the field of vacuum electronics. When charged particles interact with metamaterial, RCR or coherent enhanced transition radiation (TR) occurs. Therefore, metamaterial can be used to construct the core components of VEDs—high-frequency structures, leading to the development of new types of VEDs. Due to the strong resonant characteristics of metamaterial resulting in the enhancement of the local field, thus metamaterial-based VEDs exhibit advantages such as high-power, high-efficiency, miniaturization, or high-gain. Compared to traditional VEDs, VEDs based on the metamaterial have significant application prospects. Currently, the research in metamaterial vacuum electronics is concentrated in the United States, China, India [71], Russia, South Korea, and so on.

The introduction of metamaterial into vacuum electronics dates back to 2007 when J. D. Wilson's group attempts to apply metamaterial in traveling-wave tubes (TWTs). However, they faced challenges in finding suitable metamaterial and could only assume the existence of the LHM, preliminarily conducting the theoretical research [72]. In 2008, Z. Duan et al. reported the mechanism of the enhanced RCR due to the electron beam excitation in a circular waveguide partially filled with anisotropic double-negative materials, providing a possible new direction for VEDs [73]. In 2009, R. Seviour's group inserted the hypothetic LHM to a bent waveguide TWT, as shown in Fig. 1.12a. This approach was used to control the dispersion characteristics of the bent waveguide TWT, providing a possible new direction for TWTs [74]. In 2010, D. Shiffler's group proposed a high-frequency structure based on a metamaterial composed of SRRs, as shown in Fig. 1.12b. Using the particle simulation software MAGIC, they studied the interaction between the electron beam and electromagnetic wave, indicating the potential application of metamaterial in VEDs [75]. However, due to operating above the cut-off frequency of the circular waveguide and the need of the medium support for the conductor ring, the device was challenging to miniaturize, and high vacuum was difficult to realize, which limited the advantages of the device.

In order to further explore all-metal metamaterial, T. M. Abuelfadl et al. proposed a circular waveguide structure loaded with metamaterial in 2011, as shown in Fig. 1.13 [76]. By introducing metal rods coupled to the circular waveguide's inner wall, the structure exhibits double-negative characteristics at frequencies below the circular waveguide's cutoff frequency. This configuration holds potential applications in high-power VEDs such as backward-wave oscillators (BWOs), TWTs, and RCR detectors. However, their analysis primarily focused on dispersion characteristics, without the discussion of the device's performance parameters.

In 2012, R. J. Temkin's group and G. Shvets's group proposed a metamaterial employing complementary metal SRR unit cells, as illustrated in Fig. 1.14. This

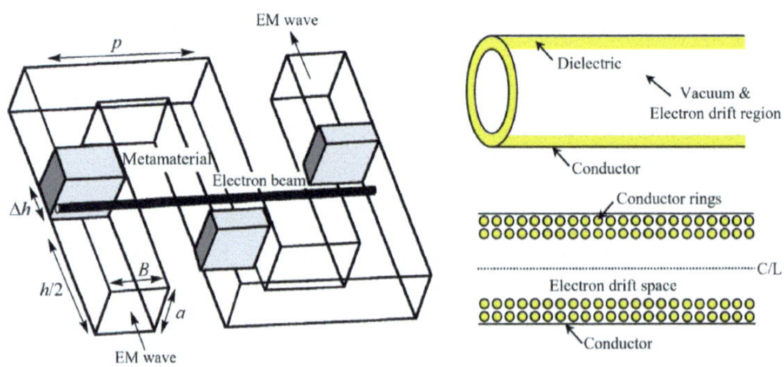

(a) (b)

Fig. 1.12 Early high-frequency structures with metamaterial. **a** Bent waveguide loaded with hypothetic metamaterial. **b** Circular waveguide loaded with the metamaterial

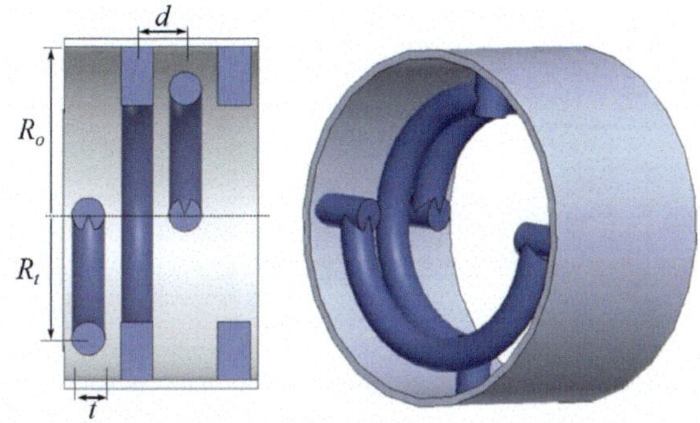

Fig. 1.13 Circular waveguide loaded with metamaterial

novel radiation mechanism offers three advantages: ① Better physical mechanisms compared to traditional BWOs; ② Low group velocity leading to the high spatial gain, reducing the threshold current and shortening the interaction length; ③ Complementary metal SRRs can be fabricated using standard planar processes, whereas traditional BWOs in the terahertz frequency range demand highly precise manufacturing, posing significant challenges. However, they assumed the existence of a high vacuum environment, making the device difficult to realize [77].

In 2012, the Air Force Office of Scientific Research (AFOSR) in the United States funded a multidisciplinary university research innovation (MURI) program. Research groups including the University of New Mexico, Massachusetts Institute of Technology, Ohio State University, University of California, Irvine, and Louisiana State University, collaborated on the study of metamaterial and their applications in

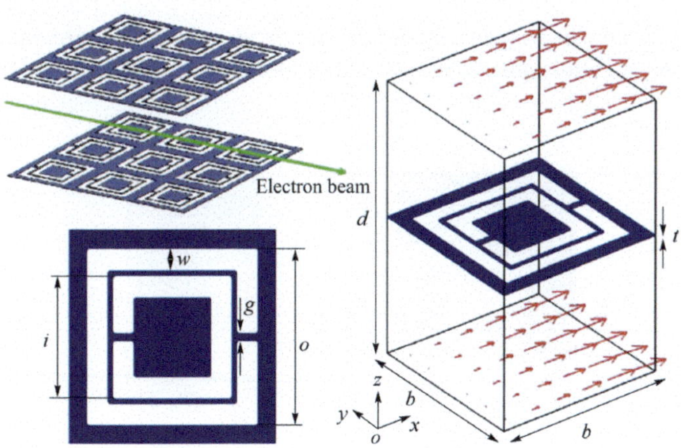

Fig. 1.14 Novel electromagnetic radiation based on metamaterial

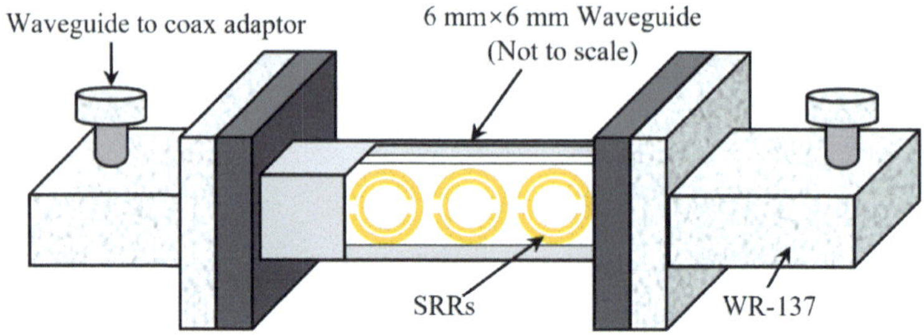

Fig. 1.15 Model of high-frequency structure with SRRs loaded into the square waveguide

high-power microwave sources. The goal was to achieve breakthroughs in the field of high-power microwave sources.

In 2013, R. J. Temkin's group reported an all-metal metamaterial suitable for VEDs, and extracted its electromagnetic parameters. However, the structure was not conducive to miniaturization. During the same year, D. Shiffler's group proposed a metamaterial high-frequency structure with an aperture resonator loaded into a square waveguide. Experimental studies on its transmission characteristics were conducted, as shown in Fig. 1.15, which marked a significant step towards practical applications of the device [78].

In 2014, based on their earlier work [77], R. J. Temkin's group developed a metamaterial-based BWO [32], as shown in Fig. 1.16. The BWO generated a pulsed output power of 5.75 MW at a frequency of 2.6 GHz, with an electron efficiency of approximately 14%.

In 2019, R. J. Temkin's group proposed a hub-shaped metamaterial unit cell, and create a novel high-power microwave radiation source employed in the wake-field particle accelerator system [79], as shown in Fig. 1.17a. Experimental results demonstrated that when a single electron cluster with parameters of 65 MeV and 45 nC passed through this structure, a pulsed output power of 25 MW was achieved at 11.4 GHz. In 2021, this group replaced the material of the metamaterial unit

Fig. 1.16 The BWO based on metamaterial

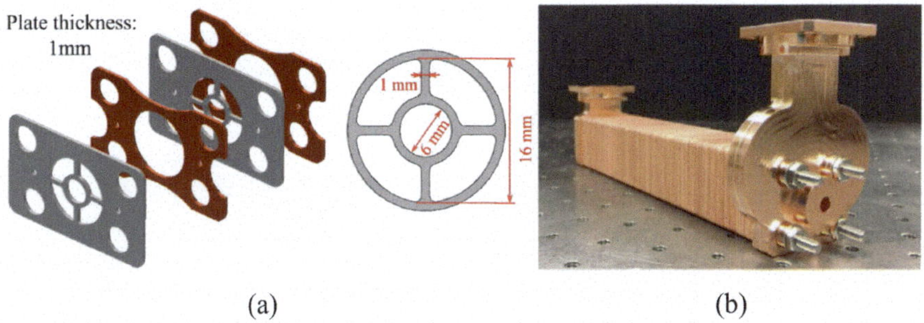

Fig. 1.17 The metamaterial unit cell and slow-wave structure. **a** Hub-shaped metamaterial unit cells. **b** Sample of a metamaterial slow-wave structure

from stainless steel to oxygen-free copper and employed Micro-Electro-Mechanical Systems (MEMS) technology to smoothen the surface of structure, which aimed to further reduce the high-frequency loss and breakdown risk, as shown in Fig. 1.17b. Experimental results indicated that the utilization of 8 electron clusters with parameters of 65 MeV and 355 nC through this structure resulted in a pulsed output power of up to 565 MW [80].

In 2016, E. Schamiloglu's group proposed a novel high-frequency structure using metamaterial, as depicted in Fig. 1.18a. Through simulations employing a short-pulse relativistic electron beam, the results indicated a pulse output power of 260 MW at 1.4 GHz, with an electron efficiency of approximately 15% [81]. In 2020, the group developed a BWO with metamaterial as shown in Fig. 1.18b. Experimental results demonstrated a pulse output power of 22 MW at 3.0 GHz, with an electron efficiency of 1.58% [33].

The extensive research has been conducted in the field of metamaterial vacuum electronics in China. F. Liu's group employed thin gold films and stacked layers of the silicon dioxide to create a hyperbolic metamaterial. This approach realized the CR with a low voltage threshold. Subsequently, they successfully utilized the magnesium fluoride to achieve hyperbolic metamaterial. By using low-energy free electrons in magnesium fluoride to generate the CR in the terahertz range, new avenues for efficient terahertz radiation were explored [82–84].

N. Bai's group proposed an integrated microstrip meander line structure composed of microstrip meander lines and compact broadband metamaterial absorbers, serving as an attenuator in a TWT. Compared to traditional carbon-coated attenuators, the metamaterial attenuator provided a flatter output power and gain [85, 86]. Additionally, they proposed a metamaterial output window for TWTs with excellent transmission characteristics [87].

M. Liu et al. utilized a double-helix metamaterial to develop a metamaterial slow-wave structure, studying a high-power, high-efficiency surface wave microwave oscillator [88, 89]. J. He's group used circular metamaterial structures to design a metamaterial slow-wave structure, optimizing spatial limiting current and coupling impedance, laying the foundation for the development of relativistic microwave

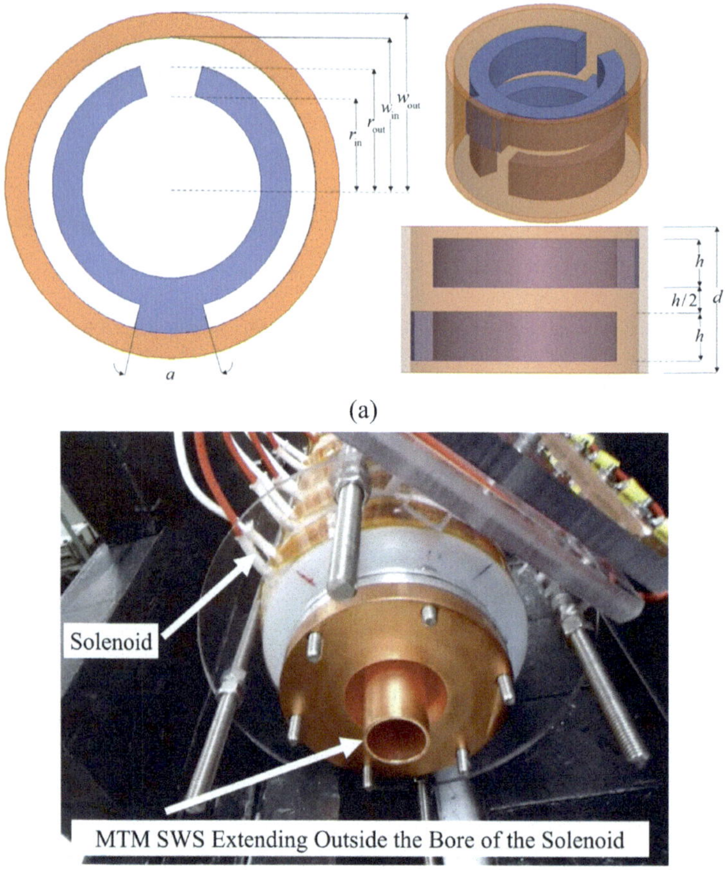

(a)

(b)

Fig. 1.18 The model of the SRR and the metamaterial-based BWO. **a** The model of the SRR. **b** The model of the metamaterial-based BWO

devices with metamaterial [90]. Research on metamaterial VEDs has also been conducted at the University of Electronic Science and Technology of China [25, 34, 91–94], which will be discussed in Chaps. 5 and 6 of this monograph.

Metamaterial, which is composed of subwavelength unit cells and arranged in a certain macroscopic sequence, have significantly expanded the fields previously covered by conventional natural materials. This remarkable feature provides metamaterial with broad prospects for various applications [95–99]. Researchers can design novel devices based on the novel characteristics of metamaterial [100–102].

In conclusion, it should be noted that the research field of metamaterial is extremely wide, encompassing optical metamaterial [103], acoustic metamaterial [104, 105], mechanical metamaterial, thermal metamaterial, and more. It is hoped that the content covered in this chapter will provide readers with new ideas for the integration of metamaterial with other fields, thereby opening doors to further cutting-edge research [106–109]. As the research on metamaterial deepens, the complexity and diversity of subwavelength unit cells become evident, which bring forth multi-dimensional, multicomponent, nonlinear, and anisotropic characteristics. The entire research field has undergone further enrichment and development. The development of metamaterial has proceeded from microwave to millimeter wave, terahertz wave, and even visible light. It has evolved from three-dimensional LHMs to nearly two-dimensional metasurfaces. The evolution also includes the transition from metal and dielectric metamaterial to all-metal metamaterial or all-dielectric metamaterial and from arrays of identical unit structures to combinations of different unit cells. Metamaterial is moving towards metasurface, intelligence, and other directions, with the potential to achieve smaller physical space, lower loss, and more flexible control of the electromagnetic wave.

References

1. Engheta N. 150 years of Maxwell's equations. Science. 2015; 349:6244–136.
2. Granatstein VL, Parker RK, Armstrong CM. Vacuum electronics at the dawn of the twenty-first century. Proceedings of the IEEE; 1999; 87:5–702.
3. Liu S. New developments in terahertz science and technology. China Basic Science. 2006 (In Chinese).
4. Veselago VG. The electrodynamics of substances with simultaneously negative values of ε and μ. Physics-Uspekhi. 1968; 10:4–509.
5. Marqués R, Martín F, Sorolla M. Metamaterials with Negative Parameters: Theory, Design, and Microwave Applications. Hoboken: John Wiley & Sons. 2011.
6. Dong YD, Itoh T. Promising future of metamaterials. IEEE Microwave Magazine. 2012; 13:2–39.
7. Ziolkowski RW. Metamaterials: The early years in the USA. EPJ Applied Metamaterials. 2014; 1:5–1.
8. Lamb H. On group-velocity. Proceedings of the London Mathematical Society. 1904; s2-1:1–473.
9. Pocklington HC. Growth of a wave-group when the group-velocity is negative. Nature. 1905; 71:1852–607.
10. Schuster A. An Introduction to the Theory of Optics. London: Edward Arnold. 1904.
11. Mandelshtam LI. Group velocity in a crystal lattice. Zhurnal Eksperimentalnoii Teoreticheskoi Fiziki. 1945; 15:9–475.
12. Malyuzhinets GD. A note on the radiation principle. Soviet Physics—Technical Physics. 1951; 21:940.
13. Sivukhin DV. The energy of electromagnetic waves in dispersive media. Optikai Spektroskopiya. 1957; 3:4–308.
14. Pafomov VE. Transition radiation and Cerenkov radiation. Soviet Physics JETP. 1959; 36:6–1321.
15. Rotman W. Plasma simulation by artificial dielectrics and parallel-plate media. IRE Transactions on Antennas and Propagation. 1962; 10:1–82.
16. Hardy WN, Whitehead LA. Split-ring resonator for use in magnetic resonance from 200-2000 MHz. Review of Scientific Instruments. 1981; 52:2–213.
17. Sievenpiper DF, Sickmiller ME, Yablonovitch E. 3D wire mesh photonic crystals. Physical Review Letters. 1996; 76:14–2480.

18. Joannopoulos JD, Johnson SG, Winn JN, et al. Photonic Crystals: Molding the Flow of Light. Second edition. Princeton: Princeton University Press. 2008.
19. Pendry JB, Holden AJ, Stewart WJ, et al. Extremely low frequency plasmons in metallic mesostructures. Physical Review Letters. 1996; 76:25–4773.
20. Pendry JB, Holden AJ, Robbins DJ, et al. Magnetism from conductors and enhanced nonlinear phenomena. IEEE Transactions on Microwave Theory and Techniques. 1999; 47:11–2075.
21. Smith DR, Kroll N. Negative refractive index in left-handed materials. Physical Review Letters. 2000; 85:14–2933.
22. Smith DR, Padilla WJ, Vier DC, et al. Composite medium with simultaneously negative permeability and permittivity. Physical Review Letters. 2000; 84:18–4184.
23. Shelby RA, Smith DR, Schultz S. Experimental verification of a negative index of refraction. Science. 2001; 292:5514–77.
24. Seddon N, Bearpark T. Observation of the inverse Doppler effect. Science. 2003; 302:5650–1537.
25. Duan ZY, Tang XF, Wang ZL, et al. Observation of the reversed Cherenkov radiation. Nature Communications. 2017; 8:1–14901.
26. Pendry JB. Negative refraction. Contemporary Physics. 2004; 45:3–191.
27. Pendry JB. Negative refraction makes a perfect lens. Physical Review Letters. 2000; 85:18–3966.
28. Liu RP, Cheng Q, Chin JY, et al. Broadband gradient index microwave quasi-optical elements based on non-resonant metamaterials. Optics Express. 2009; 17:23–21030.
29. Cui TJ, Qi MQ, Wan X, et al. Coding metamaterials, digital metamaterials and programmable metamaterials. Light: Science & Applications. 2014; 3:e218.
30. Pendry JB, Schurig D, Smith DR. Controlling electromagnetic fields. Science. 2006; 312:5781–1780.
31. Enoch S, Tayeb G, Sabouroux P, et al. A metamaterial for directive emission. Physical Review Letters. 2002; 89:21–213902.
32. Hummelt JS, Lu XY, Xu HR, et al. High power microwave generation from a metamaterial waveguide. 2016 IEEE International Vacuum Electronics Conference (IVEC). Monterey, CA, USA. 2016.
33. de Alleluia AB, Abdelshafy AF, Ragulis P, et al. Experimental testing of a 3-D-printed metamaterial slow wave structure for high-power microwave generation. IEEE Transactions on Plasma Science. 2020; 48:12–4356.
34. Wang YS, Duan ZY, Tang XF, et al. All-metal metamaterial slow-wave structure for high-power sources with high efficiency. Applied Physics Letters. 2015; 107:15–153502.
35. Valanju PM, Walser RM, Valanju AP. Wave refraction in negative-index media: Always positive and very inhomogeneous. Physical Review Letters. 2002; 88:18–187401.
36. Garcia N, Nieto-Vesperinas M. Is there an experimental verification of a negative index of refraction yet? Optics Letters. 2002; 27:11–885.
37. Garcia N, Nieto-Vesperinas M. Left-handed materials do not make a perfect lens. Physical Review Letters. 2002; 88:20–207403.
38. Pacheco J Jr, Grzegorczyk TM, Wu BI, et al. Power propagation in homogeneous isotropic frequency-dispersive left-handed media. Physical Review Letters. 2002; 89:25–257401.
39. Pendry JB, Smith DR. Comment on "Wave refraction in negative-index media: Always positive and very inhomogeneous". Physical Review Letters. 2003; 90:2–029703.
40. Ziolkowski RW. Pulsed and CW Gaussian beam interactions with double negative metamaterial slabs. Optics Express. 2003; 11:7–662.
41. Foteinopoulou S, Economou EN, Soukoulis CM. Refraction in media with a negative refractive index. Physical Review Letters. 2003; 90:10–107402.
42. Parazzoli CG, Greegor RB, Li K, et al. Experimental verification and simulation of negative index of refraction using Snell's law. Physical Review Letters. 2003; 90:10–107401.
43. Houck AA, Brock JB, Chuang IL. Experimental observations of a left-handed material that obeys Snell's law. Physical Review Letters. 2003; 90:13–137401.

44. Huangfu JT, Ran LX, Chen HS, et al. Experimental confirmation of negative refractive index of a metamaterial composed of Ω-like metallic patterns. Applied Physics Letters. 2004; 84:9–1537.
45. Yao J, Liu ZW, Liu YM, et al. Optical negative refraction in bulk metamaterials of nanowires. Science. 2008; 321:5891–930.
46. Munk BA. Metamaterials: Critique and Alternatives. Hoboken: John Wiley & Sons. 2009.
47. Kong JA. Electromagnetic Wave Theory. Hoboken: John Wiley & Sons. 1986.
48. Ritchie RH. Plasma losses by fast electrons in thin films. Physical Review. 1957; 106:5–874.
49. Ramakrishna SA. Physics of negative refractive index materials. Reports on Progress in Physics. 2005; 68:2–449.
50. Choi JW, Seo CH. Microstrip square open-loop multiple split-ring resonator for low-phase-noise VCO. IEEE Transactions on Microwave Theory and Techniques. 2008; 56:12–3245.
51. Zhang S, Fan WJ, Panoiu NC, et al. Experimental demonstration of near-infrared negative-index metamaterials. Physical Review Letters. 2005; 95:13–137404.
52. Soukoulis CM, Linden S, Wegener M. Negative refractive index at optical wavelengths. Science. 2007; 315:5808–47.
53. Liu RP, Cheng Q, Hand T, et al. Experimental demonstration of electromagnetic tunneling through an epsilon-near-zero metamaterial at microwave frequencies. Physical Review Letters. 2008; 100:2–023903.
54. Maas R, Parsons J, Engheta N, et al. Experimental realization of an epsilon-near-zero metamaterial at visible wavelengths. Nature Photonics. 2013; 7:11–907.
55. Belov PA, Slobozhanyuk AP, Filonov DS, et al. Broadband isotropic μ-near-zero metamaterials. Applied Physics Letters. 2013; 103:21–211903.
56. Silveirinha M, Engheta N. Design of matched zero-index metamaterials using nonmagnetic inclusions in epsilon-near-zero media. Physical Review B. 2007; 75:7–075119.
57. Choi MH, Lee SH, Kim YS, et al. A terahertz metamaterial with unnaturally high refractive index. Nature. 2011; 470:7334–369.
58. Chen JB, Wang Y, Jia BH, et al. Observation of the inverse Doppler effect in negative-index materials at optical frequencies. Nature Photonics. 2011; 5:4–239.
59. Duan ZY, Wu BI, Lu J, et al. Reversed Cherenkov radiation in a waveguide filled with anisotropic double-negative metamaterials. Journal of Applied Physics. 2008; 104:6–063303.
60. Chen HS, Chen M. Flipping photons backward: Reversed Cherenkov radiation. Materials Today. 2011; 14:1–34.
61. Grbic A, Eleftheriades GV. Experimental verification of backward-wave radiation from a negative refractive index metamaterial. Journal of Applied Physics. 2002; 92:10–5930.
62. Wang SB, Ng J, Xiao M, et al. Electromagnetic stress at the boundary: Photon pressure or tension? Science Advances. 2016; 2:3–e1501485.
63. Liu ZW, Lee HS, Xiong Y, et al. Far-field optical hyperlens magnifying sub-diffraction-limited objects. Science. 2007; 315:5819–1686.
64. Xu BB, Li HM, Gao SL, et al. Metalens-integrated compact imaging devices for wide-field microscopy. Advanced Photonics, 2020, 2:6–066004.
65. Liu RP, Ji C, Mock JJ, et al. Broadband ground-plane cloak. Science. 2009; 323:5912–366.
66. Qian C, Zheng B, Shen YC, et al. Deep-learning-enabled self-adaptive microwave cloak without human intervention. Nature Photonics. 2020; 14:6–383.
67. Lee Y, Hao Y. Characterization of microstrip patch antennas on metamaterial substrates loaded with complementary split-ring resonators. Microwave and Optical Technology Letters. 2008; 50:8–2131.
68. Nasser SSS, Chen ZN. Low-profile bilayer metasurface integrated antenna for cellular base stations. Sixth Asia-Pacific Conference on Antennas and Propagation (APCAP). Xi'an, China; 2017.
69. Yoo MY, Lim SJ. Polarization-independent and ultrawideband metamaterial absorber using a hexagonal artificial impedance surface and a resistor-capacitor layer. IEEE Transactions on Antennas and Propagation. 2014; 62:5–2652.

70. Xu YQ, Zhou PH, Zhang HB, et al. A wide-angle planar metamaterial absorber based on split ring resonator coupling. Journal of Applied Physics. 2011; 110:4–044102.
71. Guha R, Bandyopadhyay AK, Varshney AK, et al. Investigations into helix slow-wave structure assisted by double-negative metamaterial. IEEE Transactions on Electron Devices. 2018; 65:11–5082.
72. Starinshak DP, Wilson JD, Chevalier CT. Investigating holey metamaterial effects in terahertz traveling-wave tube amplifier. NASA/TP-2007–214701. 2007.
73. Duan ZY, Wu BI, Lu J, et al. Cherenkov radiation in anisotropic double-negative metamaterials. Optics Express. 2008; 16:22–18479.
74. Tan YS, Seviour R. Wave energy amplification in a metamaterial-based traveling-wave structure. Europhysics Letters. 2009; 87:3–34005.
75. Shiffler D, Luginsland J, French DM, et al. A Cerenkov-like maser based on a metamaterial structure. IEEE Transactions on Plasma Science. 2010; 38:6–1462.
76. Abuelfadl TM. Composite right/left-handed circular meta-waveguide. Applied Physics A. 2011; 103:759.
77. Shapiro MA, Trendafilov S, Urzhumov Y, et al. Active negative-index metamaterial powered by an electron beam. Physical Review B. 2012; 86:8–085132.
78. Shiffler D, Seviour R, Luchinskaya E, et al. Study of split-ring resonators as a metamaterial for high-power microwave power transmission and the role of defects. IEEE Transactions on Plasma Science. 2013; 41:6–1679.
79. Lu XY, Shapiro MA, Mastovsky I, et al. Generation of high-power, reversed-Cherenkov wakefield radiation in a metamaterial structure. Physical Review Letters. 2019; 122:1–014801.
80. Picard J, Mastovsky I, Shapiro MA, et al. Generation of 565 MW of X-band power using a metamaterial power extractor for structure-based wakefield acceleration. Physical Review Accelerators and Beams. 2022; 25:5–051301.
81. Yurt SC, Fuks MI, Prasad S, et al. Design of a metamaterial slow wave structure for an O-type high power microwave generator. Physics of Plasmas. 2016; 23:12–123115.
82. Liu F, Xiao L, Ye Y, et al. Integrated Cherenkov radiation emitter eliminating the electron velocity threshold. Nature Photonics. 2017; 11:5–289.
83. Li JY, Liu F, Lin YC, et al. Simulation study of terahertz Cherenkov radiation in magnesium fluoride hyperbolic material. Vacuum Electronics. 2020; 3:16 (In Chinese).
84. Lin YC, Liu F, Huang YD. Cherenkov radiation based on metamaterials. Acta Physica Sinica. 2020; 69:15–154103 (In Chinese).
85. Bai NF, Shen JX, Shen CS, et al. Metamaterial absorber structure and its application in vacuum electron devices. Vacuum Electronics. 2020; 3:25 (In Chinese).
86. Bai NF, Feng C, Liu YT, et al. Integrated microstrip meander line traveling wave tube based on metamaterial absorber. IEEE Transactions on Electron Devices. 2017; 64:7–2949.
87. Bai NF, Shen JX, Fan HH, et al. A broad bandwidth metamaterial pillbox window for W-band traveling-wave tubes. IEEE Electron Devices Letters. 2021; 42:8–1228.
88. Liu MQ, Schamiloglu E, Wang C, et al. PIC simulation of the coherent Cerenkov-Cyclotron radiation excited by a high-power electron beam in a crossed-elliptical metamaterial oscillator at S-band. IEEE Transactions on Plasma Science. 2021; 49:11–3351.
89. Liu MQ, Liu CL, Feng JJ. Study on surface wave microwave oscillator based on double-helix metamaterials. Vacuum Electronics. 2020; 3:20 (In Chinese).
90. Kong XT, He JT, Dai OZX, et al. Discussion on a novel metamaterial slow-wave structure suitable for relativistic microwave devices. Vacuum Electronics. 2020; 3:32 (In Chinese).
91. Lyu ZF, Luo HY, Wang X, et al. Compact reversed Cherenkov radiation oscillator with high efficiency. Applied Physics Letters. 2022; 120:5–053501.
92. Wang X, Li SF, Zhang XM, et al. Novel S-band metamaterial extended interaction klystron. IEEE Electron Device Letters. 2020; 41:10–1580.
93. Tang XF, Duan ZY, Ma XW, et al. Dual band metamaterial Cherenkov oscillator with a waveguide coupler. IEEE Transactions on Electron Devices. 2017; 64:5–2376.
94. Hu M, Zhong RB, Gong S, et al. Tunable free-electron-driven terahertz diffraction radiation source. IEEE Transactions on Electron Devices. 2018; 65:3–1151.

95. Smith DR, Pendry JB, Wiltshire MCK. Metamaterials and negative refractive index. Science. 2004; 305: 5685–788.
96. Solymar L, Shamonina E. Waves in Metamaterials. New York: Oxford University Press. 2009.
97. Eleftheriades GV, Balmain KG. Negative-Refraction Metamaterials: Fundamental Principles and Applications. Hoboken: John Wiley & Sons. 2005.
98. Sarychev AK, Shalaev VM. Electrodynamics of Metamaterials. Singapore: World Scientific. 2007.
99. Capolino F. Theory and Phenomena of Metamaterials. Boca Raton: CRC Press. 2017.
100. Hao Y, Mittra R. FDTD Modeling of Metamaterials: Theory and Applications. Norwood: Artech House. 2008.
101. Cui TJ, Smith DR, Liu RP. Metamaterials: Theory, Design, and Applications. New York: Springer. 2010.
102. Engheta N, Ziolkowski RW. Metamaterials: Physics and Engineering Explorations. Hoboken: John Wiley & Sons. 2006.
103. Luo XG, Subwavelength Electromagnetics. Beijing: Science Press. 2017 (In Chinese).
104. Zhu RR, Ma C, Zheng B, et al. Bifunctional acoustic metamaterial lens designed with coordinate transformation. Applied Physics Letters. 2017; 110:11–113503.
105. Lu MH, Zhang C, Feng L, et al. Negative birefraction of acoustic waves in a sonic crystal. Nature Materials. 2007; 6:10–744.
106. Shalaev VM. Optical negative-index metamaterials. Nature Photonics. 2007; 1:1–41.
107. Soukoulis CM, Kafesaki M, Economou EN. Negative-index materials: New frontiers in optics. Advanced Materials. 2006; 18:15–1941.
108. Tong LM. Frontiers in Nanophotonics Research. Shanghai: Shanghai Jiao Tong University Press. 2014 (In Chinese).
109. Li JC, Huang YQ. Time-Domain Finite Element Methods for Maxwell's Equations in Metamaterials. Berlin: Springer. 2013.

Chapter 2
Implementation and Characterization of Metamaterials

For metamaterials, to date, one of the most important research directions is how to construct them effectively since natural materials with double negative properties have not been found. The development of metamaterials can be traced back to EMT. EMT refers to an analytical method of using a continuous and uniform medium to replace non-uniform composite media, making it easier to investigate physical properties. This method was initially used to study the physical properties of composite media, including gaseous, liquid, solid, and mixed states of matters. The core idea of EMT is averaging, which defines the local physical properties of composite media through a process of averaging to create a uniform macroscopic medium. This research idea developed rapidly after it was proposed, and various related theories and approximation methods were proposed, leading to the rapid development of EMT [1]. EMT can be applied to study various physical parameters of materials, including effective permittivity, effective permeability, effective diffusion constant, effective refractive index, effective stiffness, effective thermal conductivity, effective modulus, and so on [2]. It has important scientific significance and widespread applications in electromagnetics, optics, mechanics, materials science, thermodynamics, acoustics, and more [3].

2.1 Implementation of Metamaterials

Since the LHM was proposed, it had not received widespread attentions in the academic community for over three decades. The primary reason was that LHMs had not been found in nature, and there was no implementation in experiment with artificial methods. It wasn't until the mid-to-late 1990s that J. B. Pendry et al. proposed methods for achieving the negative effective permittivity and the negative effective permeability, which marked the beginning of the LHM research. Later

Z. Duan, *Metamaterial-Based Electromagnetic Radiations and Applications*,
https://doi.org/10.1007/978-981-97-8108-9_2

on, LHMs experienced a rapid development, and multiple methods for achieving double negative property emerged, which are mainly categorized into three classes:

1. Classical combination method and its extension: This approach involves using specially designed artificial unit cells in appropriate combinations to achieve the single negative or double negative property within specific frequency bands.
2. Natural material composite method: In this approach, different natural materials are composited through physical and chemical methods to achieve single negative or double negative properties.
3. Transmission line method: Unlike the first two methods, this approach utilizes transmission line theory to achieve the reversal of the phase velocity and group velocity, thereby realizing the left-handed property.

2.1.1 Classical Combination Method and Its Extension

The double negative property within a specific frequency band was initially realized by combining a negative effective permittivity material with a negative effective permeability material. Based on the method proposed by J. B. Pendry et al., D. R. Smith et al. first realized an LHM using the classical combination method in 2000. This method creatively employed different subwavelength unit cells to separately achieve the electric and magnetic resonance response, which avoids the challenge of simultaneously achieving the electric and magnetic resonance responses using only a single metamaterial unit cell. Based on this idea of achieving the double negative property, numerous different unit cells were proposed, leading to the unprecedented development in electromagnetic metamaterials [2]. The classical combination method using metal wire arrays and SRR arrays has been described in detail in Chap. 1.

Metamaterials designed by classical combination method and its extension are most commonly found in the microwave frequency range. N. I. Landy et al. used this method to propose a metamaterial with a double negative property within a frequency range centered at 11.5 GHz, as shown in Fig. 2.1 [4]. The unit cell consists of two parts: the electric double SRR and the cut wire, as shown in Fig. 2.1a and b, respectively. The electric response was achieved by using the electric double SRR which was combined by two separate electric SRRs. The magnetic response was achieved by using the cut wire and the electric double SRR, as shown in Fig. 2.1c. The electric double SRR and the cut wire are located on opposite sides of a dielectric substrate. The electromagnetic response can be flexibly tuned by adjusting their dimensions and the thickness of the dielectric substrate. This metamaterial was applied to absorbers, and the experimental results show that the absorption peak of this metamaterial at 11.5 GHz can reach 88%, exhibiting a strong narrowband absorption characteristic.

In addition to the microwave frequency range, there has been a significant amount of the research on classical combination method and its extension in the terahertz and optical frequency range. X. Zhang's group proposed a three-dimensional metamaterial composed of a silver nanowire array with a diameter of 60 nm embedded in

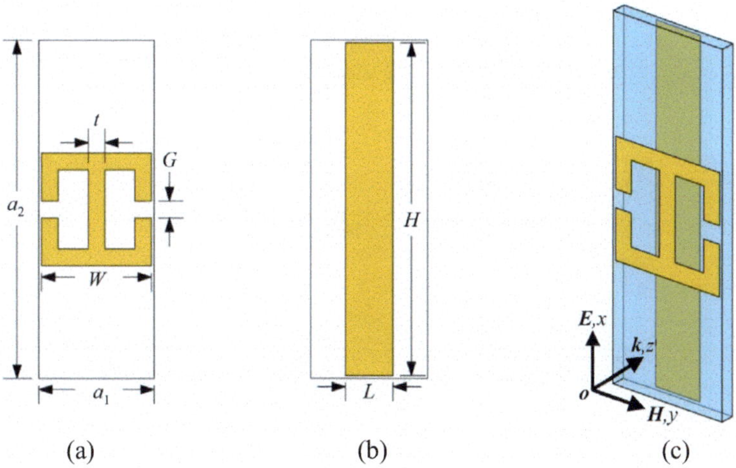

Fig. 2.1 Implementation of metamaterial using classical combination method and its extension. **a** Electric double SRR. **b** Cut wire. **c** Metamaterial unit cell

an alumina matrix, as shown in Fig. 2.2 [5]. Experimental results confirm that this three-dimensional metamaterial exhibits a negative effective refractive index in the visible light frequency range.

In addition to passive metamaterials mentioned earlier, T. Cui's group introduced active devices (such as diodes) and digital coding techniques into passive metamaterials, proposing the information metamaterials. This concept breaks free from the constraints of EMT. By combining biased diodes with specific substrate structures, they created information metamaterial unit cells that allow the real-time control of electromagnetic waves through the spatial arrangement of different digital coding sequences. In 2014, the group proposed a method using two metamaterial unit cells

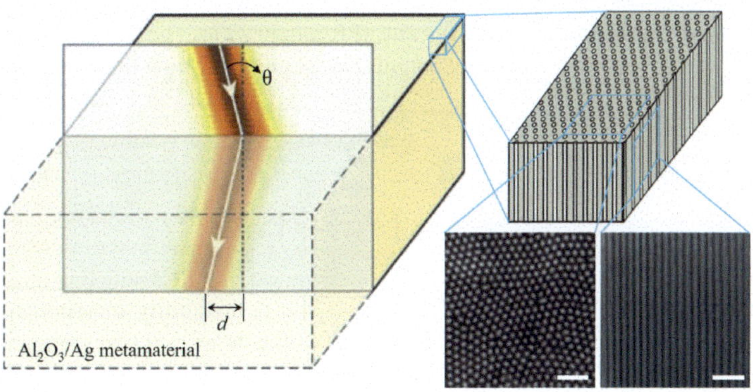

Fig. 2.2 Three-dimensional metamaterial composed of a silver nanowire array and an alumina matrix

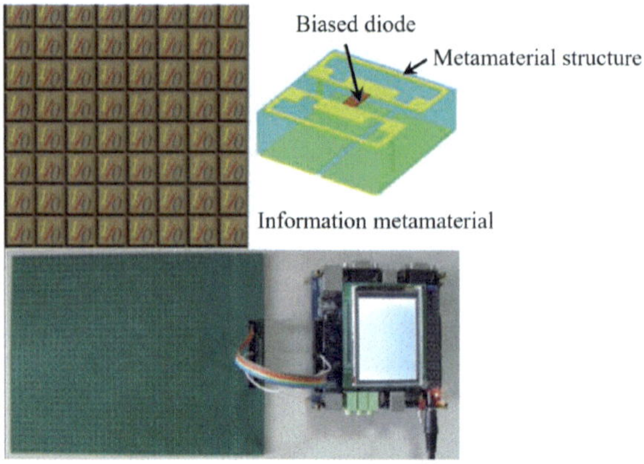

Fig. 2.3 Information metamaterial

with a 180° phase difference in electromagnetic response as digital 0 and 1 units to construct 1-bit digital coding sequences, as shown in Fig. 2.3 [6]. Due to the diversity in the spatial arrangement of 0 and 1 units, various electromagnetic control functions can be achieved.

It's important to note that metamaterials which are realized by using the classical combination method cannot be suitable for vacuum electronics. This is because the introduction of dielectric substrates is unfavorable for maintaining a high vacuum environment. Additionally, VEDs require an electron beam channel for the effective interaction between the electron beam and electromagnetic waves. Therefore, an alternative approach is needed to construct the metamaterials used in VEDs.

In 2002, R. Marqués et al. first proposed that a hollow metallic waveguide can be regarded as a one-dimensional plasma. Under the excitation of electromagnetic waves, the collective oscillation behavior of the plasma can exhibit macroscopic characteristics. According to EMT, for the TE mode, a hollow metallic waveguide can be considered as an effective electromagnetic medium with a negative effective permittivity when operating below the cutoff frequency [7]. The dispersion relations for the TE mode transmission in rectangular waveguides follow the equations:

$$k = \omega\sqrt{\mu_0 \varepsilon_{\text{eff}}}, \tag{2.1}$$

$$\varepsilon_{\text{eff}} = \varepsilon_0 \left(1 - \frac{\omega_{e0}^2}{\omega^2}\right), \tag{2.2}$$

where ε_0 is the permittivity in vacuum, μ_0 is the permeability in vacuum, ε_{eff} is the effective permittivity, and ω_{e0} corresponds to the cutoff angular frequency of a specific TE mode.

From (2.2), it can be seen that the metal wire arrays can provide the negative effective permittivity when the operating frequency is lower than the oscillation frequency of the plasma. R. Marqués et al. conducted the simulation and the experiment and found that a hollow rectangular waveguide loaded with SRRs (as shown in Fig. 2.4)

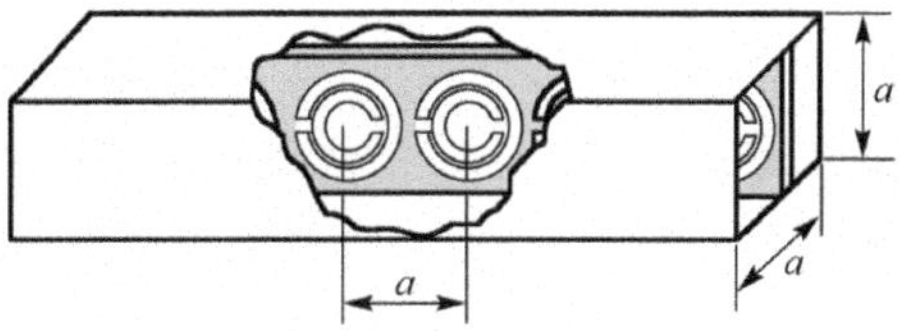

Fig. 2.4 Metamaterial constructed by loading SRRs into a hollow rectangular waveguide

can still propagate quasi-TE waves even below the cutoff frequency. It indicates that a hollow rectangular waveguide, operating below the TE mode cutoff frequency, can be regarded as a metamaterial with the negative effective permittivity. It is called the negative permittivity material here.

Building upon the work of R. Marqués et al., J. Esteban et al. further analyzed the propagation characteristics of waveguides filled with anisotropic media. They found that a hollow rectangular waveguide loaded with metamaterial unit cells could also propagate quasi-TM modes and suggested that a hollow rectangular waveguide operating below the TM mode cutoff frequency could be equivalent to a negative permeability material. The structure is shown in Fig. 2.5 [8]. The effective permeability can be predicted using the Drude model:

$$\mu_{\text{eff}} = \mu_0\left(1 - \frac{\omega_{\text{m0}}^2}{\omega^2}\right), \tag{2.3}$$

where μ_{eff} represents the effective permeability and ω_{m0} corresponds to the cutoff angular frequency of a specific TM mode of the hollow rectangular waveguide. Since it is difficult for metamaterials with a crossed-strip array to form an electron beam channel, they are not suitable for VEDs [8]. However, the research laid the theoretical and experimental foundations for developing novel metamaterials suitable for VEDs.

To address this problem, it is possible to design metamaterials with the negative permittivity separately, which have a suitable electron beam channel. These metamaterials can be combined with rectangular waveguides or circular waveguides to achieve the double negative property, serving as the high-frequency structure for

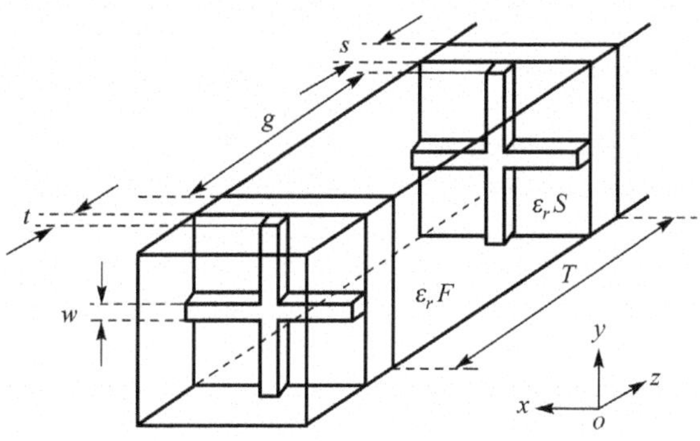

Fig. 2.5 Rectangular waveguide loaded with a crossed-strip array

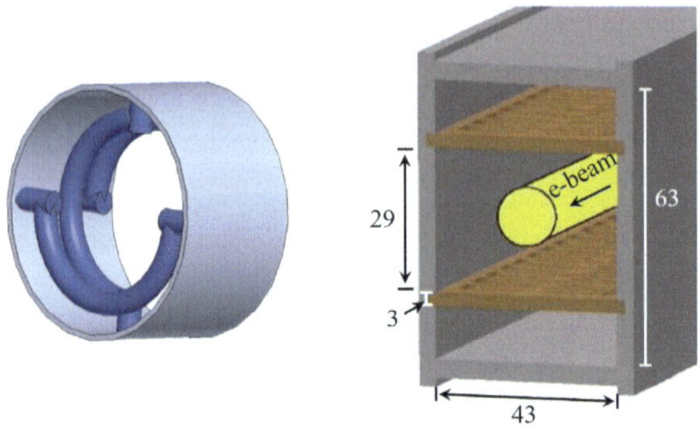

Fig. 2.6 Metamaterials suitable for VEDs. **a** Metamaterial composed of the circular waveguide loaded with the SRR. **b** Metamaterial composed of the rectangular waveguide loaded with the SRR

VEDs [9]. Based on this approach, several research groups have conducted relevant studies and proposed various metamaterials [10–12], as shown in Fig. 2.6. This implementation method will be discussed in detail in Chaps. 4 and 5 of this monograph.

2.1.2 Natural Material Composite Method

The natural material composite method is a new approach to forming metamaterials. The difference between this method and the classic combination method introduced above is that the natural material composite method is no need to design fundamental unit cells for combination. However, this method still utilizes the fundamental concept of arranging artificial atoms by regarding different natural materials, such as metals and ceramics, as these artificial atoms. Thus, the single negative or double negative characteristic is achieved through the combinations of these different natural materials.

J. Zhou's group has conducted the extensive research on the formation of metamaterials using natural materials [13]. They have made the significant progress in the field of dielectric-based and intrinsic-type metamaterials, the tunability, electromagnetic coupling, and design of metamaterials [14, 15]. In 2008, they reported a three-dimensional isotropic ceramic metamaterial [16], as shown in Fig. 2.7a. The design of this metamaterial unit cell is based on the Mie resonance theory, which suggests that if the dielectric material itself has low dielectric loss, it can exhibit the extraordinary electromagnetic response near its resonance frequency. The barium strontium titanate ceramic cubic resonator unit cells were used as artificial atoms and placed in a polymer framework, creating a three-dimensional isotropic metamaterial. Experimental results show that non-magnetic dielectric particles exhibit a strong magnetic response under the electromagnetic wave excitation. Combining such non-magnetic dielectric particles with subwavelength structures with the negative effective permittivity, such as metal wire arrays, can implement metamaterials with the double negative property. R. Fan's group used random composites to achieve a double negative property. They used a facile wet impregnation method to combine different natural dielectric materials with metallic materials in specific configurations [17], as shown in Fig. 2.7b. It is revealed that radio-frequency plasma oscillations lead to the negative effective permittivity [18]. In 2019, H. Peng's group introduced a method for creating metamaterials by combining ferromagnetic microwires and carbon fiber materials [19], as shown in Fig. 2.7c. They used the magnetic response of ferromagnetic microwires and the plasmonic response of carbon fiber arrays. Thus, a metamaterial was formed by compositing these two materials. Experimental results demonstrate that this metamaterial exhibits a double negative property from 1 to 6 GHz.

2.1.3 Transmission Line Method

In addition, metamaterials can also be realized by using the transmission line theory. On the basis of the traditional transmission line, the concept of series capacitors and parallel inductors is used to achieve the reversal of the phase velocity and the group velocity through microstrip lines or coplanar waveguides. Such one-dimensional or quasi-one-dimensional transmission line-type metamaterials are typically referred to as metalines [20].

Metalines can be employed to design novel passive devices, including new types of power dividers, antennas, couplers, filters, and more. G. V. Eleftheriades's group developed a new transmission line network by loading series capacitors and parallel inductors in a two-dimensional transmission line network. Under certain conditions, this network exhibits both negative effective permittivity and negative effective permeability [21]. Subsequently, T. Itoh's group also employed this method to realize metamaterials [22]. Transmission lines with negative characteristics are referred to as left-handed transmission lines, while traditional transmission lines are called right-handed transmission lines. Since left-handed transmission lines are constructed based on right-handed transmission lines, using combinations of capacitors and inductors, they only exhibit left-handed characteristics in a specific frequency range. Therefore, these transmission lines are called composite right/left-handed transmission lines, and their effective circuit model is depicted in Fig. 2.8. When electromagnetic waves propagate through these composite right/left-handed transmission lines, they exhibit left-handed characteristics in the low-frequency range and right-handed characteristics in the high-frequency range. In 2004, T. Itoh's group introduced the dual composite right/left-handed transmission line [23] and conducted the in-depth research on the effective circuit model of dual composite right/left-handed transmission lines. They found that these lines exhibit right-handed characteristics in the low-frequency range and left-handed characteristics in the high-frequency range, which are opposite to composite right/left-handed transmission lines.

This kind of metamaterial has several typical applications. In 2006, S. Xu's group proposed a novel balun and bandpass filter structure consisting of the composite

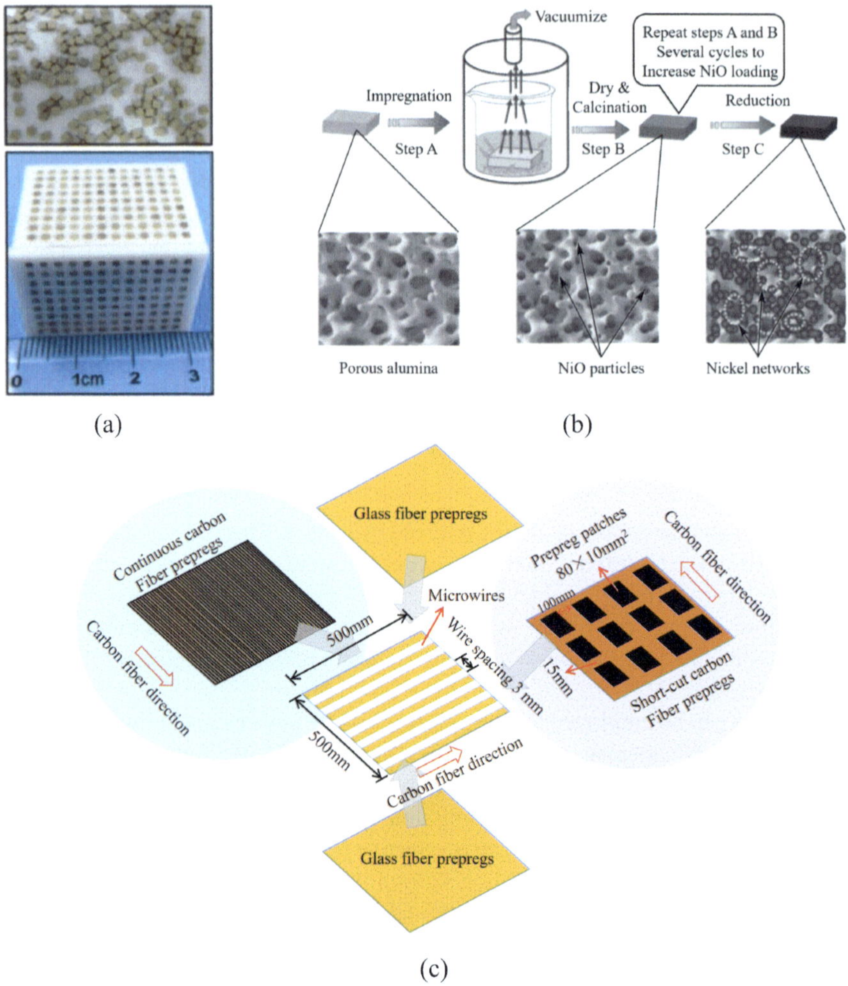

Fig. 2.7 Metamaterials prepared by using the natural material composite method. **a** The sample of ceramic metamaterials. **b** The flow chart of metamaterials prepared by the facile wet impregnation method. **c** Schematic illustration of metamaterials containing 3 mm spaced parallel Fe-based microwires

right/left-handed transmission line [24]. This device was designed to achieve a signal splitting with the equal amplitude and the reverse phase, which is suitable for dipole antennas. Meanwhile, G. A. E. Vandenbosch's group utilized composite right/left-handed transmission lines to create a reconfigurable dual-band antenna with the pattern diversity, as depicted in Fig. 2.9 [25]. By switching the dispersion curve of the composite right/left-handed transmission-line-inspired structure, the antenna can realize both a broadside pattern and an omnidirectional pattern in both the 2.4835–2.5 GHz industrial scientific medical band and the 3.4–3.5 GHz WiMAX band.

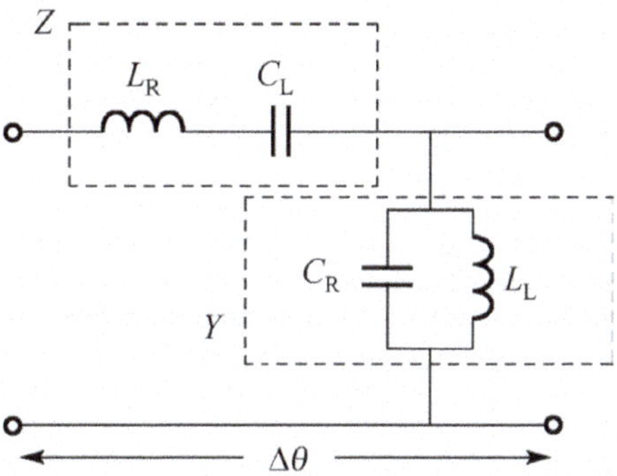

Fig. 2.8 Effective circuit model of composite right/left-handed transmission lines

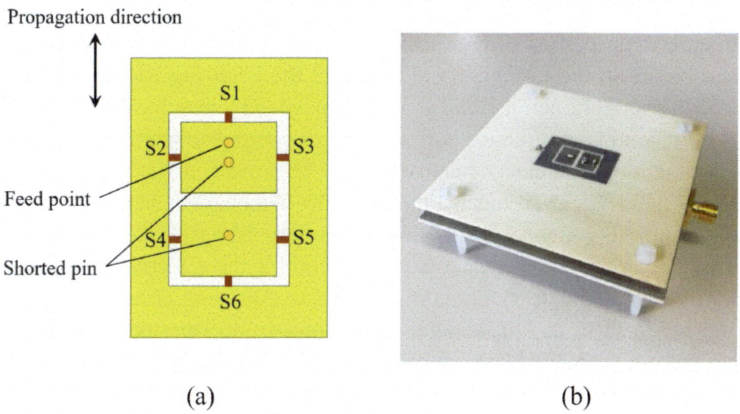

Fig. 2.9 Dual-band antenna with the pattern diversity based on composite right/left-handed transmission lines. **a** Structure diagram. **b** The sample

Another method to implement metalines is to use surface plasmon polariton (SPP) transmission lines [26]. Frequencies of the SPPs are typically very high, which can be lowered by introducing periodic structures on metal surfaces. Artificial SPP transmission lines represent a novel class of metamaterials that can finely control surface waves in the microwave, terahertz, and even optics, with configurations similar to planar circuits. These artificial SPP transmission lines usually consist of single or multiple metal wires and have also evolved into comb-shaped metal strip structures [27]. T. Cui's group used artificial SPP transmission lines to achieve a broadband microwave 3 dB power divider [28]. This concept can be extended to the terahertz and beyond.

2.2 Effective Medium Theory

EMT can be traced back to the study of the Lorentz local field. In the mid-nineteenth century, H. A. Lorentz proposed the Lorentz local field when the electrodynamic properties of macroscopic media were studied. He argued that the electromagnetic characteristics of macroscopic media must be related to the electromagnetic properties of microscopic media. Lorentz discussed the relationship of the physical quantities on the microscopic scale and the macroscopic scale and suggested that there must be a certain process of averaging involved in relating microscopic and macroscopic physical quantities. R. Clausius and O. F. Mossotti proposed the Clausius–Mossotti relation [29, 30], which describes the relationship between the polarizability of molecules and the permittivity of a medium, further clarifying the physical connection between microscopic and macroscopic scales.

At the beginning of the twentieth century, researchers started to focus on the electromagnetic properties of composite media. Various analytical methods were developed based on the Lorentz local field and the Clausius–Mossotti relation. In 1904, the Maxwell–Garnett method was introduced [31], which calculates the effective permittivity and the effective permeability of composite media by performing a specific averaging of the electromagnetic field. This method was a core component of the early EMT. However, it was limited to idealized cases where the constituents did not overlap. To enhance the applicability of the Maxwell–Garnett method, D. A. G. Bruggeman proposed a calculation method for the effective permittivity and permeability of composite media composed of overlapping constituents, which is commonly referred to as the Bruggeman method [32]. This method significantly expanded the scope of EMT. K. Lichtenecker further introduced the logarithmic mixture formula, which relates the permittivities and permeabilities of individual media to the effective permittivity and effective permeability of the composite medium when the components are uniformly mixed [33].

Initially, EMT primarily focused on the electromagnetic properties of composite media with different constituents. In the mid to late twentieth century, with the rapid development of applied electromagnetics and materials science, different periodic structures with specific shapes have emerged. For instance, in 1994, C. L. Holloway's group proposed wedge and pyramid structures for the electromagnetic wave absorption [34] and provided an initial analysis based on EMT. In 1999, F. C. Smith and colleagues introduced honeycomb-shaped absorber structures [35] and argued that EMT was not suitable for electrically large honeycomb-shaped unit cells. In 2000, EMT was used to describe honeycomb-shaped structures as an anisotropic homogeneous medium, which is more suitable for studying the electromagnetic properties of honeycomb-shaped structures [36]. These studies extended EMT from composite materials to specific electromagnetic structures, as depicted in Fig. 2.10.

Meanwhile, metamaterials also evolved in tandem. Generally, metamaterials are considered to be non-uniform and anisotropic. To accurately characterize their electromagnetic properties, Maxwell's equations must be rigorously solved. However, this process should determine some parameters such as local distributions of the

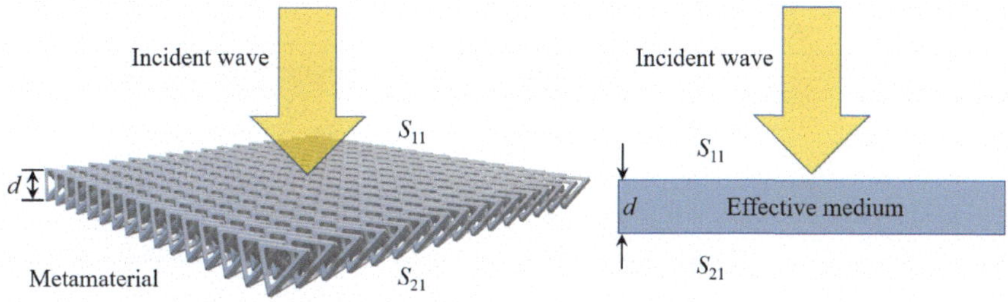

Fig. 2.10 Schematic diagram of EMT

electromagnetic field, charge density and current density, which are closely related to structural shapes, dimensions, and arrangements. Moreover, their boundary conditions are extremely complex. Especially for asymmetric unit cells, it is more difficult to obtain a solution. Therefore, one of the main scientific challenges that researchers faced at that time was which theory to use for studying metamaterials.

In 2002, D. R. Smith et al. introduced EMT into metamaterials [37]. They applied the theory of effective media to study the electromagnetic parameters of metamaterials, such as effective permittivity and effective permeability. Metamaterials were approximated as continuous and homogeneous media based on EMT. In this scenario, the electromagnetic properties of this effective medium were determined by the local distributions of the electromagnetic field, charge density, and current density within the metamaterial. Ideally, the effective homogeneous medium's response to incident electromagnetic waves was expected to be identical to the response of the original metamaterial.

Figure 2.11 presents S-parameters models of different electromagnetic media, representing homogeneous media (left), non-uniform asymmetric media (middle), and non-uniform symmetric media (right) [2]. These three types of media have different S-parameter responses to the same incident wave. For homogeneous and symmetric media: $S_{11} = S_{22}$, $S_{12} = S_{21}$; whereas for other media, $S_{11} \neq S_{22}$, or $S_{12} \neq S_{21}$. Therefore, when a metamaterial is considered as a homogeneous medium, the effective permittivity and permeability obtained from the S-parameters are approximate. It is important to note that this characterization represents the overall response of the metamaterial to electromagnetic waves.

Internationally, several research groups have used EMT to study the electromagnetic properties of metamaterials. D. R. Smith et al. proposed an S-parameters retrieval method. Using S-parameters, they derived formulas for the effective permittivity and permeability of metamaterials [2, 37]. J. A. Kong's group focused on improving the real part of the effective refractive index n in the S-parameters retrieval method in 2004 [38]. They used a mathematical iterative algorithm to expand the expression for the real part of n using a Taylor series and determined the real part of n based on the mathematical continuity. In 2005, they proposed a method for retrieving the effective electromagnetic parameters of bi-anisotropic metamaterials.

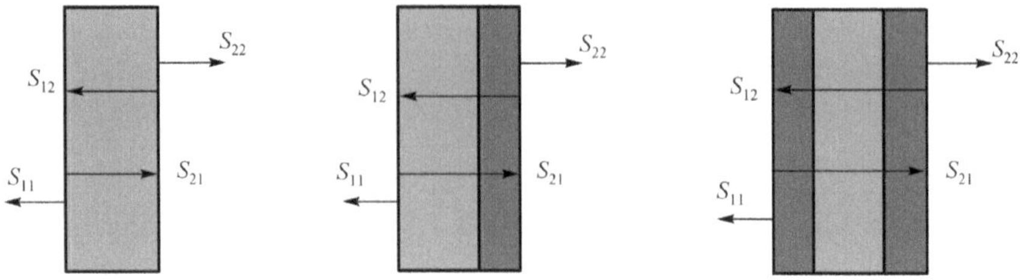

Fig. 2.11 S-parameters for different electromagnetic media

Vertically polarized and parallel polarized plane waves from three orthogonal directions were used to obtain six sets of S-parameters. Thus, effective electromagnetic parameters in different directions were obtained [39]. V. Milosevic et al. used EMT to propose a method for retrieving effective electromagnetic parameters for metamaterial unit cells that were asymmetric in the wave vector direction. They equated the asymmetric metamaterial unit cell to a bi-anisotropic medium and studied its electromagnetic properties [40]. D. Cohen and colleagues extended the conditions of the vertical incidence in the S-parameters retrieval method to the oblique incidence, as shown in Fig. 2.12 [41].

Many studies have shown that metamaterials with unit cell's dimensions smaller than one-tenth of the operating wavelength can be characterized with the effective permittivity and permeability [2]. EMT provides a unique physical approach to analyzing the electromagnetic properties of metamaterials. Although it is an approximation, it has a significant practical value.

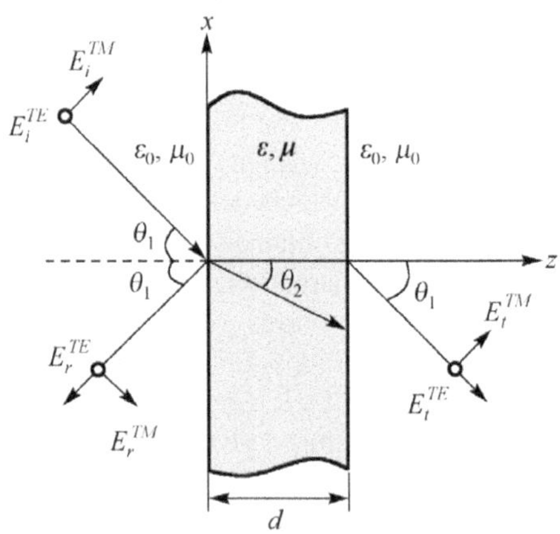

Fig. 2.12 S-parameters retrieval method for the conditions of the oblique incidence

2.3 Methods of Retrieving Effective Electromagnetic Parameters

Metamaterials are artificial subwavelength unit cell arrays, which are distinct from traditional natural materials. It is more accurate to consider them as a new structure rather than a new material. To characterize the electromagnetic properties of metamaterials, it is necessary to retrieve their effective electromagnetic parameters, such as the effective permittivity and effective permeability. For natural materials, these electromagnetic properties can be measured experimentally using traditional methods such as the free-space method [42], cavity perturbation method [43], coaxial discontinuity method [44], and waveguide method [45]. However, for metamaterials, the practical implementation of these methods is quite complex, with numerous affecting factors, making it inconvenient to rapidly retrieve their electromagnetic properties. Currently, the effective electromagnetic parameters of metamaterials are obtained by simulation, which provides a theoretical basis for designing new metamaterials.

There are four main methods for retrieving the effective electromagnetic parameters of metamaterials: First, the S-parameters retrieval method [37, 46]; Second, the electromagnetic field averaging method [47]; Third, the quasimode theory [48]; Fourth, the electromagnetic wave propagation retrieval method [49]. The first method uses the S-parameters of the medium to retrieve the effective electromagnetic parameters. Its advantage lies in quickly obtaining the S-parameters of metamaterials through the simulation and subsequently retrieving their effective electromagnetic parameters using corresponding inversion algorithms. The second method is theoretically simple but the process of averaging the electromagnetic fields is quite complex, especially for asymmetric metamaterials. Additionally, different field averaging methods significantly affect the results of the parameter retrieval. The third method can be described as embedding the metamaterial into a reference medium with the freely adjustable permittivity and permeability. Green's functions are applied to study the physical quantities such as the reference medium's self-energy, density of states, and mean free path of electromagnetic waves propagating in the medium. Then, the parameters of the reference medium are changed to maximize the density of states function to determine the effective permittivity and permeability of the metamaterial. The fourth method can obtain effective electromagnetic parameters by analyzing the transmission characteristics of the electromagnetic waves. In this approach, analytical results are used to approximate the simulated electromagnetic fields within the metamaterial. The effective electromagnetic parameters are determined by analyzing the changes in the amplitude and phase of the electromagnetic waves after passing through the metamaterial.

The S-parameters retrieval method proposed by D. R. Smith et al. is simple, feasible, and provides relatively reliable results, making it widely employed [50]. With this method, the S-parameters of metamaterials can be conveniently obtained through the simulation or experimental measurements. Next, the effective electromagnetic parameters can be calculated using the corresponding mathematics.

2.3.1 Basic Principle for Retrieving S-Parameters

Assume that the metamaterial can be equivalent to a continuous, symmetric, and homogeneous electromagnetic medium, its one-dimensional transmission matrix is defined as follows [2]:

$$\boldsymbol{F}' = \boldsymbol{T}\boldsymbol{F}, \tag{2.4}$$

where $\boldsymbol{T}$ represents the transmission matrix of this effective medium, $\boldsymbol{F}$ is the electromagnetic field incident on incident wave side of the effective medium, $\boldsymbol{F}'$ is the electromagnetic field propagated to the outgoing wave side of the effective medium, and let:

$$\boldsymbol{F} = \begin{pmatrix} \boldsymbol{E}_1 \\ \boldsymbol{H}_1 \end{pmatrix}, \boldsymbol{F}' = \begin{pmatrix} \boldsymbol{E}_2 \\ \boldsymbol{H}_2 \end{pmatrix}, \tag{2.5}$$

where $\boldsymbol{E}_1$ and $\boldsymbol{H}_1$ are the electric and magnetic fields on the incident wave side of the effective medium, respectively. $\boldsymbol{E}_2$ and $\boldsymbol{H}_2$ are the electric and magnetic fields on the outgoing wave side of the effective medium, respectively.

Suppose a continuous, symmetric, and homogeneous effective medium has a thickness d in the direction of electromagnetic wave propagation (z-direction), which is the size of the metamaterial in the direction of the electromagnetic wave propagation, as shown in Fig. 2.10. For the effective medium, the expression of the electromagnetic fields in the direction of the electromagnetic wave propagation can be derived from Maxwell's equations as follows:

$$\begin{cases} E(z) = \cos(\omega\sqrt{\mu\varepsilon}z)E(0) + \mathrm{j}\sqrt{\frac{\mu}{\varepsilon}}\sin(\omega\sqrt{\mu\varepsilon}z)H(0) \\ H(z) = \mathrm{j}\sqrt{\frac{\varepsilon}{\mu}}\sin(\omega\sqrt{\mu\varepsilon}z)E(0) + \cos(\omega\sqrt{\mu\varepsilon}z)H(0) \end{cases}. \tag{2.6}$$

Assume that it is a time-harmonic electromagnetic wave with the form $\mathrm{e}^{-\mathrm{j}\omega t}$, to simplify the transmission matrix, let $H' = -\mathrm{j}\omega\mu_0 H$, which results in the transmission matrix $\boldsymbol{T}$:

$$\boldsymbol{T} = \begin{bmatrix} \cos(nkd) & -\frac{z_w}{k}\sin(nkd) \\ \frac{k}{z_w}\sin(nkd) & \cos(nkd) \end{bmatrix}, \tag{2.7}$$

where n is the effective refractive index of the metamaterial, k is the wave number of the electromagnetic wave in a vacuum, and z_w is the relative wave impedance of this effective medium. The relationship between the relative effective permittivity ε_{eff} and the relative effective permeability μ_{eff} with respect to n and z_w is given:

$$\varepsilon_{\text{eff}} = \frac{n}{z_w}, \quad \mu_{\text{eff}} = nz_w. \tag{2.8}$$

Note that n and z_w discussed here are both complex numbers. From (2.8), it can be seen that the problem of solving the effective electromagnetic parameters is transformed into the problem of solving n and z_w. The next step is to use S-parameters to solve n and z_w. The relationship between the S-parameters matrix and the T-parameter matrix can be obtained from the transmission matrix theory:

$$\begin{cases} S_{21} = \dfrac{2}{T_{11}+T_{22}+\left(\mathrm{j}kT_{12}+\frac{T_{21}}{\mathrm{j}k}\right)} \\ S_{11} = \dfrac{T_{11}-T_{22}+\left(\mathrm{j}kT_{12}-\frac{T_{21}}{\mathrm{j}k}\right)}{T_{11}+T_{22}+\left(\mathrm{j}kT_{12}+\frac{T_{21}}{\mathrm{j}k}\right)} \\ S_{22} = \dfrac{T_{22}-T_{11}+\left(\mathrm{j}kT_{12}-\frac{T_{21}}{\mathrm{j}k}\right)}{T_{11}+T_{22}+\left(\mathrm{j}kT_{12}+\frac{T_{21}}{\mathrm{j}k}\right)} \\ S_{12} = \dfrac{2\det(\boldsymbol{T})}{T_{11}+T_{22}+\left(\mathrm{j}kT_{12}+\frac{T_{21}}{\mathrm{j}k}\right)} \end{cases}. \tag{2.9}$$

Since the effective medium is assumed to be a homogeneous medium, it follows that $T_{11} = T_{22}$, and det $(\boldsymbol{T}) = 1$. For the convenience of later calculation, let $T_s = T_{11} = T_{22}$. Since the S-parameters matrix is symmetric, we have:

$$\begin{cases} S_{21} = S_{12} = \dfrac{1}{T_s+\frac{1}{2}\left(\mathrm{j}kT_{12}+\frac{T_{21}}{\mathrm{j}k}\right)} \\ S_{11} = S_{22} = \dfrac{\frac{1}{2}\left(\mathrm{j}kT_{12}-\frac{T_{21}}{\mathrm{j}k}\right)}{T_s+\frac{1}{2}\left(\mathrm{j}kT_{12}+\frac{T_{21}}{\mathrm{j}k}\right)} \end{cases}. \tag{2.10}$$

Substituting (2.7) into (2.10), we can obtain:

$$\begin{cases} S_{21} = S_{12} = \dfrac{1}{\cos(nkd)-\frac{\mathrm{j}}{2}\left(z_w+\frac{1}{z_w}\right)\sin(nkd)} \\ S_{11} = S_{22} = \dfrac{\frac{\mathrm{j}}{2}\left(\frac{1}{z_w}-z_w\right)\sin(nkd)}{\cos(nkd)-\frac{\mathrm{j}}{2}\left(z_w+\frac{1}{z_w}\right)\sin(nkd)} \end{cases}. \tag{2.11}$$

By solving the above equations inversely, we can obtain n and z_w, which are expressed in terms of S-parameters:

$$\begin{cases} n = \dfrac{1}{kd}\arccos\left[\dfrac{1}{2S_{21}}\left(1 - S_{11}^2 + S_{21}^2\right)\right] \\ z_w = \sqrt{\dfrac{(1 + S_{11})^2 - S_{21}^2}{(1 - S_{11})^2 - S_{21}^2}} \end{cases}. \tag{2.12}$$

In this way, using (2.8) and (2.12), we can determine ε_{eff} and μ_{eff}:

$$\begin{cases} \varepsilon_{\text{eff}} = \dfrac{\arccos\left[\frac{1}{2S_{21}}\left(1 - S_{11}^2 + S_{21}^2\right)\right]}{kd\sqrt{\frac{(1+S_{11})^2 - S_{21}^2}{(1-S_{11})^2 - S_{21}^2}}} \\ \mu_{\text{eff}} = \dfrac{1}{kd}\arccos\left[\dfrac{1}{2S_{21}}\left(1 - S_{11}^2 + S_{21}^2\right)\right]\sqrt{\dfrac{(1+S_{11})^2 - S_{21}^2}{(1-S_{11})^2 - S_{21}^2}} \end{cases}. \tag{2.13}$$

By solving the (2.13), we can obtain ε_{eff} and μ_{eff}. Based on the theoretical analysis above, the problem of solving the effective electromagnetic parameters of metamaterials can be transformed into the problem of solving S-parameters. From (2.12), it can be observed that obtaining n requires solving the inverse cosine function. Since the inverse cosine function solution has multiple values, it is difficult to uniquely determine the real part of n, and multiple values arise from the periodicity of trigonometric functions, meaning multiple n values correspond to the same S-parameters. Thus, n can take the following form:

$$n = n_0 + \frac{2\pi m}{kd}, \tag{2.14}$$

where n_0 is any solution that satisfies (2.12), and m is an integer.

Due to the complex solutions for n, the complex effective refractive index $N_{\text{eff}}(\omega)$ can be defined as $N_{\text{eff}}(\omega) = n_{\text{eff}}(\omega) + \text{j}k_{\text{eff}}(\omega)$. Here, to simplify the computation, let $R_{01} = (Z_{\text{eff}} - 1)/(Z_{\text{eff}} + 1)$, where $k_{\text{eff}}(\omega)$ is the extinction coefficient, and $n_{\text{eff}}(\omega)$ is the effective refractive index. In this context, $N_{\text{eff}}(\omega)$ and Z_{eff} correspond to n and z_w mentioned in the S-parameters retrieval method. Therefore, (2.11) can be transformed into the following equations:

$$\begin{cases} S_{11} = \frac{R_{01}\left(1 - e^{j2N_{\text{eff}}kd}\right)}{1 - R_{01}^2 e^{j2N_{\text{eff}}kd}} \\ S_{21} = \frac{\left(1 - R_{01}^2\right)e^{jN_{\text{eff}}kd}}{1 - R_{01}^2 e^{j2N_{\text{eff}}kd}} \end{cases}. \tag{2.15}$$

Similarly, from (2.15), we can obtain:

$$Z_{\text{eff}} = \pm\sqrt{\frac{(1+S_{11})^2 - S_{21}^2}{(1-S_{11})^2 - S_{21}^2}}, \tag{2.16}$$

$$e^{jN_{\text{eff}}kd} = \frac{S_{21}}{1 - S_{11}R_{01}}, \tag{2.17}$$

where, due to $\left|e^{jN_{\text{eff}}kd}\right| \le 1$, the sign of the relative wave impedance Z_{eff} can be determined in (2.16). Thus, Z_{eff} can be uniquely determined by the S-parameters. According to (2.14), the complex effective refractive index can be expressed as:

$$N_{\text{eff}} = \frac{1}{kd}\left\{\text{Im}\left[\ln\left(e^{jN_{\text{eff}}kd}\right)\right] + 2m\pi - \text{j}\text{Re}\left[\ln\left(e^{jN_{\text{eff}}kd}\right)\right]\right\}. \tag{2.18}$$

In this equation, multiple values of m lead to uncertainty in the effective refractive index. The real part (called effective refractive index) and the imaginary part (called extinction coefficient) of the complex effective refractive index are denoted as follows:

$$n_{\text{eff}} = \frac{\text{Im}\left[\ln\left(e^{jN_{\text{eff}}kd}\right)\right]}{kd} + \frac{2m\pi}{kd} = n_{\text{eff}}^{0} + \frac{2m\pi}{kd}, \tag{2.19}$$

$$k_{\text{eff}} = \frac{-\text{Re}\left[\ln\left(e^{jN_{\text{eff}}kd}\right)\right]}{kd}. \tag{2.20}$$

n_{eff}^{0} is the effective refractive index corresponding to the principal branch of the logarithmic function. It can be seen from (2.20) that the imaginary part of the complex effective refractive index is independent of m.

Since the Kramers–Kronig relations can relate the real and imaginary parts of a complex variable [51], to avoid the uncertainty in m, the Kramers–Kronig relations are employed to determine the real part based on the imaginary part, thereby determining the value of m. This approach ensures the continuity of the effective refractive index with frequency, uniquely determining the real part of the effective refractive index, i.e., determining the complex effective refractive index N_{eff}. Here, the definition of its real part using the Kramers–Kronig relations is listed:

$$n_{\text{eff}}\left(\omega'\right) = 1 + \frac{2}{\pi} P \int_{0}^{\infty} \frac{\omega k_{\text{eff}}(\omega)}{\omega^2 - \omega'^2} \mathrm{d}\omega, \tag{2.21}$$

where P represents the principal value of the inverse integral, also known as the Cauchy principal value. In the cases where the integral range extends from 0 to ∞, which is physically unrealizable, thus an approximate algorithm with integral truncation should be used.

Using the trapezoidal rule for (2.21), the integral is divided into two parts:

$$\begin{cases} \Psi_{i,j} = \frac{\omega_j k_{\text{eff}}\left(\omega_j\right)}{\omega_j^2 - \omega_i^2} + \frac{\omega_j + k_{\text{eff}}\left(\omega_{j+1}\right)}{\omega_{j+1}^2 - \omega_i^2} \\ n_{\text{eff}}(\omega_i) = 1 + \frac{\Delta\omega}{\pi}\left(\sum\limits_{j=1}^{i-2} \Psi_{i,j} + \sum\limits_{j=i+1}^{N-1} \Psi_{i,j}\right) \end{cases}. \tag{2.22}$$

According to (2.19), we obtain:

$$m = \text{Round}\left[\left(n_{\text{eff}} - n_{\text{eff}}{}^{0}\right)\frac{kd}{2\pi}\right], \tag{2.23}$$

where the Round () function denotes the nearest integer to the variable on the number axis. Once m is determined, the refractive index n is also uniquely determined. At this point, both the effective refractive index and the wave impedance are determined, and they can be substituted back into (2.8) to uniquely determine the effective permittivity and permeability of the metamaterial.

It should be noted that the method based on EMT is obviously an approximate characterization of metamaterials. EMT requires that the dimensions of the metamaterial unit cell are smaller than one-tenth of the operating wavelength. However, in practically realizable metamaterials, the dimensions of unit cells typically lie between one-sixth and one-half of the operating wavelength, and it is difficult to achieve unit cell dimensions smaller than one-tenth of the operating wavelength. Therefore, EMT inevitably leads to an approximation in the S-parameters retrieval method. To obtain more accurate electromagnetic characteristics, full-wave simulations and experimental studies for metamaterials are required. Although the retrieved electromagnetic parameters are approximate, they can still provide the valuable information for the initial design of metamaterials to some extent.

2.3.2 S-Parameters Retrieval in Free Space

In the S-parameters retrieval method in free space, it is assumed that the metamaterial is placed in free space with electromagnetic waves vertically incident upon it. Its effective electromagnetic parameters can be obtained using the S-parameter retrieval method introduced above. Based on the Babinet principle, Z. Duan et al. creatively proposed a complementary electric SRR (CeSRR) unit cell, as shown in Fig. 2.13, and its size parameters are listed in Table 2.1 [52].

Assuming a plane wave propagating in the y-direction with polarization in the z-direction is vertically incident upon the CeSRR surface, and neglecting the Ohmic loss of the metal, then the relative permittivity and relative permeability have only real parts. The HFSS Driven-Modal Solver was used to obtain the S-parameters of this CeSRR, with their amplitude and phase shown in Fig. 2.14. It can be observed

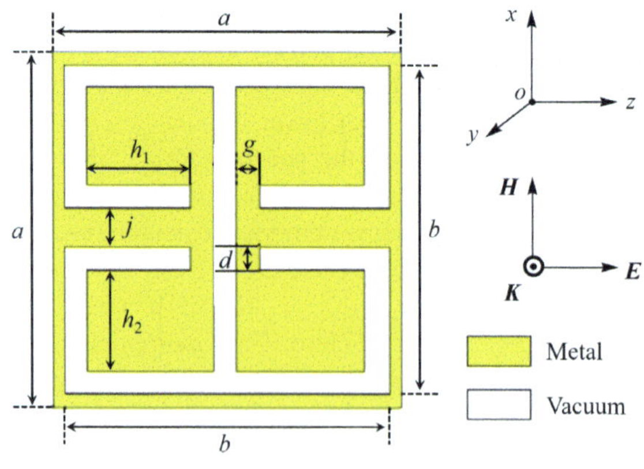

Fig. 2.13 A Planar metamaterial unit cell

Table 2.1 Size parameters of the planar metamaterial unit cell

Parameter	Value/mm
a	14.5
b	13.5
d	1
h_1	4.25
h_2	4
g	1
j	1.5

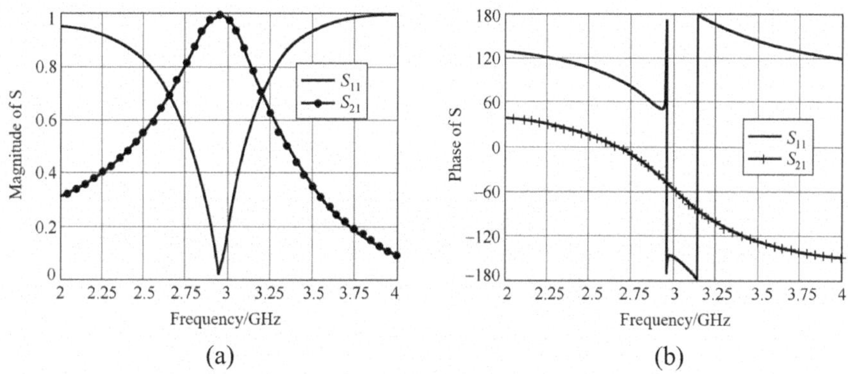

Fig. 2.14 S-parameters *vs.* frequency. **a** Magnitude. **b** Phase

that the CeSRR exhibits a strong electric resonance near 3 GHz and demonstrates good transmission characteristics.

Using the S-parameters retrieval method, the effective electromagnetic parameters ε_{zz} and μ_{xx} of this CeSRR are obtained, as shown in Fig. 2.15a. It can be seen that the planar CeSRR exhibits a negative effective permittivity in the frequency range of 2–2.82 GHz. The effective permittivity of the CeSRR results from the Drude response (the effective plasma frequency is 2.82 GHz) and the Lorentz response (the resonance frequency around 3.35 GHz) [53]. Due to the excellent symmetry of this structure, the internal symmetric ring current eliminates the magnetic response, and the effective relative permeability μ_{xx} is close to 1 in the frequency range of 2.4 GHz [52]. By changing the polarization direction of the incident wave to E_x (as shown in Fig. 2.13), another set of S-parameters is determined, resulting in effective permittivity ε_{xx} and effective permeability μ_{zz}, as shown in Fig. 2.15b. It can be observed that ε_{xx} is negative in the frequency range of 2.4 GHz, while μ_{zz} remains approximately equal to 1 in the same frequency range.

According to Sect. 2.2.1, CeSRR unit cells are periodically arranged along the z-direction with a period of a and are loaded in a hollow square waveguide with a

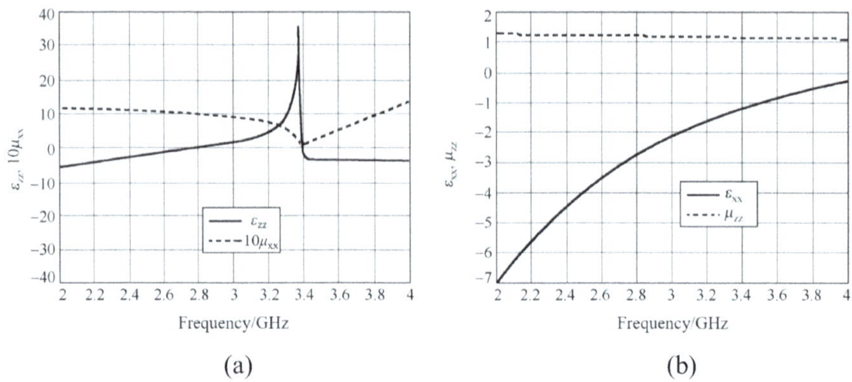

Fig. 2.15 The retrieved effective electromagnetic parameters under two different polarization conditions: **a** z-direction and **b** x-direction

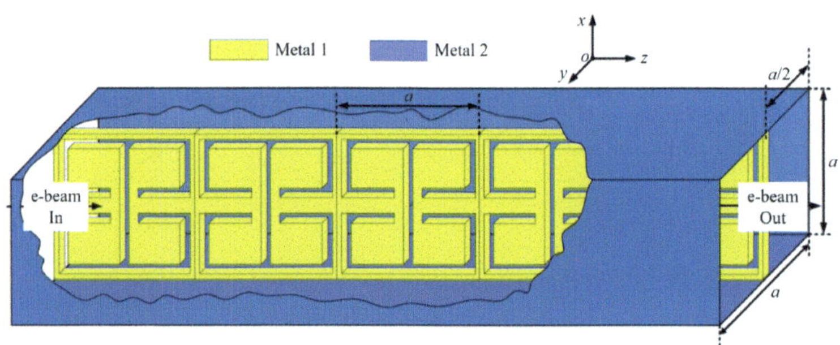

Fig. 2.16 MSWS constructed with CeSRRs

cross-sectional size of $a \times a$. Thus, a new metamaterial slow-wave structure (MSWS) is formed, as shown in Fig. 2.16.

Simulations are conducted for this structure by using the HFSS eigenmode solver. Figure 2.17a shows the distribution of longitudinal electric field E_z within one period of the MSWS. Figure 2.17b shows the amplitude of E_z in the y-direction at 3.07 GHz. It can be observed that the electric field intensity is stronger near the CeSRR. The magnetic field is attenuated near the CeSRR and is zero at the waveguide wall. At $y = 0$ and $y = a$, the electric field magnitude is very small, and the magnetic field magnitude is zero. The waveguide wall has little effect on the field distribution. Simulations of the dispersion characteristics using HFSS reveal that changing the boundary conditions from electric boundaries to magnetic boundaries at the waveguide wall does not alter the dispersion characteristics.

Assuming that the electromagnetic field does not depend on the position in y-direction and that the effective permittivity $\boldsymbol{\varepsilon}$ is anisotropic while the effective permeability μ is isotropic, their expressions are given as follows:

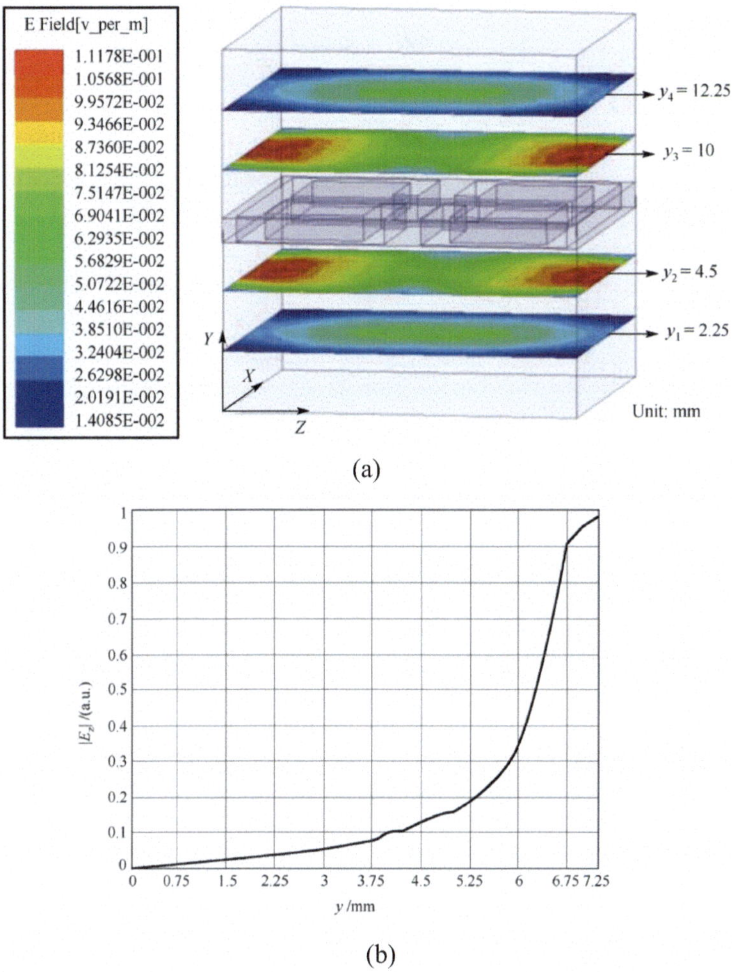

Fig. 2.17 **a** The distribution of electric field E_z within one period of the MSWS. **b** The amplitude of electric field E_z in the y-direction

$$\boldsymbol{\varepsilon} = \varepsilon_0 \begin{pmatrix} \varepsilon_{xx} & 0 & 0 \\ 0 & 1 & 0 \\ 0 & 0 & \varepsilon_{zz} \end{pmatrix}, \ \mu = \mu_0, \tag{2.24}$$

where ε_0 and μ_0 are the permittivity and permeability of free space, respectively.

Assuming a time factor of $e^{j\omega t}$ (where ω is the angular frequency), the two curl equations of Maxwell's equations can be written as:

$$\nabla \times \boldsymbol{E} = -j\omega\mu \cdot \boldsymbol{H}, \tag{2.25}$$

$$\nabla \times \boldsymbol{H} = \mathrm{j}\omega \boldsymbol{\varepsilon} \cdot \boldsymbol{E}. \tag{2.26}$$

Also, assuming that the electromagnetic wave propagates along the z-direction with a relationship to z as $\mathrm{e}^{-\mathrm{j}\beta z}$, where β is the phase constant. Substituting (2.24) into (2.25) and using the separation of variables method, the scalar wave equation for the E_z component can be obtained:

$$\left(\frac{\beta^2}{\omega^2 \varepsilon_{xx}/c_0^2 - \beta^2} + 1\right)\frac{\partial^2 E_z}{\partial x^2} + \frac{\omega^2}{c_0^2}\varepsilon_{zz} E_z = 0. \tag{2.27}$$

Using the boundary conditions, one can derive:

$$E_z(x, z)|_{x=0,a} = 0, \tag{2.28}$$

which leads to the dispersion equation for this MSWS:

$$\frac{(\pi/a)^2}{\varepsilon_{zz}} + \frac{\beta^2}{\varepsilon_{xx}} = \frac{\omega^2}{c_0^2}, \tag{2.29}$$

where ε_{zz} and ε_{xx} are functions of ω.

For this MSWS, the dispersion characteristics were obtained using the HFSS eigenmode solver, as shown in the Simulation curve in Fig. 2.18. Additionally, the dispersion relationship obtained from (2.28) is shown in the Theory curve in Fig. 2.18. It can be observed that the overall trends predicted by both curves are quite similar, which indicates that the dispersion relationship of this MSWS represents the backward wave (opposite phase and group velocities), which is caused by negative ε_{xx} and positive ε_{zz}.

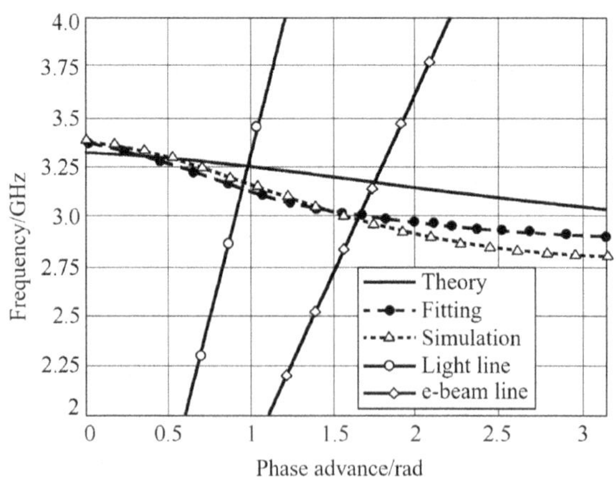

Fig. 2.18 Dispersion curves obtained via theory and simulation

It's important to note that the frequency range of the double negative property of this metamaterial does not exactly coincide with the passband of the MSWS. Two main factors contribute to this: (1) The propagation mode in the MSWS composed of CeSRR arrays loaded in a rectangular waveguide is not a standard TM mode but a quasi-TM mode. Therefore, it cannot accurately reflect the electric response of CeSRRs. (2) The S-parameters retrieval method in free space characterizes the effective electromagnetic parameters using a single CeSRR as a substitute for the effective electromagnetic parameters of the metamaterial composed of CeSRRs. It does not account for the mutual coupling between adjacent CeSRRs and the influence of the rectangular waveguide. Furthermore, most existing metamaterials are anisotropic subwavelength electromagnetic structures and their effective permittivity and permeability can change with the change of direction of the excitation electric field or excitation magnetic field. Typically, it needs to achieve the negative effective permittivity or negative effective permeability in a specific direction. Therefore, the effective permittivity tensor and the effective permeability tensor are used to describe the anisotropic metamaterials. In such cases, S-parameters in different directions can be obtained by changing the directions of the applied electric field and magnetic field. The effective permittivity and the effective permeability in different directions can be obtained by using the aforementioned S-parameters retrieval method in free space.

In addition to the study of anisotropic metamaterials, some researchers have focused on the electromagnetic properties and retrieval of effective electromagnetic parameters of bi-anisotropic metamaterials. Bi-anisotropic metamaterials are characterized by both the electric displacement vector and the magnetic induction intensity varying with the intensities of the electric field and magnetic field. They exhibit polarization and magnetization phenomena under the influence of the electric field or magnetic field [54]. I. O. Vardiambasis et al. conducted the theoretical research on the propagation characteristics of electromagnetic waves in bi-anisotropic metamaterials [55]. In 2005, J. A. Kong's group employed the S-parameters retrieval method to retrieve the effective electromagnetic parameters of bi-anisotropic metamaterials [39]. The bi-anisotropic metamaterial consisting of SRRs was used to retrieve the effective electromagnetic parameters [56].

2.4 Overview of Metasurfaces

2.4.1 Basic Concept and Development

Metamaterials, as a new type of artificially electromagnetic structure, possess unique electromagnetic properties and have widespread applications in various research fields [57–60]. However, the development of three-dimensional metamaterials has been hindered by their structural complexity, manufacturing challenges, large volume, high loss, narrow bandwidth, and so on. To address these issues and reduce the thickness and complexity of metamaterials, metasurfaces have emerged as a solution [61].

Metasurfaces can be seen as the two-dimensional metamaterials, which consist of artificial subwavelength unit cells arranged periodically or non-periodically on a plane or surface. Metasurfaces enable the realization of attractive electromagnetic responses on a two-dimensional scale, which are difficult to achieve with traditional natural or composite materials. Similar to three-dimensional metamaterials, metasurfaces can manipulate electromagnetic wave properties. Metasurfaces have the thinner thickness, smaller volume, and lighter weight, which are more conducive to the miniaturization and integration with devices. Additionally, metasurfaces can avoid the challenges associated with the fabrication and assembly of three-dimensional metamaterials. More importantly, three-dimensional metamaterials, especially all-metal metamaterials, have high loss for terahertz and optical waves, limiting their development in higher frequency range. In contrast, metasurfaces, with simpler and thinner structure, exhibit lower electromagnetic loss in the higher frequency range. At present, metasurfaces have become one of the mainstream development directions of metamaterials, with broad prospects in scientific and engineering fields such as acoustics, mechanics, electromagnetism, optics, and thermodynamics.

In fact, the research on artificial electromagnetic surfaces dates back several decades. In 1962, K. M. Siegel proposed the idea of using two-dimensional periodic surface structures for the frequency selection while studying radar echoes [62]. From 1971 to 1983, B. A. Munk extensively investigated the transmission, reflection, and impedance properties of two-dimensional periodic surface structures [63–65]. In 1999, E. Yablonovitch introduced a high-impedance surface (HIS) with a mushroom-like structure [66]. This structure can serve as a bandgap structure of a two-dimensional artificial electromagnetic surface, and allow for the modulation of its dispersion curve by varying the geometric dimensions of the mushroom-like unit cell, as shown in Fig. 2.19.

In 2003, C. L. Holloway et al. used the metafilm [67] and D. F. Sievenpiper et al. employed the metasurface to describe two-dimensional metamaterial [68]. The

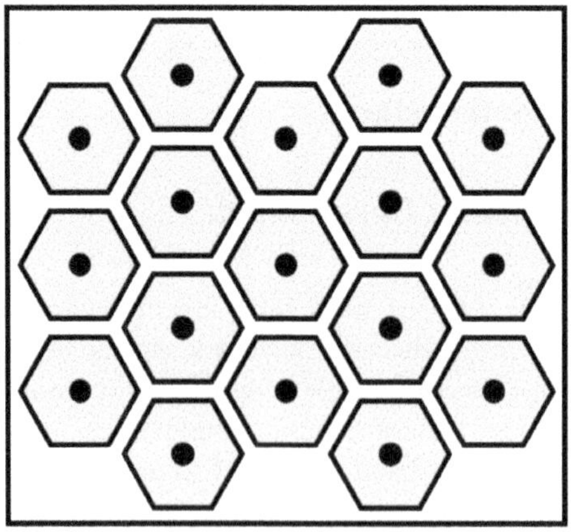

Fig. 2.19 High impedance surface

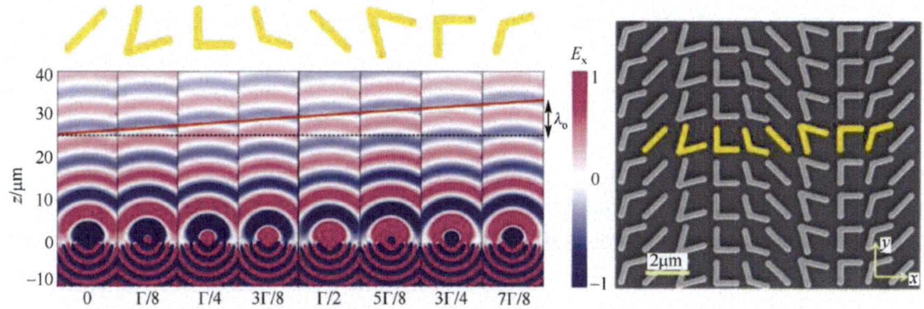

Fig. 2.20 Metasurface based on the generalized Snell's law

Clausius–Mossotti relationship from EMT is used to approximate the electromagnetic properties of these two-dimensional metamaterials [29, 30]. In 2009, C. L. Holloway et al. employed the metasurface term to describe two-dimensional metamaterials [69]. In 2011, F. Capasso's group used the metasurface term and introduced the generalized Snell's law, as depicted in Fig. 2.20 [70]. Since then, the metasurface term had been accepted in the academic community.

Currently, metasurfaces can be classified into plasmonic metasurfaces based on metallic materials, dielectric metasurfaces based on high refractive index materials, and other novel metasurfaces based on their material type [71–73]. Plasmonic metasurfaces are based on metallic meta-atoms, and their electromagnetic response is driven by the particle plasmon resonances supported by metallic meta-atoms. Dielectric metasurfaces, whose structures are constructed of high refractive index dielectric materials, such as silicon, germanium, or tellurium, can support the electric and magnetic dipole responses based on the Mie resonance. In addition, many novel metasurfaces have been proposed, such as metasurfaces based on novel thin film materials and semiconductor materials [74]. Most of metasurfaces have been used to manipulate electromagnetic waves. According to the different electromagnetic parameters, metasurfaces can be employed for amplitude modulation, phase modulation, polarization modulation, and multi-dimensional modulation. In recent years, to achieve higher degree of freedom in controlling electromagnetic waves, new concepts and design techniques for active reconfigurable metasurfaces and nonlinear metasurfaces have been proposed. By adjusting the structure of metasurfaces and the properties of substrate material in real time, the electromagnetic response of the metasurface can be dynamically controlled. The reconfigurable metasurfaces include electrically reconfigurable metasurfaces, optically reconfigurable metasurfaces, thermally reconfigurable metasurfaces, mechanically reconfigurable metasurfaces, power reconfigurable metasurfaces, and so on [75].

2.4.2 Implementation and Characterization of Metasurfaces

The retrieval of effective electromagnetic parameters is an important method of analyzing and characterizing metamaterials, and is beneficial to their design, improvement, and application. Although EMT can lead to the approximation of the retrieval of effective electromagnetic parameters, it can accurately reflect the negative characteristics of metamaterials. For three-dimensional metamaterials, EMT is used to determine the effective permittivity and permeability, which describe their electromagnetic characteristics. However, since the thickness of the metasurface is much smaller than the operating wavelength, it can be approximated as a planar structure. Effective electromagnetic parameters are no longer suitable for characterizing metasurfaces. Instead, the surface impedance and similar concepts are commonly used to characterize metasurface [76].

The metasurface unit cells are typically composed of metal or dielectric materials, and their design methods are similar to three-dimensional metamaterials. Metasurface has a specific surface impedance. According to the effective circuit theory, the surface impedance is associated with the effective inductance L and the effective capacitance C, resulting in a resonant frequency of $\sqrt{1/LC}$. Near the resonant frequency, the metasurface exhibit the extremely high surface impedance, which implies that the transverse magnetic field is very weak while the transverse electric field is very strong. Their electromagnetic characteristics are analogous to a magnetic conductor. When the frequency of incident wave is far from the resonant frequency, the characteristics of the metasurface revert to an electric conductor [66]. The surface impedance properties enable the metasurface to have the significantly different reflection and transmission phases at different frequencies. Metasurfaces can modulate electromagnetic waves by controlling the phase of reflected and propagated waves. Therefore, the surface impedance can be used to characterize the electromagnetic properties of the metasurface in specific cases.

In recent years, researchers have proposed various methods including generalized averaged transition conditions [67], Babinet principle [77], generalized Snell's law [69], and Huygens principle [78], to study metasurfaces. Since metasurfaces represent a new research direction, both theoretical and experimental studies are going on. Metasurfaces can achieve many functions by changing the shapes, sizes, arrangements, and materials of unit cells. At present, there is still a gap between the theoretical design and practical application of metasurfaces. The present research methods have many limitations and are difficult to adapt all types of metamaterials. Artificial intelligence (AI) [79] is an emerging, powerful and effective computing tool. With its powerful ability to process information, AI has been rapidly integrated into the research of many academic fields in recent years. When AI is applied to the design of metasurfaces, more features can be retrieved from traditional numerical calculations and can significantly improve the design efficiency. To date, many researchers have incorporated AI into their metasurface research and achieved many exciting results.

Different implementation and characterization methods have provided in-depth research on metasurfaces. As a frontier research field, metasurfaces intersect multiple disciplines such as physics, chemistry, material science, electronics, and information technology, and have produced many astonishing phenomena [80–82].

2.4.3 Application Prospectives of Metasurfaces

Since 2011, the metasurface research has developed significantly and spanned microwave, terahertz, visible light, and other frequency ranges [83]. Metasurfaces are widely used to control the amplitude, phase, polarization, and other characteristics of electromagnetic waves. Metasurfaces have potential applications in vortex beams [84], holography [85], filters [86], absorbers [87], planar lenses [88], cloaking [83].

The amplitude of the electromagnetic wave is controlled by reflecting, absorbing or propagating the incident electromagnetic wave energy for a metasurface. For example, X. Liu et al. proposed a terahertz all dielectric metasurface absorber based on the hybrid dielectric waveguide resonance. The metasurface geometry was changed in order to overlap electric and magnetic dipole resonances at the same frequency, thus achieving the efficient absorption of electromagnetic waves [89].

The phase of electromagnetic wave is closely related to the size, shape, rotation mode, and constituent materials of the metasurface unit cell. In recent years, researchers have successively proposed many metasurfaces such as Pancharatnam-Berry (P-B) metasurfaces, reflective-type metasurfaces, and Huygens' metasurfaces to control the phase. L. Zhou's group designed a gradient-index metasurface with the H-type unit cell, as shown in Fig. 2.21a. This metasurface can efficiently convert the propagating waves in space into the surface waves [90]. In addition to electromagnetic mode conversion properties, the phase modulation capability of metasurface has brought new possibilities to holography. D. R. Smith's group used a metasurface to achieve the holography in the infrared wavelength range, as shown in Fig. 2.21b. The refractive indices of this metasurface in different regions are varied by changing the sizes of the unit cell. As a result, planar light waves accumulate different transmission phases, leading to the construction of a new wavefront. Thus, holography is achieved [91].

The polarization of electromagnetic waves can be achieved by designing specific metasurface unit cells, and the phase difference of the propagated or reflected electromagnetic waves in the orthogonal direction can be realized [81]. H. T. Chen's group proposed a metasurface for the linear polarization conversion and anomalous refraction [92]. The polarization direction of the x-polarized vertical incident waves is changed to y-direction after passing through multiple layers of metasurface, and their propagation direction is deflected by a certain angle, allowing for polarization control of incident waves.

The metasurface research has made a significant progress in theory and application. A variety of metasurface implementation methods have been proposed. Based on

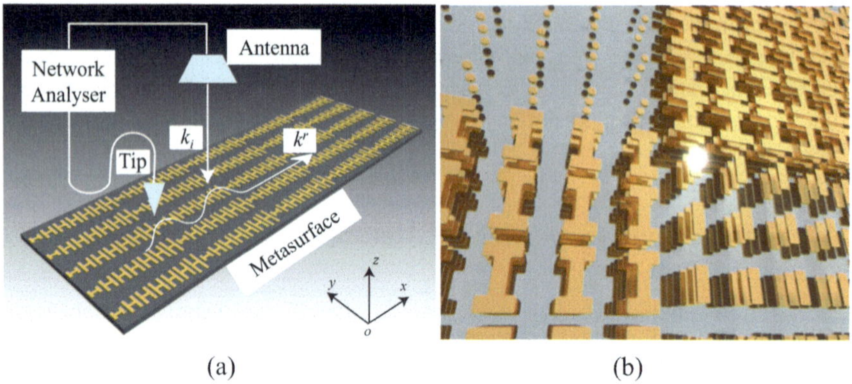

Fig. 2.21 Typical examples of metasurface applications. **a** Gradient-index metasurface. **b** Holography metasurface

the breakthrough of metasurfaces, new microwave, terahertz, and optical frequency band devices have been developed. For various metasurfaces, especially active reconfigurable metasurfaces, some of them have achieved the independent control of each metasurface unit cell. In the meantime, cutting-edge technologies such as AI make metasurfaces to move towards intelligent and digital directions.

References

1. Choy TC. Effective Medium Theory: Principles and Applications. Second Edition. Oxford: Oxford University Press. 2015.
2. Smith DR, Vier DC, Koschny T, et al. Electromagnetic parameter retrieval from inhomogeneous metamaterials. Physical Review E. 2005; 71:3–036617.
3. Peng HX, Zhou J, Cui TJ, et al. Metamaterial. Beijing: China Railway Publishing House. 2020 (In Chinese).
4. Landy NI, Sajuyigbe S, Mock JJ, et al. Perfect metamaterial absorber. Physical Review Letters. 2008; 100:20–207402.
5. Yao J, Liu ZW, Liu YM, et al. Optical negative refraction in bulk metamaterials of nanowires. Science. 2008; 321:5891–930.
6. Cui TJ, Qi MQ, Wan X, et al. Coding metamaterials, digital metamaterials and programmable metamaterials. Light: Science & Applications. 2014; 3:e218.
7. Marqués R, Martel J, Mesa F, et al. Left-handed-media simulation and transmission of EM waves in subwavelength split-ring-resonator-loaded metallic waveguides. Physical Review Letters. 2002; 89:18–183901.
8. Esteban J, Camacho-Peñalosa C, Page JE, et al. Simulation of negative permittivity and negative permeability by means of evanescent waveguide modes: Theory and experiment. IEEE Transactions on Microwave Theory and Techniques. 2005; 53:4–1506.
9. Wang YS, Duan ZY, Tang XF, et al. All-metal metamaterial slow-wave structure for high-power sources with high efficiency. Applied Physics Letters. 2015; 107:15–153502.
10. Hummelt JS, Lewis SM, Shapiro MA, et al. Design of a metamaterial-based backward-wave oscillator. IEEE Transactions on Plasma Science. 2014; 42:4–930.

11. Abuelfadl TM. Composite right/left-handed circular meta-waveguide. Applied Physics A. 2011; 103:3–759.
12. Shapiro MA, Trendafilov S, Urzhumov Y, et al. Active negative-index metamaterial powered by an electron beam. Physical Review B. 2012; 86:8–085132.
13. Zhou J. Some thoughts on the integration of metamaterials and natural materials. New Materials Industry. 2014; 9:5–8 (In Chinese).
14. Dong GY, Bi K, Zhou J. Zero phase delay in metamaterials. Scientia Sinica Physica, Mechanica & Astronomica. 2014; 44:4–406 (In Chinese).
15. Sun JB, Litchinitser NM, Zhou J. Indefinite by nature: From ultraviolet to terahertz. ACS Photonics. 2014; 1:4–293.
16. Zhao Q, Kang L, Du B, et al. Experimental demonstration of isotropic negative permeability in a three-dimensional dielectric composite. Physical Review Letters. 2008; 101:2–027402.
17. Shi ZC, Fan RH, Zhang ZD, et al. Random composites of nickel networks supported by porous alumina toward double negative materials. Advanced Materials. 2012; 24:17–2349.
18. Shi ZC, Fan RH, Zhang ZD, et al. Experimental and theoretical investigation on the high frequency dielectric properties of Ag/Al_2O_3 composites. Applied Physics Letters. 2011; 99:3–032903.
19. Luo Y, Estevez D, Scarpa F, et al. Microwave properties of metacomposites containing carbon fibers and ferromagnetic microwires. Research. 2019; 2019:3239879.
20. Nakano H. Low-Profile Natural and Metamaterial Antennas: Analysis Methods and Applications. New York: John Wiley & Sons. 2016.
21. Grbic A, Eleftheriades GV. Experimental verification of backward-ware radiation from a negative refractive index metamaterial. Journal of Applied Physics. 2002; 92:10–5930.
22. Caloz C, Itoh T. Application of the transmission line theory of left-handed (LH) materials to the realization of a microstrip "LH line". IEEE Antennas and Propagation Society International Symposium. San Antonio, TX, USA. 2002.
23. Lai A, Itoh T, Caloz C. Composite right/left-handed transmission line metamaterials. IEEE Microwave Magazine. 2004; 5:3–34.
24. Xu SJ, Zhang ZX. Novel balun and bandpass filter structure consisting of composite right/left-handed transmission line. International Journal of Microwave and Optical Technology. 2006; 1:2–458.
25. Zhang JH, Yan S, Vandenbosch GAE. Realization of dual band pattern diversity with a CRLH-TL inspired reconfigurable metamaterial. IEEE Transactions on Antennas and Propagation. 2018; 66:10–5130.
26. Zhou YJ, Jiang Q, Cui TJ. Bidirectional bending splitter of designer surface plasmons. Applied Physics Letters. 2011; 99:11–111904.
27. Shen XP, Cui TJ, Martin-Cano D, et al. Conformal surface plasmons propagating on ultrathin and flexible films. Proceedings of the National Academy of Sciences. 2013; 110:1–40.
28. Gao X, Zhou L, Yu XY, et al. Ultra-wideband surface plasmonic Y-splitter. Optics Express. 2015; 23:18–23270.
29. Clausius R. Abhandlungen über die Mechanische Wärmetheorie. Braunschweig: Friedrich Vieweg und Sohn. 1864.
30. Mossotti OF. Discussione analitica sull'influenza che l'azione di un mezzo dielettrico ha sulla distribuzione dell'elettricità alla superficie di più corpi elettrici disseminati in esso. Memorie di Mathematica e di Fisica della Società Italiana della Scienza. 1850; 24:49–74.
31. Maxwell GJC. Colours in metal glasses and in metallic film. Philosophical Transactions of the Royal Society of London, Series A. 1904; 203:359–371.
32. Bruggeman DAG. Berechnung verschiedener physikalischer Konstanten von heterogenen Substanzen. I. Dielektrizitätskonstanten und Leitfähigkeiten der Mischkörper aus isotropen Substanzen. Annalen der Physik. 1935; 416:7–636.
33. Simpkin R. Derivation of Lichtenecker's logarithmic mixture formula from Maxwell's equations. IEEE Transactions on Microwave Theory and Techniques. 2010; 58:3–545.
34. Kuester EF, Holloway CL. A low-frequency model for wedge or pyramid absorber arrays—I: Theory. IEEE Transactions on Electromagnetic Compatibility. 1994; 36:4–300.

35. Smith FC. Effective permittivity of dielectric honeycombs. IEE Proceedings-Microwaves, Antennas and Propagation. 1999; 146:1–55.
36. Smith FC, Scarpa F, Chambers B. The electromagnetic properties of re-entrant dielectric honeycombs. IEEE Microwave and Guided Wave Letters. 2000; 10:11–451.
37. Smith DR, Schultz S, Markoš P, et al. Determination of effective permittivity and permeability of metamaterials from reflection and transmission coefficients. Physical Review B. 2002; 65:19–195104.
38. Chen XD, Grzegorczyk TM, Wu BI, et al. Robust method to retrieve the constitutive effective parameters of metamaterials. Physical Review E. 2004; 70:1–016608.
39. Chen XD, Wu BI, Kong JA, et al. Retrieval of the effective constitutive parameters of bianisotropic metamaterials. Physical Review E. 2005; 71:4–046610.
40. Milosevic V, Jokanovic B, Bojanic R. Effective electromagnetic parameters of metamaterial transmission line loaded with asymmetric unit cells. IEEE Transactions on Microwave Theory and Techniques. 2013; 61:8–2761.
41. Cohen D, Shavit R. Bi-anisotropic metamaterials effective constitutive parameters retrieval using oblique incidence S-parameters method. IEEE Transactions on Antennas and Propagation. 2015; 63:5–2071.
42. Ghodgaonkar DK, Varadan VV, Varadan VK. Free-space measurement of complex permittivity and complex permeability of magnetic materials at microwave frequencies. IEEE Transactions on Instrumentation and Measurement. 1990; 39:2–387.
43. Chen LF, Ong CK, Tan BTG. Cavity perturbation technique for the measurement of permittivity tensor of uniaxially anisotropic dielectrics. IEEE Transactions on Instrumentation and Measurement. 1999; 48:6–1023.
44. Belhadj-Tahar NE, Fourrier-Lamer A. Broad-band simultaneous measurement of the complex permittivity tensor for uniaxial materials using a coaxial discontinuity. IEEE Transactions on Microwave Theory and Techniques. 1991; 39:10–1718.
45. Damaskos NJ, Mack RB, Maffett AL, et al. The inverse problem for biaxial materials. IEEE Transactions on Microwave Theory and Techniques. 1984; 32:4–400.
46. Smith DR, Vier DC, Kroll N, et al. Direct calculation of permeability and permittivity for a left-handed metamaterial. Applied Physics Letters. 2000; 77:14–2246.
47. Smith DR, Pendry JB. Homogenization of metamaterials by field averaging. Journal of the Optical Society of America B. 2006; 23:3–391.
48. Sun SL, Chui ST, Zhou L. Effective-medium properties of metamaterials: A quasimode theory. Physical Review E. 2009; 79:6–066604.
49. Andryieuski A, Malureanu R, Lavrinenko AV. Wave propagation retrieval method for metamaterials: Unambiguous restoration of effective parameters. Physical Review B. 2009; 80:19–193101.
50. Smith DR. Analytic expressions for the constitutive parameters of magnetoelectric metamaterials. Physical Review E. 2010; 81:3–036605.
51. Szabó Z, Park GH, Hedge R, et al. A unique extraction of metamaterial parameters based on Kramers–Kronig relationship. IEEE Transactions on Microwave Theory and Techniques. 2010; 58:10–2646.
52. Duan ZY, Hummelt JS, Shapiro MA, et al. Sub-wavelength waveguide loaded by a complementary electric metamaterial for vacuum electron devices. Physics of Plasmas. 2014; 21:10–103301.
53. Chen HT, O'Hara JF, Taylor AJ, et al. Complementary planar terahertz metamaterials. Optics Express. 2007; 15:3–1084.
54. Kong JA. Electromagnetic Wave Theory. Cambridge: EMW Publishing. 2005.
55. Vardiambasis IO, Tsalamengas JL, Kostogiannis K. Propagation of EM waves in composite bianisotropic cylindrical structures. IEEE Transactions on Microwave Theory and Techniques. 2003; 51:3–761.
56. Marqués R, Mesa F, Martel J, et al. Comparative analysis of edge- and broadside- coupled split ring resonators for metamaterial design: Theory and experiments. IEEE Transactions on Antennas and Propagation. 2003; 51:10–2572.

57. Zheludev NI, Kivshar YS. From metamaterials to metadevices. Nature Materials. 2012; 11:11–917.
58. Cai WS, Chettiar UK, Kildishev AV, et al. Nonmagnetic cloak with minimized scattering. Applied Physics Letters. 2007; 91:11–111105.
59. Pendry JB, Schurig D, Smith DR. Controlling electromagnetic fields. Science. 2006; 312:5781–1780.
60. Soukoulis CM, Wegener M. Optical metamaterials: More bulky and less lossy. Science. 2010; 330:6011–1633.
61. Holloway CL, Kuester EF, Gordon JA, et al. An overview of the theory and applications of metasurfaces: The two-dimensional equivalents of metamaterials. IEEE Antennas and Propagation Magazine. 2012; 54:2–10.
62. Siegel KM, Kennaugh EM, Moffatt DL. Radar cross section of a cone sphere. Proceedings of the IEEE. 1963; 51:1–231.
63. Munk BA, Kouyoumjian RG, Peters L Jr. Reflection properties of periodic surfaces of loaded dipoles. IEEE Transactions on Antennas and Propagation. 1971; 19:5–612.
64. Munk BA, Luebbers RJ, Fulton RD. Transmission through a two-layer array of loaded slots. IEEE Transactions on Antennas and Propagation. 1974; 22:6–804.
65. Larson CJ, Munk BA. The broad-band scattering response of periodic arrays. IEEE Transactions on Antennas and Propagation. 1983; 31:2–261.
66. Sievenpiper D, Zhang LJ, Broas RFJ, et al. High-impedance electromagnetic surfaces with a forbidden frequency band. IEEE Transactions on Microwave Theory and Techniques. 1999; 47:11–2059.
67. Kuester EF, Mohamed MA, Piket-May M, et al. Averaged transition conditions for electromagnetic fields at a metafilm. IEEE Transactions on Antennas and Propagation. 2003; 51:10–2641.
68. Sievenpiper DF, Schaffner JH, Song HJ, et al. Two-dimensional beam steering using an electrically tunable impedance surface. IEEE Transactions on Antennas and Propagation. 2003; 51:10–2713.
69. Holloway CL, Dienstfrey A, Kuester EF, et al. A discussion on the interpretation and characterization of metafilms/metasurfaces: The two-dimensional equivalent of metamaterials. Metamaterials. 2009; 3:2–100.
70. Yu NF, Genevet P, Kats MA, et al. Light propagation with phase discontinuities: Generalized laws of reflection and refraction. Science. 2011; 334:6054–333.
71. Kamali SM, Arbabi E, Arbabi A, et al. A review of dielectric optical metasurfaces for wavefront control. Nanophotonics. 2018; 7:6–1041.
72. Meinzer N, Barnes WL, Hooper IR. Plasmonic meta-atoms and metasurfaces. Nature Photonics. 2014; 8:12–889.
73. Wang J, Du J. Plasmonic and dielectric metasurfaces: Design, fabrication and applications. Applied Sciences. 2016; 6:9–239.
74. Zahra S, Ma L, Wang W, et al. Electromagnetic metasurfaces and reconfigurable metasurfaces: A review. Frontiers in Physics. 2021; 8:593411.
75. Shaltout AM, Kinsey N, Kim JB, et al. Development of optical metasurfaces: emerging concepts and new materials. Proceedings of the IEEE. 2016; 104:12–2270.
76. Fong BH, Colburn JS, Ottusch JJ, et al. Scalar and tensor holographic artificial impedance surfaces. IEEE Transactions on Antennas and Propagation. 2010; 58:10–3212.
77. Falcone F, Lopetegi T, Laso MAG, et al. Babinet principle applied to the design of metasurfaces and metamaterials. Physical Review Letters. 2004; 93:19–197401.
78. Pfeiffer C, Grbic A. Metamaterial Huygens' surfaces: Tailoring wave fronts with reflectionless sheets. Physical Review Letters. 2013; 110:19–197401.
79. Russell SJ, Norvig P. Artificial Intelligence: A Modern Approach (4th ed.). Hoboken: Pearson. 2021.
80. Yu NF, Capasso F. Flat optics with designer metasurfaces. Nature Materials. 2014; 13:2–139.
81. Zhang L, Mei ST, Huang K, et al. Advances in full control of electromagnetic waves with metasurfaces. Advanced Optical Materials. 2016; 4:6–818.

82. Kildishev AV, Boltasseva A, Shalaev VM. Planar photonics with metasurfaces. Science. 2013; 339:6125–1232009.
83. Glybovski SB, Tretyakov SA, Belov PA, et al. Metasurfaces: From microwaves to visible. Physics Reports. 2016; 634:1.
84. Ren HR, Li XP, Zhang QM, et al. On-chip noninterference angular momentum multiplexing of broadband light. Science. 2016; 352:6287–805.
85. Wen DD, Yue FY, Li GX, et al. Helicity multiplexed broadband metasurface holograms. Nature Communications. 2015; 6:8241.
86. Shen F, Kang Q, Wang J, et al. Dielectric metasurface-based high-efficiency mid-infrared optical filter. Nanomaterials. 2018; 8:11–938.
87. Guo W, Liu Y, Han T, et al. Ultra-broadband infrared metasurface absorber. Optics Express. 2016; 24:18–20586.
88. Chen XZ, Huang LL, Mühlenbernd H, et al. Dual-polarity plasmonic metalens for visible light. Nature Communications. 2012; 3:1198.
89. Liu XY, Fan KB, Shadrivov IV, et al. Experimental realization of a terahertz all-dielectric metasurface absorber. Optics Express. 2017; 25:1–191.
90. Sun SL, He Q, Xiao SY, et al. Gradient-index meta-surfaces as a bridge linking propagating waves and surface waves. Nature Materials. 2012; 11:5–426.
91. Larouche S, Tsai YJ, Tyler T, et al. Infrared metamaterial phase holograms. Nature Materials. 2012; 11:450.
92. Grady NK, Heyes JE, Chowdhury DR, et al. Terahertz metamaterials for linear polarization conversion and anomalous refraction. Science. 2013; 340:6138–1304.

Chapter 3
Basic Theory of Reversed Cherenkov Radiation

CR refers to a type of electromagnetic radiation excited by charged particles. CR occurs when the velocity of the charged particles exceeds the phase velocity of electromagnetic waves in the surrounding environment. The wavefront of CR forms a cone surface, propagating in the direction of the charged particles' motion. The theoretical predictions of this electromagnetic phenomenon can be traced back to the theoretical predictions by O. Heaviside in 1888 [1] and A. Sommerfeld in 1904 [2]. In 1910, M. Curie observed a faint blue glow in a highly concentrated solution of radium, but the physical mechanism was not deeply investigated at that time [3]. In 1926, L. Mallet discovered that the luminescent radiation emitted when radium irradiates water has a continuous spectrum [4]. At that time, this faint blue glow was mistakenly understood as fluorescence and was not recognized as a new physical phenomenon, hence no in-depth research was conducted. In 1934, under the guidance of S. I. Vavilov, P. A. Cherenkov observed a faint blue glow in water where the radioactive substances are immersed. After the research, P. A. Cherenkov discovered that this glow was not fluorescence, but a new physical phenomenon, as shown in Fig. 3.1, which is known as CR, named after him [5]. In 1937, I. M. Frank and I. Y. Tamm were the first to provide a perfect physical explanation of CR using macroscopic electromagnetic theory [6]. Subsequently, extensive research was conducted internationally, especially in the Soviet Union [7, 8]. P. A. Cherenkov, I. M. Frank, and I. Y. Tamm were awarded the Nobel Prize in Physics in 1958 for the discovery and explanation of the CR phenomenon.

CR has important applications in many fields such as high-energy particle physics, optics, cosmic ray physics, and vacuum electronics [9]. For example, in high-energy particle physics, it has been successfully used to detect the charged particles with different velocities, leading to the discovery of antiprotons [10] and J/ψ particles [11], for which the Nobel Prizes in Physics were awarded in 1959 and 1976, respectively. In the field of vacuum electronics, R. Kompfner invented the TWT in 1943 [12], and B. Epsztein invented the BWO in 1951 [13]. These devices are based on CR [14–16].

Z. Duan, *Metamaterial-Based Electromagnetic Radiations and Applications*,
https://doi.org/10.1007/978-981-97-8108-9_3

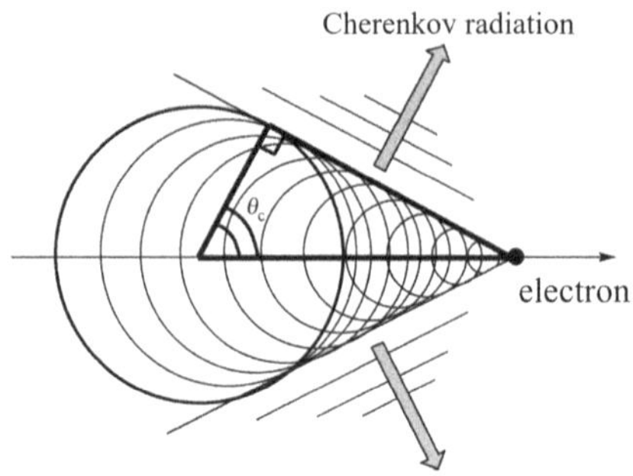

Fig. 3.1 Schematic diagram of CR

Microwave VEDs have been widely used in radar, radio navigation, communications, electronic countermeasures, and other fields [17–21].

3.1 Reversed Cherenkov Radiation in Unbounded Anisotropic Double Negative Metamaterial

In 1967, V. G. Veselago theoretically predicted the phenomenon of RCR in LHMs. Due to the simultaneous negative permittivity and negative permeability of LHMs [also known as double negative metamaterials (DNMs)], RCR exhibits a novel physical form, which is characterized by the wave vector of the radiated wave and the Poynting vector being opposite (in isotropic DNMs) or approximately opposite (in anisotropic DNMs). In other words, the direction of phase velocity is opposite or approximately opposite to that of group velocity v_g. Therefore, this type of electromagnetic radiation in LHMs is referred to as RCR.

However, to date, naturally occurring materials with a negative permittivity and a negative permeability have not been found. As described in Chap. 1, it was not until 1996 and 1999 that J. B. Pendry and colleagues proposed electromagnetic structures like metallic wire arrays and SRRs, which provided a method to achieve negative effective permittivity and negative effective permeability, respectively, in the microwave frequency range. In 2000, D. R. Smith et al. reported the realization of left-handed characteristics using a combination of metallic wire arrays and SRRs [22], marking the beginning of research in metamaterials. For more information on the novel electromagnetic properties and realization of metamaterials, readers can refer to Sects. 1.3 and 2.1 in this monograph.

Research on RCR can be traced back to around 2002, when J. Lu et al. first studied RCR excited by charged particles in unbounded isotropic LHMs [23]. Subsequently, Z. Duan, S. Antipov, and others systematically studied RCR excited by charged particles in unbounded, semi-unbounded, and waveguides filled with anisotropic DNMs, exploring new methods to enhance RCR [24–30]. Moreover, Y. O. Averkov, S. N. Galyamin, A. V. Kats, and others conducted theoretical studies on RCR and TR which is produced when charged particles pass through the boundary of DNMs

[31–34]. These research findings provide a theoretical foundation for experimental studies and potential applications of RCR [35–38].

3.1.1 Single-Particle Model

In 2003, J. Lu et al. studied a single-particle model in an unbounded space filled with isotropic LHMs [23]. However, practically realizable LHMs are generally anisotropic. Therefore, an unbounded anisotropic DNM, whose electromagnetic properties are characterized by a permittivity tensor and a permeability tensor, is considered. The elements of the permittivity tensor are described using the Drude model, while the elements of the permeability tensor are described using the Lorentz model.

In the cylindrical coordinate system (ρ, θ, z), the permittivity tensor can be represented as:

$$\boldsymbol{\varepsilon} = \varepsilon_0 \begin{bmatrix} \varepsilon_{r\rho} & 0 & 0 \\ 0 & \varepsilon_{r\theta} & 0 \\ 0 & 0 & \varepsilon_{rz} \end{bmatrix} = \begin{bmatrix} \varepsilon_{\rho} & 0 & 0 \\ 0 & \varepsilon_{\theta} & 0 \\ 0 & 0 & \varepsilon_{z} \end{bmatrix}, \tag{3.1}$$

$$\varepsilon_{r\rho}(\omega) = 1 - \frac{\omega_{p\rho}^2}{\omega^2 + \mathrm{j}\gamma_{e\rho}\omega}, \tag{3.2}$$

$$\varepsilon_{r\theta}(\omega) = 1 - \frac{\omega_{p\theta}^2}{\omega^2 + \mathrm{j}\gamma_{e\theta}\omega}, \tag{3.3}$$

$$\varepsilon_{rz}(\omega) = 1 - \frac{\omega_{pz}^2}{\omega^2 + \mathrm{j}\gamma_{ez}\omega}, \tag{3.4}$$

where ω is the angular frequency of the electromagnetic wave, $\omega_{p\rho}$, $\omega_{p\theta}$, and ω_{pz} are the equivalent plasma frequencies in the ρ, θ, and z directions, respectively, and $\gamma_{e\rho}$, $\gamma_{e\theta}$, and γ_{ez} represent the collision frequencies associated with the electric loss in the DNMs.

The permeability tensor can be represented as:

$$\boldsymbol{\mu} = \mu_0 \begin{bmatrix} \mu_{r\rho} & 0 & 0 \\ 0 & \mu_{r\theta} & 0 \\ 0 & 0 & \mu_{rz} \end{bmatrix} = \begin{bmatrix} \mu_{\rho} & 0 & 0 \\ 0 & \mu_{\theta} & 0 \\ 0 & 0 & \mu_{z} \end{bmatrix}, \tag{3.5}$$

$$\mu_{r\rho}(\omega) = 1 - \frac{F_{\rho}\omega^2}{\omega^2 - \omega_{0\rho}^2 + \mathrm{j}\gamma_{m\rho}\omega}, \tag{3.6}$$

$$\mu_{r\theta}(\omega) = 1 - \frac{F_{\theta}\omega^2}{\omega^2 - \omega_{0\theta}^2 + \mathrm{j}\gamma_{m\theta}\omega}, \tag{3.7}$$

$$\mu_{rz}(\omega) = 1 - \frac{F_z\omega^2}{\omega^2 - \omega_{0z}^2 + \mathrm{j}\gamma_{mz}\omega}, \tag{3.8}$$

In the permeability tensor, $\gamma_{m\rho}$, $\gamma_{m\theta}$, and γ_{mz} are the collision frequencies associated with the magnetic loss in the DNMs, $\omega_{0\rho}$, $\omega_{0\theta}$, and ω_{0z} are the magnetic resonance frequencies, F_ρ, F_θ, and F_z are the filling factors. It should be noted that when the conditions $\varepsilon_{r\rho}(\omega) = \varepsilon_{r\theta}(\omega) = \varepsilon_{rz}(\omega)$ and $\mu_{r\rho}(\omega) = \mu_{r\theta}(\omega) = \mu_{rz}(\omega)$ are satisfied, the anisotropic DNMs can degenerate into isotropic DNMs.

To simplify the analysis, here let's use $\mathrm{e}^{-\mathrm{j}\omega t}$, and suppose a charged particle moves along the $+z$ axis with a constant velocity $\boldsymbol{v} = \hat{z}v$. Then, the current density of the moving charged particle is:

$$\boldsymbol{J}(\boldsymbol{r}, t) = \hat{z}qv\delta(z - vt)\delta(x)\delta(y), \tag{3.9}$$

where q is the charge of the particle, and $\delta(x)$ is the Dirac function. In the cylindrical coordinate system, the current density can be written as [39]:

$$\boldsymbol{J}(\boldsymbol{r}, t) = \hat{z}qv\delta(z - vt)\frac{\delta(\rho)}{2\pi\rho}. \tag{3.10}$$

Transforming (3.10) into the frequency domain yields:

$$\boldsymbol{J}(\boldsymbol{r}, \omega) = \hat{z}\frac{q}{4\pi^2\rho}e^{\mathrm{j}\omega z/v}\delta(\rho). \tag{3.11}$$

According to Ampère's law, one can derive:

$$\boldsymbol{D} = \frac{\mathrm{j}}{\omega}\nabla \times \left(\boldsymbol{\mu}^{-1} \cdot \boldsymbol{B}\right) - \boldsymbol{J}(\boldsymbol{r}, \omega), \tag{3.12}$$

where $\boldsymbol{B} = \nabla \times \boldsymbol{A}$, and $\boldsymbol{A} = \hat{z}A_z$ is the vector potential. Therefore, by substituting (3.12) and $\boldsymbol{B} = \nabla \times \boldsymbol{A}$ into the Faraday's law, one can obtain the vector wave equation:

$$\nabla \times [\boldsymbol{\mu}^{-1} \cdot \nabla \times (\hat{z}A_z)] - \boldsymbol{J} = \omega^2\boldsymbol{\varepsilon} \cdot (\hat{z}A_z) + \mathrm{j}\omega\boldsymbol{\varepsilon} \cdot \nabla\varphi. \tag{3.13}$$

where φ represents the scalar potential. Subsequently, by decomposing the vector wave Eq. (3.13) into three scalar wave equations, with the given expression $A_z = g(\rho)\mu_\theta q/(2\pi)\exp(\mathrm{j}\omega z/v)$, one can obtain the following scalar wave equation for $g(\rho)$:

$$\left[\frac{1}{\rho}\frac{\partial}{\partial\rho}\left(\rho\frac{\partial}{\partial\rho}\right) + k_\rho^2\right]g(\rho) = -\frac{\delta(\rho)}{2\pi\rho}, \tag{3.14}$$

where $k_\rho = -\sqrt{\omega^2\varepsilon_z\mu_\theta - \varepsilon_z/\varepsilon_\rho\omega^2/v^2}$ represents the radial wave number. Equation (3.14) has the following solution:

$$g(\rho) = -\frac{\mathrm{j}}{4} H_0^{(2)}(k_\rho \rho), \quad (\rho \neq 0), \tag{3.15}$$

where $H_0^{(2)}(k_\rho \rho)$ is the second kind of Hankel function.

On the one hand, if $\mathrm{Re}(\varepsilon_{r\rho}\mu_{r\theta}) > 1/\beta^2$, CR will occur. Therefore, the conditions for CR to occur in DNMs are:

$$\mathrm{Re}(\varepsilon_{r\rho}\mu_{r\theta}) > 1/\beta^2, \mathrm{Re}(\varepsilon_{r\rho}) < 0, \mathrm{Re}(\mu_{r\theta}) < 0, \mathrm{Re}(\varepsilon_{rz}) < 0, \tag{3.16}$$

where $\beta = \upsilon/c_0$, and Re() denotes the real part operator.

When the conditions for CR are met ($\rho \neq 0$), its magnetic field intensity can be expressed as:

$$\boldsymbol{H}_\theta(\boldsymbol{r}, \omega) = -\hat{\boldsymbol{\theta}} \frac{\mathrm{j}qk_\rho}{8\pi} \mathrm{e}^{\mathrm{j}\omega z/\upsilon} H_1^{(2)}(k_\rho \rho). \tag{3.17}$$

From (3.12), one can derive the electric field component:

$$\boldsymbol{E}_z(\boldsymbol{r}, \omega) = \hat{\boldsymbol{z}} \frac{qk_\rho}{8\pi\varepsilon_z\omega} \mathrm{e}^{\mathrm{j}\omega z/\upsilon} \left\{ \frac{1}{\rho} H_1^{(2)}(k_\rho \rho) + \frac{k_\rho}{2} [H_0^{(2)}(k_\rho \rho) - H_2^{(2)}(k_\rho \rho)] \right\}, \tag{3.18}$$

$$\boldsymbol{E}_\rho(\boldsymbol{r}, \omega) = -\hat{\boldsymbol{\rho}} \frac{\mathrm{j}qk_\rho}{8\pi\varepsilon_\rho \upsilon} \mathrm{e}^{\mathrm{j}\omega z/\upsilon} H_1^{(2)}(k_\rho \rho). \tag{3.19}$$

When anisotropic DNMs degenerate into isotropic ones, the expressions for the field components are the same as those in [23]. Following a similar definition [9], the phase angle θ_{CR} is the angle between the phase velocity $\boldsymbol{\upsilon}_p$ and the particle velocity $\boldsymbol{\upsilon}$, as shown in Fig. 3.1. Its expression is given as:

$$\cos\theta_{CR} = -\frac{1}{\mathrm{Re}\left\{\sqrt{\varepsilon_z\mu_\theta + (1 - \varepsilon_z/\varepsilon_\rho)/\upsilon^2}\right\}\upsilon}. \tag{3.20}$$

It should be noted that the phase velocity, in the direction of the wave vector $\boldsymbol{k}$, is the same as that of $\boldsymbol{k} = \hat{\boldsymbol{\rho}}\mathrm{Re}(k_\rho) + \hat{\boldsymbol{z}}k_z$ ($k_z = \omega/\upsilon$). When anisotropic DNMs degenerate into isotropic ones, (3.20) degenerates into $\cos\theta_{CR} = 1/(\beta\mathrm{Re}(n))$ (in the presence of loss) [40], and $\cos\theta_{CR} = 1/(-\left|\sqrt{\varepsilon_r\mu_r}\right|\beta)$ (in the absence of loss) [31], where $n = -\sqrt{\varepsilon_{r\rho}\mu_{r\theta}}$ represents the complex refractive index. Clearly, the phase angle θ_{CR} depends largely on the frequency of the observed RCR. Similarly defined, $\theta_{s\upsilon}$ is the angle between the time-averaged Poynting vector <*S*> and the velocity $\boldsymbol{\upsilon}$ of the charged particle, denoted as $\theta_{s\upsilon} = \pi/2 - \arctan(\langle \boldsymbol{S}_z\rangle / \langle \boldsymbol{S}_\rho\rangle)$, referred to as the radiation angle (or group angle). As shown in Fig. 3.2, $\theta_{s\upsilon}$ is obtuse, but in ordinary electromagnetic media, $\theta_{s\upsilon}$ is acute. Due to $\theta_{CR} \neq \theta_{s\upsilon}$, <*S*> and $\boldsymbol{k}$ are not completely opposite in anisotropic DNMs. This feature is entirely different from the case in isotropic DNMs. This is because in isotropic DNMs, the vectors <*S*> and $\boldsymbol{k}$ are completely opposite in the far field [31]. Therefore, for a DNM, such electromagnetic

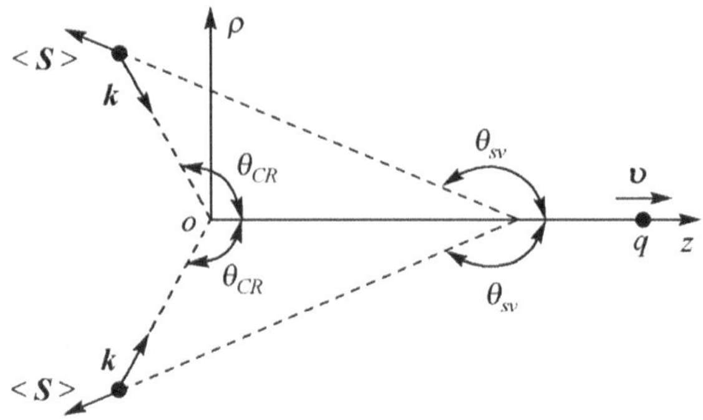

Fig. 3.2 Diagram of RCR in DNMs

radiation is referred to as RCR. Equation (3.20) describes the relationship among the phase angle θ_{CR}, the velocity of the charged particles, and the electromagnetic media. Thus, different phase angles can be obtained by changing the velocity of the charged particles or selecting different electromagnetic media.

Furthermore, the energy loss per unit path length is determined by the retardation force exerted on a moving charge by the field generated by the moving charge, as discussed in [7]:

$$\begin{aligned}\frac{dW}{dz} &= qE_z\big|_{z\to \upsilon t,\,\rho\to 0}\\ &= \frac{q^2}{4\pi}\mathrm{Re}\left\{\int_0^\infty d\omega\frac{k_\rho}{\varepsilon_z\omega}\left[\frac{1}{\rho_0}H_1^{(2)}(k_\rho\rho_0)+\frac{k_\rho}{2}[H_0^{(2)}(k_\rho\rho_0)-H_2^{(2)}(k_\rho\rho_0)]\right]\right\}\end{aligned} \tag{3.21}$$

where ρ_0 represents the minimum average distance from the field source. It's worth noting that the integration range in (3.21) is determined by (3.16).

On the other hand, if $\mathrm{Re}(\varepsilon_{r\rho}\mu_{r\theta}) < 1/\beta^2$, the moving charged particles excite evanescent waves, and CR does not occur. Here, this case is not discussed.

3.1.2 Numerical Calculations

Based on the theoretical analysis described above, numerical calculations are employed to study the radiation spectral density per unit path length and the total radiation energy. First, characteristic models represented by (3.1) and (3.5) are selected. Based on the experimental results from [41], the following values are chosen: $\omega_{p\rho} = 2\pi \times 10^{10}$ rad/s, $\omega_{pz} = 2\pi \times 7 \times 10^9$ rad/s, $\omega_{0\theta} = 2\pi \times 4 \times 10^9$ rad/s, $F_\theta = 0.56$, $\gamma_{e\rho} = \gamma_{ez} = \gamma_{m\theta} = \gamma = 10^7$ rad/s. In addition, $\rho_0 = 10^{-6}$ m and $\upsilon = 2.4 \times 10^8$ m/s. As a result, the real and imaginary parts of $\varepsilon_{r\rho}$, ε_{rz}, and $\mu_{r\theta}$ are shown in Figs. 3.3 and 3.4, respectively. Clearly, $\mathrm{Re}(\varepsilon_{r\rho})$, $\mathrm{Re}(\varepsilon_{rz})$, and $\mathrm{Re}(\mu_{r\theta})$ are all negative within the frequency range of 4.002–6 GHz. Furthermore, the electric loss decreases with frequency, while the magnetic loss is relatively high near the resonant frequency. Here, $\omega_{0\theta}$ and F_θ are held constant to investigate the difference in CR for the two cases of anisotropic and isotropic LHMs.

Next, the conditions for CR, phase angle, and radiation angle are studied. Using the parameters mentioned above, calculations are performed for the conditions of CR, as well as the phase and radiation angles, as shown in Fig. 3.5. In Fig. 3.5a, n_1, n_2 and n_3 represent the complex refractive indices in anisotropic DNMs, with corresponding effective plasma frequencies in the $\hat{\rho}$ direction of $\omega_{p\rho} = 2\pi \times 6 \times 10^9$ rad/s, $\omega_{p\rho} = 2\pi \times 7 \times 10^9$ rad/s, $\omega_{p\rho} = 2\pi \times 10^{10}$ rad/s, respectively. The conditions for CR are met in the radiation frequency range when $\text{Re}(n^2)$ is greater than $1/\beta^2$,

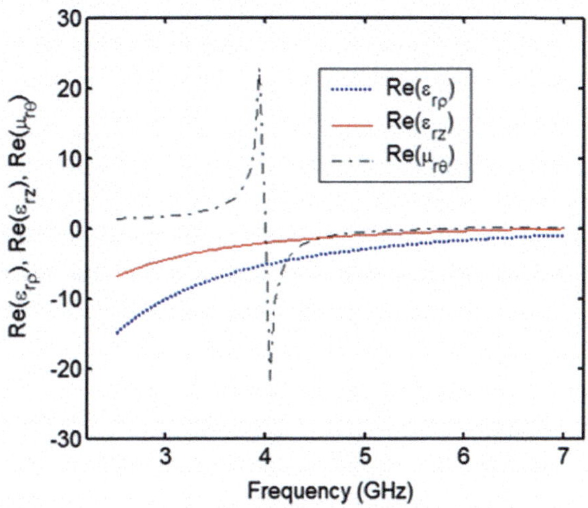

Fig. 3.3 Real parts of relative permittivity and permeability versus frequency

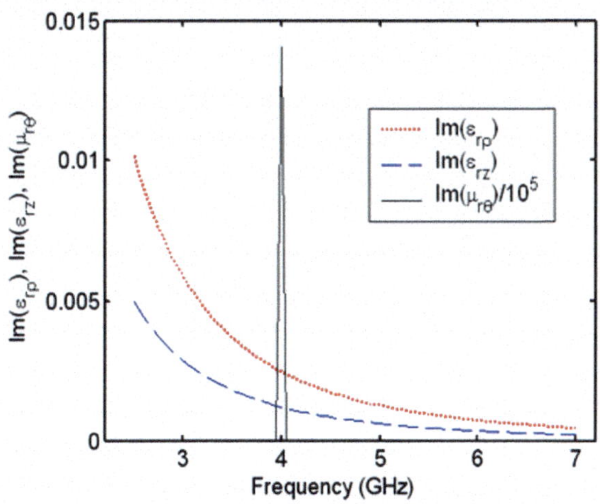

Fig. 3.4 Imaginary parts of relative permittivity and permeability versus frequency

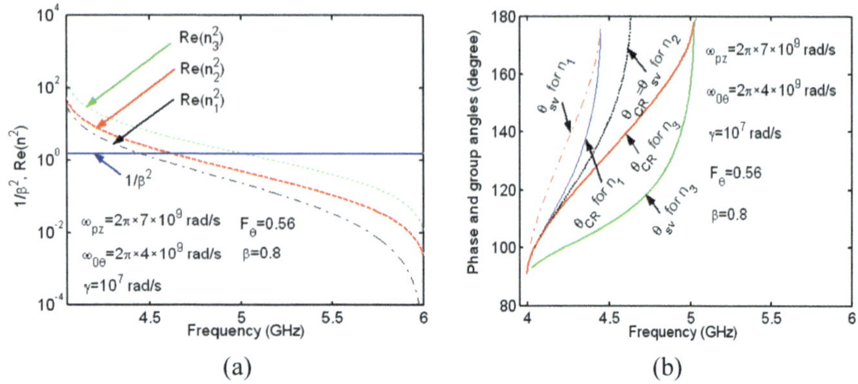

Fig. 3.5 **a** Frequency dependence of CR condition for different refractive indices; **b** Frequency dependence of phase and group angles for different refractive indices

as indicated by the solid line. It can be seen from Fig. 3.5a that $\mathrm{Re}(n^2)$ decreases with increasing operating frequency and increases with the increase of the effective plasma frequency $\omega_{p\rho}$ in the $\hat{\rho}$ direction at certain operating frequencies. Therefore, the frequency range that satisfies RCR increases with the increase of the angular frequency $\omega_{p\rho}$. This is because that at a given operating frequency, the absolute value of $\mathrm{Re}(\varepsilon_{r\rho})$ increases with $\omega_{p\rho}$, which means a decrease in phase velocity, making it easier to satisfy the CR conditions. It is worth noting that the frequency ranges in which the electromagnetic media with complex refractive indices n_1, n_2, and n_3 satisfy the conditions for RCR are from 4.002 to 4.450 GHz, from 4.002 to 4.628 GHz, and from 4.002 to 5.024 GHz, respectively.

Subsequently, the relationship between the phase angle and the radiation angle with respect to the effective plasma frequency and operating frequency in the $\hat{\rho}$ direction is shown in Fig. 3.5b. It is assumed that the radiation angle θ_{sv} is obtuse ($90° < \theta_{sv} < 180°$), corresponding to RCR. For different refractive indices, θ_{CR} and θ_{sv} increase with increasing operating frequency. For isotropic DNMs with a complex refractive index of n_2, under the condition of a given operating frequency, θ_{CR} only completely equals θ_{sv} in the far-field region, where the directions of vectors $<\boldsymbol{S}>$ and $\boldsymbol{k}$ are opposite. However, for anisotropic DNMs with complex refractive indices of n_1 and n_3, θ_{CR} is generally equal to θ_{sv} under the given operating frequency. This indicates that in anisotropic DNMs, the directions of the wave vectors and time-averaged Poynting vectors are not opposite. As for particle detectors based on CR, the phase angle θ_{CR} can be used to directly measure the velocity of charged particles, while the radiation angle θ_{sv} can be measured to determine the radiation pattern of RCR. Therefore, both the phase angle and radiation angle should be used to describe the radiation characteristics of RCR in DNMs. Clearly, these properties of RCR are different from CR. Additionally, when the operating frequency is kept constant, both the phase angle and radiation angle decrease with increasing the absolute value of the real part of the refractive index. This characteristic describes the dependence of the phase angle and radiation angle of DNMs on material properties, which is useful for particle detection and identification in particle physics.

Finally, the radiation spectral density and the total radiation energy per unit path length of anisotropic and isotropic DNMs are separately investigated. The values of $\omega_{p\rho} = 2\pi \times 10 \times 10^9$ rad/s and $\omega_{pz} = 2\pi \times 7 \times 10^9$ rad/s are selected for an anisotropic DNM. To consider the loss, the loss factors ($\gamma_{e\rho} = \gamma_{ez} = \gamma_{m\theta} = \gamma$) are increased accordingly. The spectral density for anisotropic DNMs is shown in Fig. 3.6, where f_i represents the spectral density as a function of frequency. When both dispersion and loss are considered, the spectral density becomes continuous [39, 42]. The spectral density of RCR at lower frequencies is much larger than at higher frequencies. As loss increases, the spectral density becomes flatter in the radiation frequency range. To study the impact of loss and different effective plasma frequencies on the total radiation energy, numerical calculations are performed, and the results are shown in Fig. 3.7. Case 1, Case 2, and Case 3 correspond to situations where $\omega_{pz} = 14\pi \times 10^9$ rad/s $< \omega_{p\rho}$, $\omega_{pz} = 20\pi \times 10^9$ rad/s $= \omega_{p\rho}$, and $\omega_{pz} = 26\pi \times 10^9$ rad/s $> \omega_{p\rho}$, respectively. Here, Case 2 implies the degeneration of anisotropic DNMs into isotropic ones. It can be observed from Fig. 3.7a that for both anisotropic and isotropic DNMs, the total radiation energy increases with decreasing loss. However, when the loss values are kept constant, the total radiation energy of anisotropic DNMs in Case 1 is greater than that of isotropic DNMs in Case 2, while the total radiation energy of anisotropic DNMs in Case 3 is less than that of isotropic DNMs in Case 2. As loss increases, the difference in total radiation energy between anisotropic and isotropic DNMs decreases, as a result of the overlapping distribution of total radiation energy. Furthermore, Fig. 3.7b illustrates that the total radiation energy is a decreasing function of the effective plasma frequency in the z-direction. This is because, at a given operating frequency, the absolute value of $\mathrm{Re}(\varepsilon_{rz})$ increases with the frequency ω_{pz} and the difference in k_ρ is small. Therefore, it is possible to determine the radiation spectral density and total radiation energy by designing appropriate DNMs. The difference in the total radiation energy between anisotropic and isotropic DNMs arises from the different electromagnetic properties of the materials.

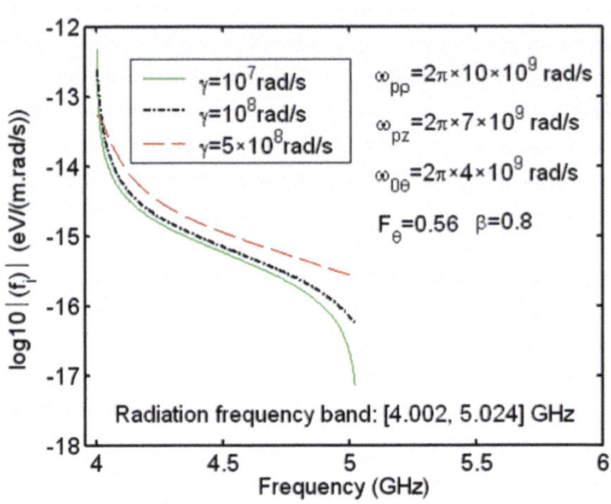

Fig. 3.6 Spectral density of CR versus frequency for different loss cases

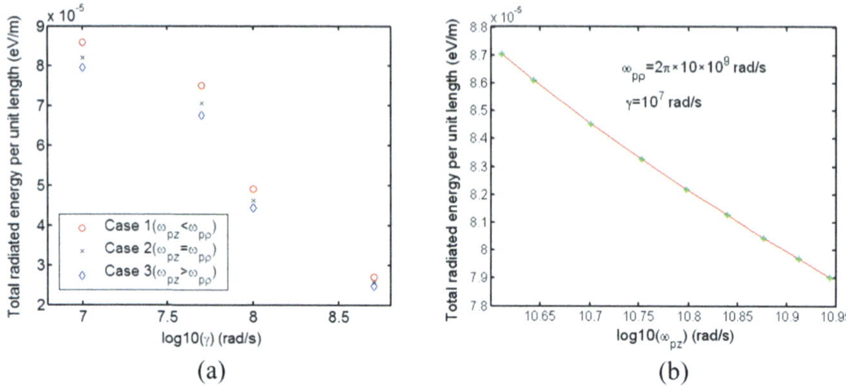

Fig. 3.7 **a** Comparison of total radiated energies among the three cases for different loss; **b** total radiated energy as a function of ω_{pz}

3.2 Reversed Cherenkov Radiation in Semi-Unbounded Double Negative Metamaterial

3.2.1 Theoretical Analysis of Single-Particle Model

In the previous section, the RCR excited by a single charged particle when filling anisotropic DNMs in unbounded space is analyzed. Based on this, the RCR excited by a single charged particle when filling isotropic DNMs in semi-unbounded space is considered.

Consider the general case of a single charged particle moving along the boundary, with the physical model shown in Fig. 3.8. The distance between the particle and the boundary is d. Half of the space is filled with isotropic DNMs, while the other half remains in vacuum environment. The equivalent electromagnetic parameters of the DNMs are described in a Cartesian coordinate system (x, y, z) as follows [26]:

$$\varepsilon_2(\omega) = 1 - \frac{\omega_p^2}{\omega^2 + \mathrm{j}\gamma_e\omega}, \tag{3.22}$$

$$\mu_2(\omega) = 1 - \frac{F\omega^2}{\omega^2 - \omega_0^2 + \mathrm{j}\gamma_m\omega}, \tag{3.23}$$

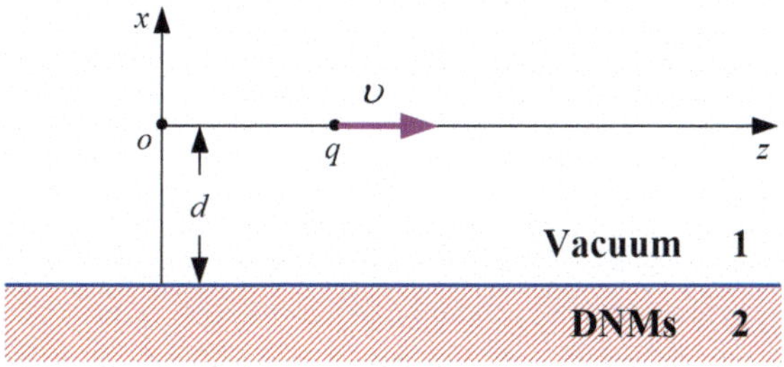

Fig. 3.8 Schematic diagram of a charged particle moving with speed υ along a semi-unbounded space filled with isotropic DNMs

where ω_p is the electric plasma frequency, γ_e is the collision frequency for electric dissipation in the material, ω_0 is the magnetic resonant frequency, γ_m is the collision frequency for magnetic loss in the material, and F is the filling factor of the SRR unit cell. The charged particle moves in the z-direction with a velocity $\boldsymbol{\upsilon}$ within the semi-unbounded space filled with isotropic DNMs. Its charge density is given by

$$\rho(\boldsymbol{r}, t) = q\delta(x)\delta(y)\delta(z - \upsilon t). \tag{3.24}$$

According to the definition of electromagnetic potentials, in Gaussian units, the electric and magnetic fields are given by

$$\boldsymbol{E} = -\frac{1}{c_0}\frac{\partial \boldsymbol{A}}{\partial t} - \nabla\varphi, \tag{3.25}$$

$$\boldsymbol{B} = \nabla \times \boldsymbol{A}, \tag{3.26}$$

In this context, a set of ($\boldsymbol{A}$, φ) satisfying the Lorenz gauge is chosen for theoretical analysis. By using Maxwell's equations and employing the electromagnetic potential method, one can derive the d'Alembertian equation. Additionally, it is assumed that $\boldsymbol{A}$ and φ have the following Fourier integral forms:

$$\boldsymbol{A}(\boldsymbol{r}, t) = \int_{-\infty}^{+\infty}\int_{-\infty}^{+\infty}\int_{-\infty}^{+\infty}\int_{-\infty}^{+\infty} \boldsymbol{A}(\boldsymbol{k}, \omega)e^{\mathrm{j}(\boldsymbol{k}\cdot\boldsymbol{r}-\omega t)}\mathrm{d}k_x\mathrm{d}k_y\mathrm{d}k_z\mathrm{d}\omega, \tag{3.27}$$

$$\rho(\boldsymbol{r}, t) = \int_{-\infty}^{+\infty}\int_{-\infty}^{+\infty}\int_{-\infty}^{+\infty}\int_{-\infty}^{+\infty} \rho(\boldsymbol{k}, \omega)e^{\mathrm{j}(\boldsymbol{k}\cdot\boldsymbol{r}-\omega t)}\mathrm{d}k_x\mathrm{d}k_y\mathrm{d}k_z\mathrm{d}\omega, \tag{3.28}$$

and:

$$\boldsymbol{A}(\boldsymbol{k},\omega)=\frac{1}{(2\pi)^4}\int\limits_{-\infty}^{+\infty}\int\limits_{-\infty}^{+\infty}\int\limits_{-\infty}^{+\infty}\int\limits_{-\infty}^{+\infty}\boldsymbol{A}(\boldsymbol{r},t)\mathrm{e}^{-\mathrm{j}(\boldsymbol{k}\cdot\boldsymbol{r}-\omega t)}\mathrm{d}x\mathrm{d}y\mathrm{d}z\mathrm{d}t, \tag{3.29}$$

$$\rho(\boldsymbol{k},\omega)=\frac{1}{(2\pi)^4}\int\limits_{-\infty}^{+\infty}\int\limits_{-\infty}^{+\infty}\int\limits_{-\infty}^{+\infty}\int\limits_{-\infty}^{+\infty}\rho(\boldsymbol{r},t)\mathrm{e}^{-\mathrm{j}(\boldsymbol{k}\cdot\boldsymbol{r}-\omega t)}\mathrm{d}x\mathrm{d}y\mathrm{d}z\mathrm{d}t, \tag{3.30}$$

Substituting (3.24) into the d'Alembertian equation and using the Fourier integrals, the expression for the vector potential $\boldsymbol{A}$ is obtained:

$$\boldsymbol{A}(\boldsymbol{r},\omega)=\hat{\boldsymbol{z}}\frac{q}{2\pi^2}\frac{1}{c_0}\int\limits_{-\infty}^{+\infty}\mathrm{d}k_y\mathrm{e}^{\mathrm{j}k_y y}\mathrm{e}^{\mathrm{j}\frac{\omega}{\upsilon}z}\int\limits_{-\infty}^{+\infty}\frac{\mu\mathrm{e}^{\mathrm{j}k_x x}}{k_x^2-k_{x1}^2}\mathrm{d}k_x=\frac{\mu\pi\mathrm{j}}{k_{x1}}\mathrm{e}^{\mathrm{j}k_{x1}|x|}, \tag{3.31}$$

where $k_{x1}=\mathrm{j}\sqrt{k_y^2+\omega^2/\upsilon^2-\varepsilon_1\mu_1\omega^2/c_0^2}$ represents the x-component of the wave vector $\boldsymbol{k}_1$. Therefore, (3.31) can be simplified for the unbounded free space case:

$$\boldsymbol{A}_0(\boldsymbol{r},\omega)=\hat{\boldsymbol{z}}A_{0z}=\hat{\boldsymbol{z}}\frac{q}{2\pi}\frac{\mathrm{j}}{c_0}\int\limits_{-\infty}^{+\infty}\frac{\mathrm{d}k_y}{k_{x1}}\mathrm{e}^{\mathrm{j}\left(k_{x1}|x|+k_y y+\frac{\omega}{\upsilon}z\right)}, \tag{3.32}$$

For the case depicted in Fig. 3.8, the part $\boldsymbol{A}_1(\boldsymbol{r},\omega)$ in vacuum can be written in the following form:

$$\boldsymbol{A}_1(\boldsymbol{r},\omega)=\mathrm{j}\int\limits_{-\infty}^{+\infty}\boldsymbol{a}_1\mathrm{d}k_y\mathrm{e}^{\mathrm{j}(2k_{x1}d+k_{x1}x+k_y y+\frac{\omega}{\upsilon}z)}, \tag{3.33}$$

where $\boldsymbol{a}_1=a_{1x}\hat{\boldsymbol{x}}+a_{1y}\hat{\boldsymbol{y}}+a_{1z}\hat{\boldsymbol{z}}$ represents the field coefficients in vacuum. Clearly, since CR does not occur, the electromagnetic wave excited by the charged particle in vacuum must be a surface wave, and this surface wave decays exponentially with distance from the interface at $x=-d$.

In the space filled with DNMs, $\boldsymbol{A}_2(\boldsymbol{r},\omega)$ can be expressed in the following form:

$$\boldsymbol{A}_2(\boldsymbol{r},\omega)=\mathrm{j}\int\limits_{-\infty}^{+\infty}\boldsymbol{a}_2\mathrm{d}k_y\mathrm{e}^{\mathrm{j}[k_{x1}d+k_{x2}(x+d)+k_y y+\frac{\omega}{\upsilon}z]}, \tag{3.34}$$

where $\boldsymbol{a}_2=a_{2x}\hat{\boldsymbol{x}}+a_{2y}\hat{\boldsymbol{y}}+a_{2z}\hat{\boldsymbol{z}}$ represents the field coefficients in the region filled with DNMs. The x-component of the wave vector $\boldsymbol{k}_2$ should be:

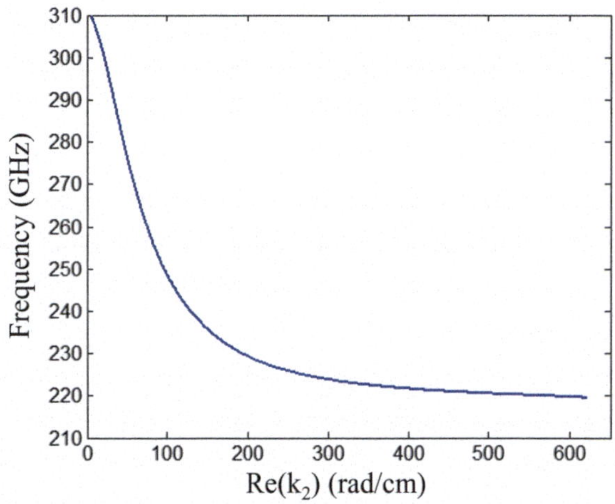

Fig. 3.9 The dispersion relation in the double-negative frequency range

$$k_{x2}^2 = \frac{\omega^2}{c_0^2}\varepsilon_2\mu_2 - \frac{\omega^2}{\upsilon^2} - k_y^2 = \frac{\omega^2}{c_0^2}[\varepsilon_2'\mu_2' - \varepsilon_2''\mu_2'' + \mathrm{j}(\varepsilon_2''\mu_2' + \varepsilon_2'\mu_2'')] - \frac{\omega^2}{\upsilon^2} - k_y^2, \tag{3.35}$$

where $\varepsilon_2' = \mathrm{Re}\,(\varepsilon_2)$, $\varepsilon_2'' = \mathrm{Im}\,(\varepsilon_2)$, $\mu_2' = \mathrm{Re}\,(\mu_2)$, and $\mu_2'' = \mathrm{Im}\,(\mu_2)$. In any passive electromagnetic media, $\varepsilon_2'' > 0$ and $\mu_2'' > 0$. Therefore, for DNMs, $\mathrm{Im}\left(k_{x2}^2\right) < 0$. If the propagating waves in DNMs are considered, that is, $-\pi/2 < \arg k_{x2}^2 < 0$. So, there are two choices here: ① $-\pi/4 < \arg k_{x2} < 0$; ② $3\pi/4 < \arg k_{x2} < \pi$. The first choice satisfies the requirement $\mathrm{Im}(k_{x2}) < 0$, while the second choice does not. Therefore, for positive frequencies, $-\pi/4 < \arg k_{x2} < 0$ implies $\mathrm{Re}\,(k_{x2}) > 0$. Both the directions of the group velocity and energy flux density are opposite to the direction of $\boldsymbol{k}_2$. Therefore, these vectors have negative projections on the x-axis. The dispersion curves in the double-negative frequency range clearly illustrate these physical characteristics, as shown in Fig. 3.9.

Additionally, the scalar potential is represented as follows:

$$\varphi_0(\boldsymbol{r}, \omega) = \frac{c_0}{\upsilon}A_{0z}(\boldsymbol{r}, \omega), \tag{3.36}$$

$$\varphi_1(\boldsymbol{r}, \omega) = 0, \tag{3.37}$$

$$\varphi_2(\boldsymbol{r}, \omega) = 0, \tag{3.38}$$

The electromagnetic field in vacuum half-space is determined by φ_0 and $\boldsymbol{A}_0 + \boldsymbol{A}_1$, while the electromagnetic field in the half-space filled with DNMs is determined by $\boldsymbol{A}_2$. Therefore, by matching the electromagnetic fields at the boundary $x = -d$, the field components in both vacuum half-space and the half-space filled with DNMs can be obtained [7].

The total field of a single charged particle in semi-unbounded space, composed of the field due to a single charged particle in unbounded space and the field generated due to the existence of the boundary, allows for the following calculation of the total radiation energy of the unit path length charge:

$$\frac{dW_t}{dz} = qE_{1z}\Bigg|_{\substack{x=y=0\\ z\to \upsilon t}} = \frac{q}{c_0^2}\int_{-\infty}^{\infty} d\omega \int_{-\infty}^{\infty} dk_y \left[\frac{q}{2\pi k_{x1}}\left(\frac{c_0^2}{\upsilon^2}-1\right) - c_0 a_{1z}\omega e^{j2k_{x1}d}\right]$$
$$= \frac{2q}{c_0^2}\mathrm{Re}\left\{\int_0^{\infty} d\omega \int_{-\infty}^{\infty} dk_y \left[\frac{q}{2\pi k_{x1}}\left(\frac{c_0^2}{\upsilon^2}-1\right) - c_0 a_{1z}\omega e^{j2k_{x1}d}\right]\right\}. \tag{3.39}$$

Since the first term within the curly braces in (3.39) does not contribute to CR [7], the energy of CR per unit path length charge can be expressed as:

$$\frac{dW}{dz} = -\frac{2q}{c_0}\mathrm{Re}\left\{\int_0^{\infty} d\omega \int_{-\infty}^{\infty} dk_y a_{1z}\omega e^{j2k_{x1}d}\right\}. \tag{3.40}$$

The upper limit of the integration region for k_y can be determined by $k_y^2 < \omega^2 / c_0^2(\varepsilon_2'\mu_2' - \varepsilon_2''\mu_2'') - \omega^2 / \upsilon^2$, where the integrand in (3.40) is referred to as the radiation spectral density.

3.2.2 Numerical Calculations of Single-Particle Model

To analyze the physical characteristics of the electromagnetic radiation described above, numerical calculations are performed using the following parameters: the operating frequency ω is in the terahertz range, electric plasma frequency $\omega_p = 2\pi \times 500 \times 10^9$ rad/s, collision frequency $\gamma_e = \gamma_m = 1 \times 10^{10}$ rad/s, magnetic resonance frequency $\omega_0 = 2\pi \times 219 \times 10^9$ rad/s, the filling factor $F = 0.5$ for the SRR, electron velocity $\upsilon = 0.2c_0$, and distance $d = 0.05$ mm. Therefore, the electrons can excite RCR in the frequency range of 219.1–226.6 GHz, which falls within the double-negative region (219.05–309.65 GHz). In comparison to the case shown in Fig. 3.10, the results in this example clearly exhibit the backward characteristics of CR in DNMs, as indicated by the solid lines with arrows, while CR in common media like distilled water shows forward characteristics, as shown by the dashed lines with arrows. This result is in good agreement with the theoretical predictions presented in Sect. 3.2.1. In unbounded isotropic DNMs, the CR angle is symmetric about the z-axis [24]. However, in this example, the surface wave excited by the electron in vacuum is not symmetric about the z-axis, as seen in Fig. 3.10, and its

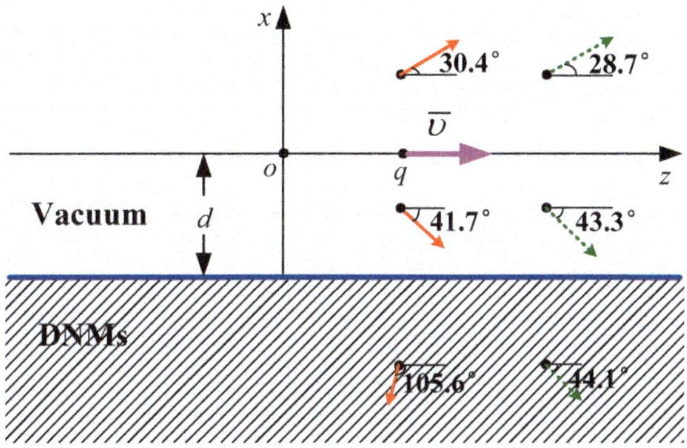

Fig. 3.10 The directions of the time-averaged Poynting vectors in vacuum and DNMs

amplitude decays exponentially as it moves away from the interface between vacuum and DNMs.

Next, the RCR spectral density is studied. Figure 3.11 indicates that this example exhibits a similar continuous distribution of spectral density as the unbounded case [24]. Subsequently, the influence of important parameters such as electron velocity and distance *d* on the RCR energy is investigated. It is evident from Fig. 3.12 that as the particle velocity increases, the radiation energy per unit length of CR increases. This is because the bandwidth of RCR will widen with increasing particle velocity, similar to the unbounded case. When the electron velocity is kept the same, the energy of RCR in the semi-unbounded case is significantly greater than that in the unbounded case, as shown in Fig. 3.12.

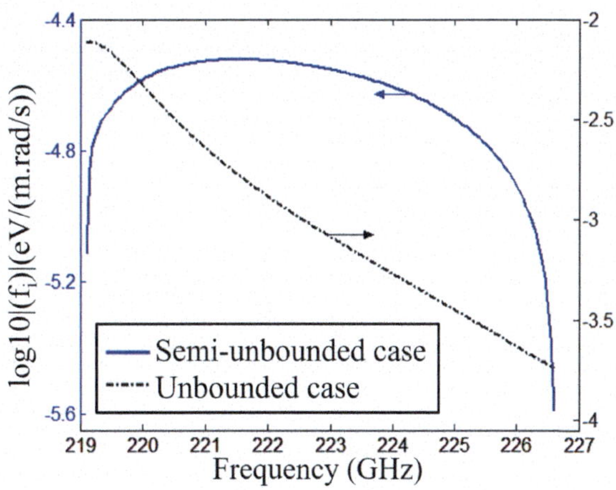

Fig. 3.11 Comparison of the spectral densities over the radiation frequency band between the semi-unbounded and unbounded cases

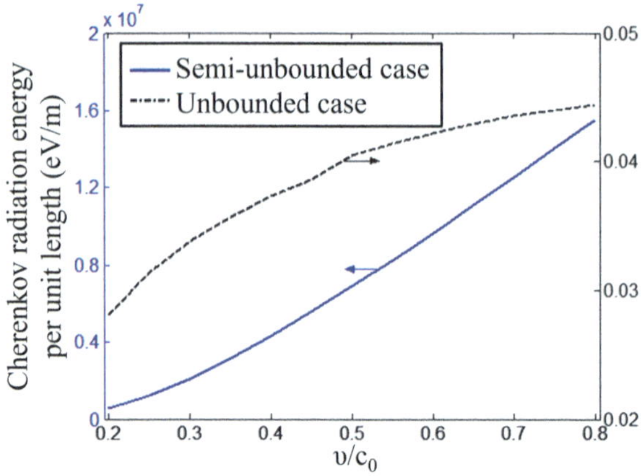

Fig. 3.12 Comparison of the CR energies as a function of the electron speed between the semi-unbounded and unbounded cases

The impact of distance *d* on RCR energy is shown in Fig. 3.13. It can be seen that the total radiation energy increases as the distance *d* decreases. This behavior is determined by the nature of surface waves. In other words, as the distance *d* increases, the electromagnetic field exhibits an exponential decay. Therefore, one can increase the velocity of the charged particle or bring the charged particle as close as possible to the interface, i.e., by reducing the distance *d*, to enhance RCR.

Finally, using the parameters and an operating frequency of 220 GHz, one can study the amplitude of surface waves in semi-unbounded vacuum. Compared to the case of ordinary electromagnetic media (i.e., the DNMs are replaced by RHMs), the

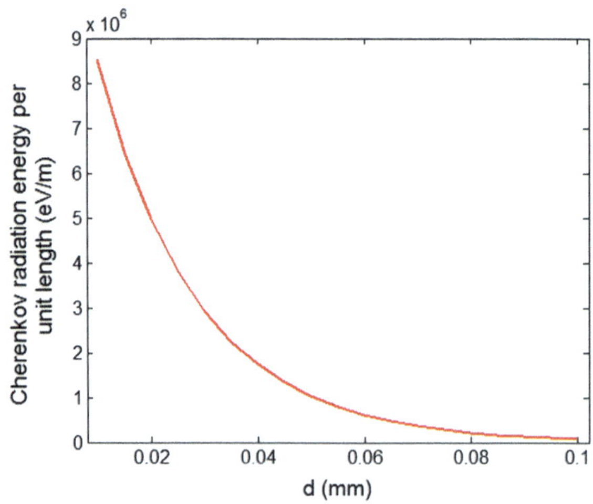

Fig. 3.13 Total radiation energy as a function of the distance *d* in the present case

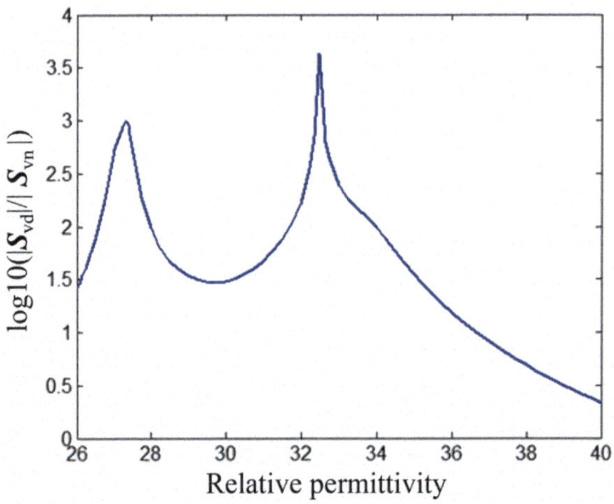

Fig. 3.14 Ratio of the amplitudes of the surface wave between the present and the normal dielectric material cases

amplitude of the time-averaged Poynting vector at $x = d/2$ for this surface wave increases by a factor of 3×10^3–5×10^3, as shown in Fig. 3.14. Here, $|\langle \boldsymbol{S}_{vd} \rangle|$ and $|\langle \boldsymbol{S}_{vn} \rangle|$ represent the amplitudes of the time-averaged Poynting vector at $x = d/2$ for semi-unbounded vacuum and RHMs, respectively. When the relative permittivity of the RHM is less than 25, the conditions for CR are not met. Compared to some RHMs (such as BaO with a relative permittivity of 34 ± 1), the use of DNMs can significantly increase the amplitude of surface waves. This method presented here is helpful for the generation of electromagnetic waves, especially in the terahertz frequency range.

3.2.3 Theoretical Analysis of Multi-Particles Model

In analyzing the RCR in semi-unbounded isotropic LHMs, a single-particle model is used. Since a single charged particle is involved, the total energy of the excited RCR is inevitably very small. To effectively enhance its radiation energy, a multi-particle model is utilized to study the RCR of the semi-unbounded case. The multi-particle model here employs a sheet beam [43–45], with dimensions of $2x_0 \times 2y_0 \times 2z_0$. The generation and focusing of the sheet beam will be discussed in Sect. 4.2 of Chap. 4. Using a Cartesian coordinate system, the coordinate origin is located at the center of the sheet beam, as shown in Fig. 3.15. Like the case in [46], if the charged particles are compressed into the sheet beam smaller than one operating wavelength, coherent fields' superposition will occur. For simplicity, it is assumed that the charged particles are uniformly distributed within the sheet beam and moving with velocity $\boldsymbol{v}$ along the $+z$ direction. To increase the total current, one of the best methods is to increase the transverse dimension $2y_0$ of the sheet beam while keeping $2x_0$ unchanged. The charge density of the sheet beam is given by the following expression:

$$\rho(\boldsymbol{r}, t) = \frac{Nq}{8x_0 y_0 z_0} X(x)\, Y(y)\, T(t, z), \tag{3.41}$$

where $\boldsymbol{r}$ is the position vector, N is the total number of charged particles in the sheet beam, $X(x)$, $Y(y)$, and $T(t, z)$ are defined as follows:

$$X(x) = Y(y) = T(t, z) = \begin{cases} 1, & (|x| < x_0, |y| < y_0, |z - vt| < z_0) \\ 0, & \text{otherwise} \end{cases}, \tag{3.42}$$

The electromagnetic properties of isotropic and homogeneous DNMs are represented by their effective electromagnetic parameters [24]:

$$\varepsilon_{r2}(\omega) = 1 - \frac{\omega_p^2}{\omega^2 + \mathrm{j}\gamma_e \omega}, \tag{3.43}$$

$$\mu_{r2}(\omega) = 1 - \frac{F\omega^2}{\omega^2 - \omega_0^2 + \mathrm{j}\gamma_m \omega}, \tag{3.44}$$

Combining arrays of SRRs that can generate negative permeability $\mu_{r2}(\omega)$ with arrays of metallic wires that can generate negative permittivity $\varepsilon_{r2}(\omega)$ can achieve isotropic DNMs [47]. Such DNMs with a thickness of several wavelengths can be realized using metasurfaces [48]. All-metal metasurfaces do not require supporting structure and do not suffer from dielectric outgassing issue, making them suitable for high-vacuum environment.

The space investigated here is divided into two regions: ① the vacuum region 1 ($x > -d$); ② the region 2 filled with DNMs ($x \leq -d$). The Fourier transform is performed using the time factor $e^{-\mathrm{j}\omega t}$. Combining the electromagnetic potentials ($\boldsymbol{A}$, φ), applying the generalized Ampere's law and Faraday's law, substituting the current density defined in (3.41) ($\boldsymbol{J} = \rho \boldsymbol{v}$) into the generalized Ampere's law, and choosing the Lorenz gauge, which is given by $\nabla \cdot \boldsymbol{A} + (\partial\varphi)/(c_0 \cdot \partial t) = 0$, one can derive the wave equation for region 1.

Similarly, due to $\boldsymbol{J} = 0$ in region 2, one can derive the corresponding wave equation. By using Green's functions to solve the non-homogeneous and homogeneous differential equations for regions 1 and 2 separately, one can obtain their general solutions. Note that the case where the RCR conditions are satisfied in the DNMs is considered [29]. Next, according to the boundary conditions at $x = -d$, one can determine the coefficients and thus obtain the solutions for the vector and scalar potentials ($\boldsymbol{A}$, φ). Based on the relationship between the fields and potentials: $\boldsymbol{B} = \nabla \times \boldsymbol{A}$ and $\boldsymbol{E} = -\nabla\varphi + \mathrm{j}\omega \boldsymbol{A}/c_0$, the field components in regions 1 and 2 will have quite complex analytical expressions that can be determined through mathematical manipulation.

Finally, the power density of the surface wave characterized by the time-averaged Poynting vector ($\langle \boldsymbol{S}_1 \rangle$) in region 1, the power density of the RCR characterized by the time-averaged Poynting vector ($\langle \boldsymbol{S}_2 \rangle$) in region 2, and the total radiation energy per unit path length of the charge (dW/dz) can be obtained. The time-averaged Poynting vector ($\langle \boldsymbol{S}_1 \rangle$) in vacuum can be expressed:

$$
\begin{aligned}
\langle \boldsymbol{S}_1 \rangle &= \frac{1}{2}\mathrm{Re}\left[\boldsymbol{E}_1(\boldsymbol{r},\omega) \times \boldsymbol{H}_1^*(\boldsymbol{r},\omega)\right] \\
&= \frac{1}{2}\mathrm{Re}\left\{\hat{\boldsymbol{x}}\left[E_{1y}(\boldsymbol{r},\omega)H_{1z}^*(\boldsymbol{r},\omega) - E_{1z}(\boldsymbol{r},\omega)H_{1y}^*(\boldsymbol{r},\omega)\right]\right\} \\
&+ \frac{1}{2}\mathrm{Re}\left\{\hat{\boldsymbol{y}}\left[E_{1z}(\boldsymbol{r},\omega)H_{1x}^*(\boldsymbol{r},\omega) - E_{1x}(\boldsymbol{r},\omega)H_{1z}^*(\boldsymbol{r},\omega)\right]\right\} \\
&+ \frac{1}{2}\mathrm{Re}\left\{\hat{\boldsymbol{z}}\left[E_{1x}(\boldsymbol{r},\omega)H_{1y}^*(\boldsymbol{r},\omega) - E_{1y}(\boldsymbol{r},\omega)H_{1x}^*(\boldsymbol{r},\omega)\right]\right\}
\end{aligned} \tag{3.45}
$$

For $x > 0$ and $-d < x < 0$, the expressions for the electromagnetic field components in vacuum region are different.

(1) when $x > 0$:

$$
\begin{aligned}
E_{1x}(\boldsymbol{r},\omega) &= \frac{c_0}{\upsilon}\int_{-\infty}^{+\infty} a_{0z}k_{x1}dk_y e^{\mathrm{j}\left(k_{x1}x + k_y y + \frac{\omega}{\upsilon}z\right)} \\
&- \frac{1}{c_0}\int_{-\infty}^{+\infty} a_{1x}\omega \mathrm{d}k_y e^{\mathrm{j}\left(2k_{x1}d + k_{x1}x + k_y y + \frac{\omega}{\upsilon}z\right)}
\end{aligned} \tag{3.46}
$$

$$
\begin{aligned}
E_{1y}(\boldsymbol{r},\omega) &= \frac{c_0}{\upsilon}\int_{-\infty}^{+\infty} a_{0z}k_y \mathrm{d}k_y e^{\mathrm{j}\left(k_{x1}x + k_y y + \frac{\omega}{\upsilon}z\right)} \\
&- \frac{1}{c_0}\int_{-\infty}^{+\infty} a_{1y}\omega \mathrm{d}k_y e^{\mathrm{j}\left(2k_{x1}d + k_{x1}x + k_y y + \frac{\omega}{\upsilon}z\right)}
\end{aligned} \tag{3.47}
$$

$$
\begin{aligned}
E_{1z}(\boldsymbol{r},\omega) &= \int_{-\infty}^{+\infty}\left(\frac{1}{\upsilon^2} - \frac{1}{c_0^2}\right)c_0\omega a_{0z}dk_y e^{\mathrm{j}\left(k_{x1}x + k_y y + \frac{\omega}{\upsilon}z\right)} \\
&- \frac{1}{c_0}\int_{-\infty}^{+\infty} a_{1z}\omega dk_y e^{\mathrm{j}\left(2k_{x1}d + k_{x1}x + k_y y + \frac{\omega}{\upsilon}z\right)}
\end{aligned} \tag{3.48}
$$

$$
H_{1x}(\boldsymbol{r},\omega) = -\int_{-\infty}^{+\infty}\left(a_{1z}k_y - a_{1y}\frac{\omega}{\upsilon}\right)\mathrm{d}k_y e^{\mathrm{j}\left(2k_{x1}d + k_{x1}x + k_y y + \frac{\omega}{\upsilon}z\right)}
$$

$$-\int_{-\infty}^{+\infty} a_{0z}k_y \mathrm{d}k_y \mathrm{e}^{\mathrm{j}\left(k_{x1}x+k_y y+\frac{\omega}{\upsilon}z\right)} \tag{3.49}$$

$$H_{1y}(\boldsymbol{r},\omega) = \int_{-\infty}^{+\infty} \left(a_{1z}k_{x1} - a_{1x}\frac{\omega}{\upsilon}\right)\mathrm{d}k_y \mathrm{e}^{\mathrm{j}\left(2k_{x1}d+k_{x1}x+k_y y+\frac{\omega}{\upsilon}z\right)}$$

$$+\int_{-\infty}^{+\infty} a_{0z}k_{x1} \mathrm{d}k_y \mathrm{e}^{\mathrm{j}\left(k_{x1}x+k_y y+\frac{\omega}{\upsilon}z\right)} \tag{3.50}$$

$$H_{1z}(\boldsymbol{r},\omega) = -\int_{-\infty}^{+\infty} (a_{1y}k_{x1} - a_{1x}k_y)\mathrm{d}k_y \mathrm{e}^{\mathrm{j}\left(2k_{x1}d+k_{x1}x+k_y y+\frac{\omega}{\upsilon}z\right)} \tag{3.51}$$

(2) when $-d < x < 0$:

$$E_{1x}(\boldsymbol{r},\omega) = -\frac{c_0}{\upsilon}\int_{-\infty}^{+\infty} a_{0z}k_{x1} \mathrm{d}k_y \mathrm{e}^{\mathrm{j}\left(-k_{x1}x+k_y y+\frac{\omega}{\upsilon}z\right)}$$

$$-\frac{1}{c_0}\int_{-\infty}^{+\infty} a_{1x}\omega \mathrm{d}k_y \mathrm{e}^{\mathrm{j}\left(2k_{x1}d+k_{x1}x+k_y y+\frac{\omega}{\upsilon}z\right)} \tag{3.52}$$

$$E_{1y}(\boldsymbol{r},\omega) = \frac{c_0}{\upsilon}\int_{-\infty}^{+\infty} a_{0z}k_y \mathrm{d}k_y \mathrm{e}^{\mathrm{j}\left(-k_{x1}x+k_y y+\frac{\omega}{\upsilon}z\right)}$$

$$-\frac{1}{c_0}\int_{-\infty}^{+\infty} a_{1y}\omega \mathrm{d}k_y \mathrm{e}^{\mathrm{j}\left(2k_{x1}d+k_{x1}x+k_y y+\frac{\omega}{\upsilon}z\right)} \tag{3.53}$$

$$E_{1z}(\boldsymbol{r},\omega) = \int_{-\infty}^{+\infty} \left(\frac{1}{\upsilon^2} - \frac{1}{c_0^2}\right)c_0\omega \mathrm{d}k_y \mathrm{e}^{\mathrm{j}\left(-k_{x1}x+k_y y+\frac{\omega}{\upsilon}z\right)}$$

$$-\frac{1}{c_0}\int_{-\infty}^{+\infty} a_{1z}\omega \mathrm{d}k_y \mathrm{e}^{\mathrm{j}\left(2k_{x1}d+k_{x1}x+k_y y+\frac{\omega}{\upsilon}z\right)} \tag{3.54}$$

$$H_{1x}(r,\omega) = -\int_{-\infty}^{+\infty} \left(a_{1z}k_y - a_{1y}\frac{\omega}{\upsilon}\right)dk_y e^{\mathrm{j}\left(2k_{x1}d+k_{x1}x+k_y y+\frac{\omega}{\upsilon}z\right)}$$

$$-\int_{-\infty}^{+\infty} a_{0z}k_y dk_y e^{\mathrm{j}\left(-k_{x1}x+k_y y+\frac{\omega}{\upsilon}z\right)} \tag{3.55}$$

$$H_{1y}(\boldsymbol{r},\omega)=\int_{-\infty}^{+\infty}\left(a_{1z}k_{x1}-a_{1x}\frac{\omega}{\upsilon}\right)\mathrm{d}k_y\mathrm{e}^{\mathrm{j}\left(2k_{x1}d+k_{x1}x+k_yy+\frac{\omega}{\upsilon}z\right)}-\int_{-\infty}^{+\infty}a_{0z}k_{x1}\mathrm{d}k_y\mathrm{e}^{\mathrm{j}\left(-k_{x1}x+k_yy+\frac{\omega}{\upsilon}z\right)} \tag{3.56}$$

$$H_{1z}(\boldsymbol{r},\omega)=-\int_{-\infty}^{+\infty}(a_{1y}k_{x1}-a_{1x}k_y)\mathrm{d}k_y\mathrm{e}^{\mathrm{j}\left(2k_{x1}d+k_{x1}x+k_yy+\frac{\omega}{\upsilon}z\right)} \tag{3.57}$$

It should be mentioned that $\boldsymbol{k}_1=k_{x1}\hat{\boldsymbol{x}}+k_y\hat{\boldsymbol{y}}+k_z\hat{\boldsymbol{z}}$ and $\boldsymbol{k}_2=k_{x2}\hat{\boldsymbol{x}}+k_y\hat{\boldsymbol{y}}+k_z\hat{\boldsymbol{z}}$ are the wave vectors in vacuum and DNM, respectively. The field coefficients in vacuum can be represented as:

$$a_{0z}=\frac{qN}{2\pi c_0}\frac{\sin(k_{x1}x_0)}{k_{x1}^2x_0}\frac{\sin(k_yy_0)}{k_yy_0}\frac{\sin\left(\frac{\omega}{\upsilon}z_0\right)}{\frac{\omega}{\upsilon}z_0} \tag{3.58}$$

$$a_{1x}=-\frac{k_y(-k_{x2}^2M_2+k_yM_3)+\frac{\omega}{\upsilon}\left(\frac{\omega}{\upsilon}M_3-k_{x2}^2M_1+\mu_2k_{x1}k_{x2}a_{0z}\right)}{(k_{x1}+\mu_2k_{x2})\left(k_y^2+\frac{\omega^2}{\upsilon^2}\right)+k_{x1}k_{x2}(k_{x2}+\mu_2k_{x1})} \tag{3.59}$$

$$a_{1y}=\frac{-(k_{x1}+\mu_2k_{x2})k_y\frac{\omega}{\upsilon}\left(\frac{\omega}{\upsilon}M_3-k_{x2}^2M_1+\mu_2k_{x1}k_{x2}a_{0z}\right)}{(k_{x1}+\mu_2k_{x2})(k_{x2}+\mu_2k_{x1})\left(k_{x2}k_y^2+k_{x2}\frac{\omega^2}{\upsilon^2}\right)+k_{x1}k_{x2}^2(k_{x2}+\mu_2k_{x1})^2}+\frac{(-k_{x2}^2M_2+k_yM_3)\left[\frac{\omega^2}{\upsilon^2}(k_{x1}+\mu_2k_{x2})+k_{x1}(k_{x2}^2+\mu_2k_{x1}k_{x2})\right]}{(k_{x1}+\mu_2k_{x2})(k_{x2}+\mu_2k_{x1})\left(k_{x2}k_y^2+k_{x2}\frac{\omega^2}{\upsilon^2}\right)+k_{x1}k_{x2}^2(k_{x2}+\mu_2k_{x1})^2}, \tag{3.60}$$

$$a_{1z}=\frac{(k_{x1}+\mu_2k_{x2})k_y^2(-k_{x2}^2M_1+\mu_2k_{x1}k_{x2}a_{0z})+k_y\frac{\omega}{\upsilon}(k_{x1}+\mu_2k_{x2})k_{x2}^2M_2}{(k_{x1}+\mu_2k_{x2})(k_{x2}+\mu_2k_{x1})\left(k_{x2}k_y^2+k_{x2}\frac{\omega^2}{\upsilon^2}\right)+k_{x1}k_{x2}^2(k_{x2}+\mu_2k_{x1})^2}+\frac{k_{x1}(k_{x2}^2+\mu_2k_{x1}k_{x2})\left(\frac{\omega}{\upsilon}M_3-k_{x2}^2M_1+\mu_2k_{x1}k_{x2}a_{0z}\right)}{(k_{x1}+\mu_2k_{x2})(k_{x2}+\mu_2k_{x1})\left(k_{x2}k_y^2+k_{x2}\frac{\omega^2}{\upsilon^2}\right)+k_{x1}k_{x2}^2(k_{x2}+\mu_2k_{x1})^2} \tag{3.61}$$

The expressions for M_1, M_2, and M_3 are listed below:

$$M_1=\left(1-\frac{c_0^2}{\upsilon^2}\right)a_{0z}, \tag{3.62}$$

$$M_2=-\frac{k_yc_0^2}{\omega\upsilon}a_{0z}, \tag{3.63}$$

$$M_3 = -\frac{\omega}{\upsilon}M_1 - k_y M_2, \tag{3.64}$$

Similarly, the time-averaged Poynting vector (<S_2>) in the DNM region can be derived. The expressions for the electromagnetic field components in the DNM region are given below:

$$E_{2x}(\boldsymbol{r},\omega) = -\frac{1}{c_0}\int_{-\infty}^{+\infty} a_{2x}\omega dk_y e^{j\left[k_{x1}d + k_{x2}(x+d) + k_y y + \frac{\omega}{\upsilon}z\right]}, \tag{3.65}$$

$$E_{2y}(\boldsymbol{r},\omega) = -\frac{1}{c_0}\int_{-\infty}^{+\infty} a_{2y}\omega \mathrm{d}k_y e^{j\left[k_{x1}d + k_{x2}(x+d) + k_y y + \frac{\omega}{\upsilon}z\right]}, \tag{3.66}$$

$$E_{2z}(\boldsymbol{r},\omega) = -\frac{1}{c_0}\int_{-\infty}^{+\infty} a_{2z}\omega \mathrm{d}k_y e^{j\left[k_{x1}d + k_{x2}(x+d) + k_y y + \frac{\omega}{\upsilon}z\right]} \tag{3.67}$$

$$H_{2x}(\boldsymbol{r},\omega) = -\frac{1}{\mu_2}\int_{-\infty}^{+\infty} \left(a_{2z}k_y - a_{2y}\frac{\omega}{\upsilon}\right)\mathrm{d}k_y e^{j\left[k_{x1}d + k_{x2}(x+d) + k_y y + \frac{\omega}{\upsilon}z\right]}, \tag{3.68}$$

$$H_{2y}(\boldsymbol{r},\omega) = \frac{1}{\mu_2}\int_{-\infty}^{+\infty} \left(a_{2z}k_{x2} - a_{2x}\frac{\omega}{\upsilon}\right)\mathrm{d}k_y e^{j\left[k_{x1}d + k_{x2}(x+d) + k_y y + \frac{\omega}{\upsilon}z\right]}, \tag{3.69}$$

$$H_{2z}(\boldsymbol{r},\omega) = -\frac{1}{\mu_2}\int_{-\infty}^{+\infty} (a_{2y}k_{x2} - a_{2x}k_y)\mathrm{d}k_y e^{j\left[k_{x1}d + k_{x2}(x+d) + k_y y + \frac{\omega}{\upsilon}z\right]} \tag{3.70}$$

The field coefficients in the DNM region can be calculated:

$$a_{2x} = \frac{k_{x1}a_{1x} + M_3}{k_{x2}}, \tag{3.71}$$

$$a_{2y} = a_{1y} + M_2, \tag{3.72}$$

$$a_{2z} = a_{1z} + M_1, \tag{3.73}$$

Finally, the total RCR energy per unit path length for the charges can be obtained from (3.74) [7]:

$$\frac{dW}{dz} = \frac{2qN}{c_0^2}\mathrm{Re}\left\{\int_0^{\infty} d\omega \int_{-\infty}^{\infty} dk_y \left[\left(\frac{c_0^2}{\upsilon^2} - 1\right)c_0 a_{0z}\omega - c_0 a_{1z}\omega e^{j(2k_{x1}d)}\right]\right\} \tag{3.74}$$

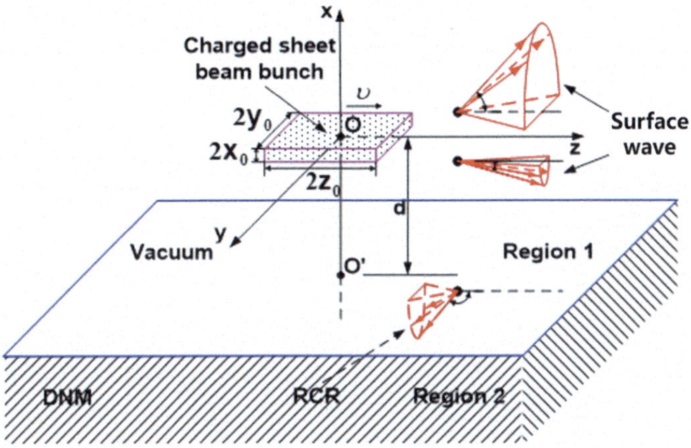

Fig. 3.15 The schematic of a sheet beam moving with speed υ in vacuum above the DNM, showing the resultant radiation patterns of the RCR and surface wave

The physical mechanism of enhancing electromagnetic radiation using a physical model based on semi-unbounded DNMs and a sheet beam is explained: at the operating frequency, the time-varying magnetic field can induce strong resonance in the SRRs, leading to two stages of electromagnetic radiation. First, RCR occurs within the DNMs. Next, the RCR couples with the interface between the DNMs and vacuum, resulting in the appearance of surface waves. In the case of the Smith-Purcell effect, the axial electric field directly generates electromagnetic radiation [49].

The radiation directions of surface waves and RCR are shown in Fig. 3.15. It can be seen that surface waves radiate forward and are not symmetric about the z-axis. However, RCR in DNMs is concentrated in half the CR angle and is symmetric in half of the azimuthal angle.

3.2.4 Numerical Calculations of Multi-Particles Model

An example provided for the theoretical analysis illustrates the enhancement of the surface wave amplitude through coupling with enhanced RCR. The following parameters are used for the numerical calculations: $\omega_p = 2\pi \times 450 \times 10^9$ rad/s, $\omega_0 = 2\pi \times 213 \times 10^9$ rad/s, $\gamma_e = \gamma_m = \gamma = 2\pi \times 1.5 \times 10^{10}$ rad/s, $F = 0.6$, $x_0 = 0.1$ mm, $y_0 = 0.4$ mm, $z_0 = 0.5$ mm, $\upsilon = 0.8c_0$, $N = 5 \times 10^{11}$, and $d = 0.5$ mm. Figure 3.16 shows the normalized amplitudes $|\langle S_1\rangle|$ at $x = -d/2$ and $|\langle S_2\rangle|$ at $x = -3d/2$ as functions of the electron velocity $\boldsymbol{\upsilon}$. From Fig. 3.16, it can be observed that when the electron velocity $\boldsymbol{\upsilon}$ is less than the phase velocity $\boldsymbol{\upsilon}_p$, $|\langle S_1\rangle|$ can be neglected, and $|\langle S_2\rangle|$ is zero. Once the RCR conditions are met, $|\langle S_2\rangle|$ exhibits a threshold effect similar to normal forward CR. Furthermore, due to the coupling between RCR and surface waves, $|\langle S_1\rangle|$ and $|\langle S_2\rangle|$ increase with increasing

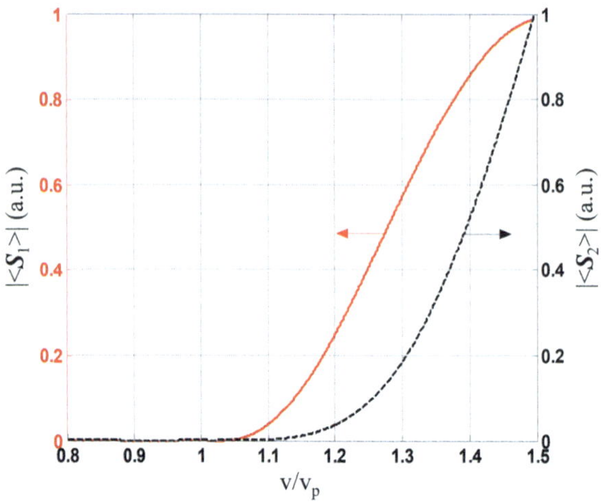

Fig. 3.16 The normalized time-averaged Poynting vectors in regions 1 and 2, respectively, as functions of the particle velocity normalized to the phase velocity

electron velocity, indicating the enhancement of surface waves. To clearly show the relationship between $|<S_1>|$ and $|<S_2>|$ with velocity, they are normalized based on the values at $\upsilon/\upsilon_p = 1.5$. The results indicate that $|<S_1>|$ and $|<S_2>|$ at $\upsilon/\upsilon_p = 1.49$ are 10^3 and 10^5 times greater, respectively, than the values at $\upsilon/\upsilon_p = 1.01$. These results demonstrate that the physical mechanism of the enhancement of surface waves is the generation of enhanced RCR in the DNMs, which couples with the vacuum interface to produce enhanced surface waves.

Next, let's discuss the key factors affecting the enhanced surface wave and RCR. By changing the parameters of the DNMs so that $|\text{Re}(\varepsilon_{r2})|$ and $|\text{Re}(\mu_{r2})|$ are greater than 1, the resonance spectra can be provided in any desired terahertz frequency range, as shown in Fig. 3.17. To enhance the resonance of the SRR, a feasible approach is to increase the filling factor F. For convenience, the amplitude of the time-averaged Poynting vector $|<S_1>|$ at the operating frequency of 230 GHz is used to characterize the amplitude of the surface wave. Figure 3.18 shows that $|<S_1>|$ at $x = -d/2$ and dW/dz both increase with the increase in F. According to the theory of negative effective permeability [46], when F increases, both the resonance frequency and loss decrease. Therefore, increasing F can enhance the amplitude of the surface wave and RCR, which is equivalent to reducing the loss in the DNMs. This is because the resonance enhancement in the DNM leads to enhanced RCR, which can more effectively couple to form an enhanced surface wave. As discussed earlier, in theory, one can freely manipulate the real part of the refractive index in the DNMs, making its absolute value much greater than 1. This results in a particle velocity threshold $\upsilon_t = c_0/\sqrt{\text{Re}(\varepsilon_{r2})\text{Re}(\mu_{r2})}$ that is much smaller than c_0. This is a significant advantage compared to RHMs and allows for the generation of enhanced RCR and surface waves using lower electron beam voltage.

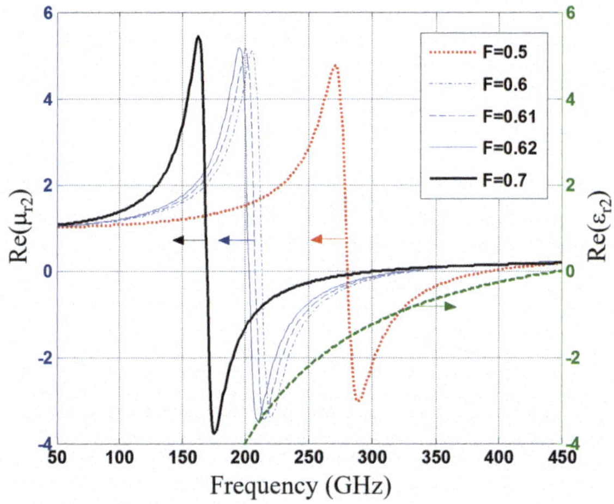

Fig. 3.17 The relative permittivity and permeability for different filling factors (F) as functions of frequency

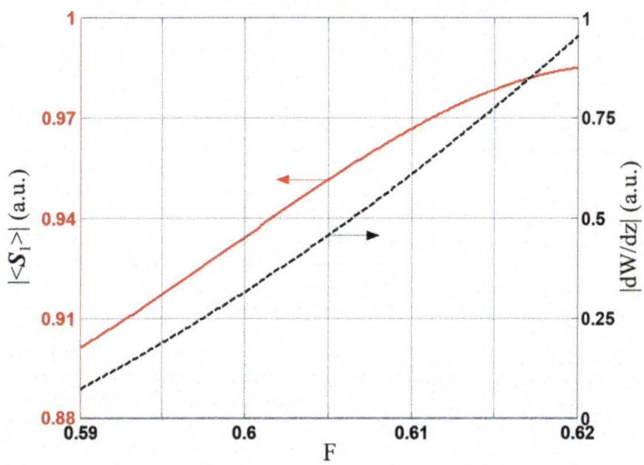

Fig. 3.18 The effects of the filling factor (F) on the amplitude of the surface wave in region 1 and on the RCR energy in region 2, respectively

Finally, let's discuss the impact of the sheet beam on the surface wave and RCR. As the number of electrons increases, both the amplitudes of the surface wave ($|<\boldsymbol{S}_1>|$ at $x = -d/2$) and the RCR increase, as shown in Fig. 3.19. Furthermore, in Fig. 3.20, when the current density of the electron beam is kept constant and the transverse size $2y_0$ is increased by a factor of 10, the amplitude of the surface wave at $x = -d/2$ and the energy of RCR increase by approximately 32 times and 78 times, respectively. To avoid the transport instability of the sheet beam, the ratio y_0/x_0 should be less than ~ 50 [30].

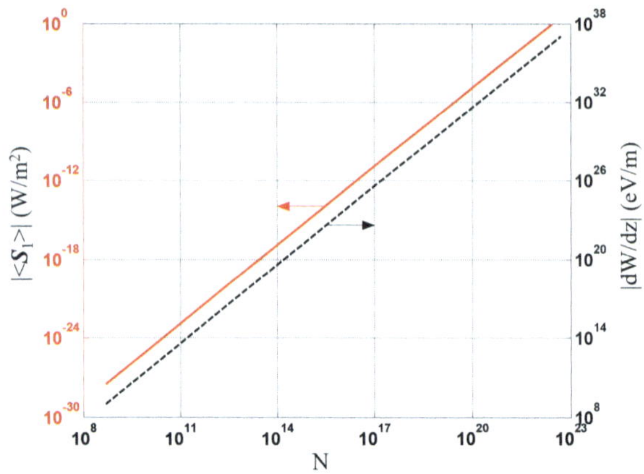

Fig. 3.19 The effects of the charged particle number (N) on the amplitude of the surface wave in region 1 and on the RCR energy in region 2, respectively

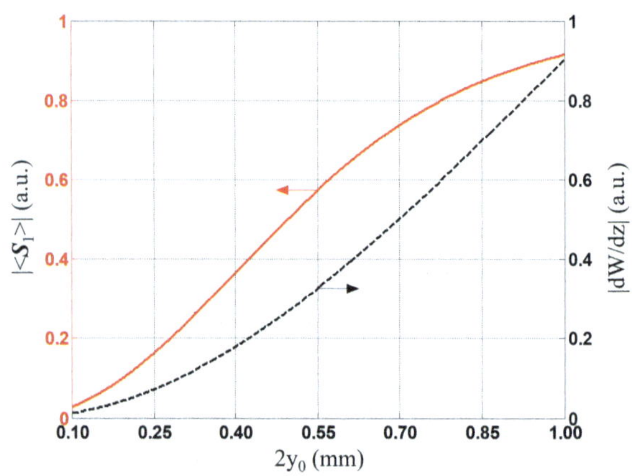

Fig. 3.20 The effects of the transverse dimension $2y_0$ on the amplitude of the surface wave in region 1 and on the RCR energy in region 2, respectively

3.3 Reversed Cherenkov Radiation in a Circular Waveguide Filled with Double Negative Metamaterial

3.3.1 *Theoretical Analysis of Single-Particle Model*

For the case of a circular waveguide filled with anisotropic DNMs in which single particle moves, the idealized cases of unbounded and semi-unbounded DNMs are not suitable for experimental verification. In this section, the excitation of RCR by a single charged particle in a circular waveguide filled with anisotropic DNMs is analyzed. The DNMs are characterized by negative permittivity and negative permeability tensors, represented by using the Drude model for permittivity and the Lorentz model for permeability, respectively [41]. Here, a represents the radius of the circular waveguide, which is assumed to be made of perfect electric conductor (PEC). As shown in Fig. 3.21, a charged particle with charge q moves uniformly along the axis of the circular waveguide with a velocity $\boldsymbol{v} = \hat{z}v$. In the cylindrical coordinate system (ρ, θ, z), the permittivity and permeability tensors for the anisotropic DNM are given below:

$$\boldsymbol{\varepsilon} = \varepsilon_0 \begin{bmatrix} \varepsilon_{r\rho} & 0 & 0 \\ 0 & \varepsilon_{r\theta} & 0 \\ 0 & 0 & \varepsilon_{rz} \end{bmatrix} = \begin{bmatrix} \varepsilon_{\rho} & 0 & 0 \\ 0 & \varepsilon_{\theta} & 0 \\ 0 & 0 & \varepsilon_{z} \end{bmatrix}, \tag{3.75}$$

$$\varepsilon_{r\rho}(\omega) = 1 - \frac{\omega_{p\rho}^2}{\omega^2 + \mathrm{j}\gamma_{e\rho}\omega}, \tag{3.76}$$

$$\varepsilon_{r\theta}(\omega) = 1 - \frac{\omega_{p\theta}^2}{\omega^2 + \mathrm{j}\gamma_{e\theta}\omega}, \tag{3.77}$$

$$\varepsilon_{rz}(\omega) = 1 - \frac{\omega_{pz}^2}{\omega^2 + \mathrm{j}\gamma_{ez}\omega}, \tag{3.78}$$

where $\omega_{p\rho}$, $\omega_{p\theta}$, and ω_{pz} are the effective plasma frequencies in the ρ, θ, and z directions, respectively, with the following relationships:

$$\boldsymbol{\mu} = \mu_0 \begin{bmatrix} \mu_{r\rho} & 0 & 0 \\ 0 & \mu_{r\theta} & 0 \\ 0 & 0 & \mu_{rz} \end{bmatrix} = \begin{bmatrix} \mu_{\rho} & 0 & 0 \\ 0 & \mu_{\theta} & 0 \\ 0 & 0 & \mu_{z} \end{bmatrix}, \tag{3.79}$$

$$\mu_{r\rho}(\omega) = 1 - \frac{F_\rho \omega^2}{\omega^2 - \omega_{0\rho}^2 + \mathrm{j}\gamma_{m\rho}\omega}, \tag{3.80}$$

$$\mu_{r\theta}(\omega) = 1 - \frac{F_\theta \omega^2}{\omega^2 - \omega_{0\theta}^2 + \mathrm{j}\gamma_{m\theta}\omega}, \tag{3.81}$$

$$\mu_{rz}(\omega) = 1 - \frac{F_z \omega^2}{\omega^2 - \omega_{0z}^2 + \mathrm{j}\gamma_{mz}\omega}, \tag{3.82}$$

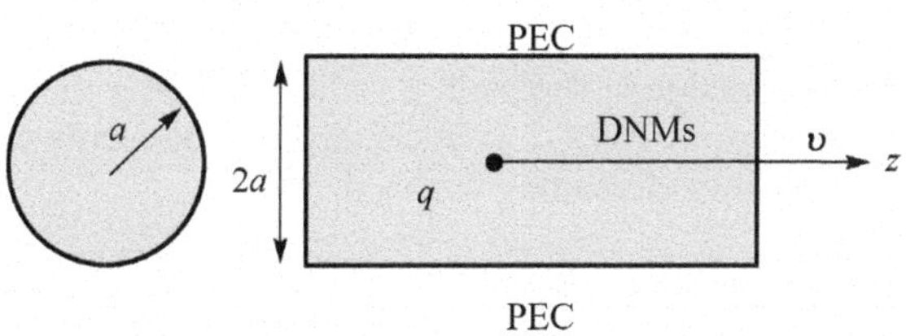

Fig. 3.21 Schematic diagram of RCR in a circular waveguide filled with anisotropic DNMs

When $\varepsilon_{r\rho}(\omega) = \varepsilon_{r\theta}(\omega) = \varepsilon_{rz}(\omega)$ and $\mu_{r\rho}(\omega) = \mu_{r\theta}(\omega) = \mu_{rz}(\omega)$, the anisotropic DNM will degenerate into an isotropic one.

On the one hand, by using the vector potential, i.e., $\boldsymbol{B} = \nabla \times \boldsymbol{A}_z$, and applying the boundary condition $E_z(\rho = a) = 0$, one can obtain the scalar Green's function $g(\rho)$ satisfying the condition for RCR ($\mathrm{Re}(\varepsilon_{r\rho}\mu_{r\theta}) > 1/\beta^2$):

$$g(\rho) = -\frac{1}{4}N_0(k_\rho \rho) + \alpha J_0(k_\rho \rho),\ 0 < \rho < a \tag{3.83}$$

where $\beta = \upsilon/c_0$, Re() represents the real part operator, $J_0(k_\rho \rho)$ and $N_0(k_\rho \rho)$ are the zeroth-order Bessel functions of the first kind and second kind respectively, $k_\rho = \sqrt{\omega^2 \varepsilon_z \mu_\theta - \varepsilon_z/\varepsilon_\rho \omega^2/\upsilon^2}$ is the radial wave number, and $\alpha = N_0(k_\rho a)/(4J_0(k_\rho a))$. Thus, the electromagnetic fields can be expressed as:

$$\boldsymbol{E}_z(\boldsymbol{r}, \omega) = \hat{\boldsymbol{z}}\frac{\mathrm{j}qk_\rho}{2\pi\omega\varepsilon_z}\mathrm{e}^{\mathrm{j}\omega z/\upsilon}\left\{\begin{array}{l}\left[-\frac{1}{4}(N_0(k_\rho\rho) - N_2(k_\rho\rho)) + \alpha(J_0(k_\rho\rho) - J_2(k_\rho\rho))\right]\frac{k_\rho}{2} + \\ \left[-\frac{1}{4}N_1(k_\rho\rho) + \alpha J_1(k_\rho\rho)\right]\frac{1}{\rho}\end{array}\right\}, \tag{3.84}$$

$$\boldsymbol{E}_\rho(\boldsymbol{r}, \omega) = \hat{\boldsymbol{\rho}}\frac{qk_\rho}{2\pi\varepsilon_\rho \upsilon}\mathrm{e}^{\mathrm{j}\omega z/\upsilon}\left[-\frac{1}{4}N_1(k_\rho\rho) + \alpha J_1(k_\rho\rho)\right], \tag{3.85}$$

$$\boldsymbol{H}_\theta(\boldsymbol{r}, \omega) = \hat{\boldsymbol{\theta}}\frac{qk_\rho}{2\pi}\mathrm{e}^{\mathrm{j}\omega z/\upsilon}\left[-\frac{1}{4}N_1(k_\rho\rho) + \alpha J_1(k_\rho\rho)\right], \tag{3.86}$$

As predicted, if the anisotropic DNMs filled in the circular waveguide degenerate into isotropic ones, the expressions for the electromagnetic fields will degenerate into those given in [40]. According to the definition [7], the formula for calculating the radiation energy per unit length of charge can be derived as follows:

$$\begin{aligned}\frac{dW}{dz} &= qE_z|_{z\to \upsilon t,\ \rho\to 0} \\ &= \frac{q^2}{\pi}\mathrm{Re}\left\{\int_0^\infty d\omega\frac{\mathrm{j}k_\rho}{\varepsilon_z\omega}\left\{\begin{array}{l}\left[-\frac{1}{4}(N_0(k_\rho\rho_0) - N_2(k_\rho\rho_0)) + \alpha(J_0(k_\rho\rho_0) - J_2(k_\rho\rho_0))\right]\frac{k_\rho}{2} + \\ \left[-\frac{1}{4}N_1(k_\rho\rho_0) + \alpha J_1(k_\rho\rho_0)\right]\frac{1}{\rho_0}\end{array}\right\}\right\}\end{aligned} \tag{3.87}$$

The formulae (3.84), (3.85), and (3.86) for the electric and magnetic fields show that the RCR can be observed when the direction of energy flow is determined by the sign of the real part of the elements of the permittivity tensor $\boldsymbol{\varepsilon}$. For example, when $\mathrm{Re}(\varepsilon_\rho) > 0$, the direction of energy flow is the same as the direction of particle motion along the z-axis, while $\mathrm{Re}(\varepsilon_\rho) < 0$, the direction of energy flow is opposite to the direction of particle motion along the z-axis.

On the other hand, when $\mathrm{Re}(\varepsilon_{r\rho}\mu_{r\theta}) < 1/\beta^2$, only evanescent waves are excited by the motion of the particle. In other words, no CR is produced.

3.3.2 Numerical Calculations of Single-Electron Model

In the microwave frequency range, DNMs can be realized using SRR arrays and metallic wire arrays. Based on experimental results from [41], the constitutive parameters for the DNMs are chosen: $\omega_{p\rho} = 2\pi \times 10^{10}$ rad/s, $\omega_{pz} = 2\pi \times 7 \times 10^9$ rad/s, $\omega_{0\theta} = 2\pi \times 4 \times 10^9$ rad/s, $F_\theta = 0.56$, and $\gamma_{e\rho} = \gamma_{ez} = \gamma_{m\theta} = \gamma = 10^7$ rad/s. Additionally, for the single electron model, $\beta = 0.8$ and $a = 1$ cm. In the frequency range studied here, all values are calculated at the same distance $\rho_0 = 1\ \mu\text{m}$. To explore the difference in the CR properties between anisotropic and isotropic DNMs, $\omega_{0\theta}$ and F_θ are kept constant.

First, the impact of electron velocity on the RCR condition is studied, considering the material properties. Based on the parameters mentioned above, the relationship between the RCR condition and frequency at different electron velocities can be obtained, as shown in Fig. 3.22. It can be observed that the operating frequency band that satisfies the RCR condition decreases as the electron velocity decreases. For example, if $1/\beta^2 = 0.2$, the operating frequency band is from 4.002 to 4.118 GHz, while if $1/\beta^2 = 0.8$, the operating frequency band is from 4.002 to 5.024 GHz.

Next, the radiation spectral density of RCR within the operating frequency band is studied, considering the loss and the equivalent plasma frequency in the z-direction. The radiation spectral density for smaller loss and larger one is shown in Fig. 3.23, where f_i represents the radiation spectral density as a function of ω. Case 1, Case 2, and Case 3 correspond to $\omega_{pz} < \omega_{p\rho}$ (Case 1), $\omega_{pz} = \omega_{p\rho}$ (Case 2), and $\omega_{pz} > \omega_{p\rho}$

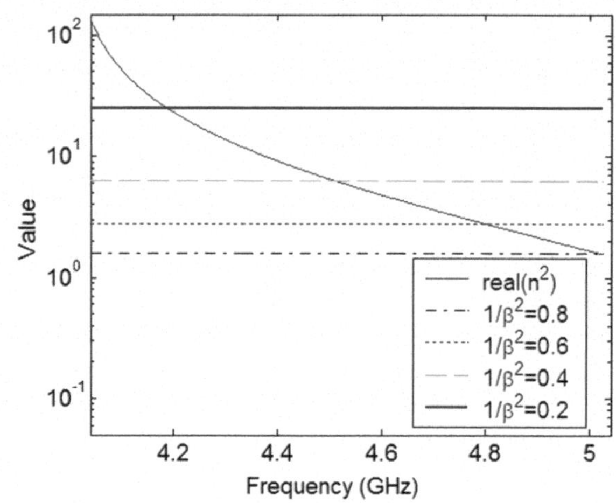

Fig. 3.22 RCR condition *versus* operating frequency with different particle velocities

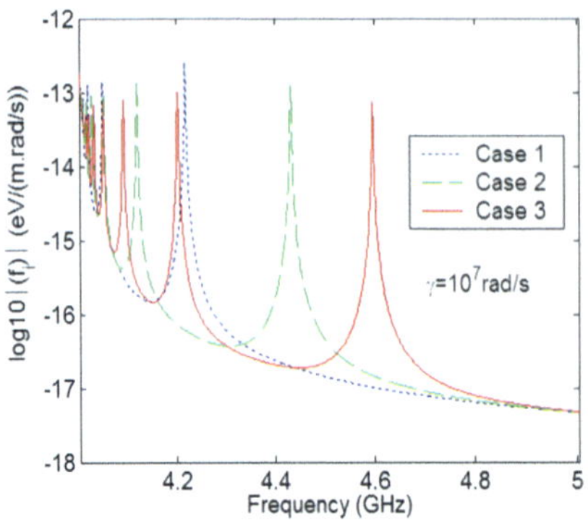

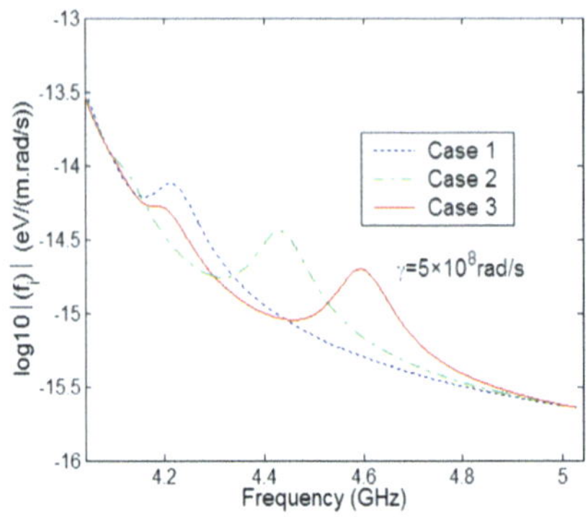

Fig. 3.23 Radiation energy spectral density over frequency for the smaller (**a**) and larger (**b**) loss

(Case 3), respectively, with $\omega_{pz} = 2\pi \times 7 \times 10^9$ rad/s for Case 1, $\omega_{pz} = 2\pi \times 10^{10}$ rad/s for Case 2, and $\omega_{pz} = 2\pi \times 13 \times 10^9$ rad/s for Case 3. Cases 1 and 3 correspond to different anisotropic DNMs, while Case 2 corresponds to an isotropic one [40]. From Fig. 3.23, it can be observed that when dispersion and loss are simultaneously considered, the radiation spectral density is continuous, and it has different peaks at different frequencies corresponding to the lossless case. For example, at certain frequencies, Case 2 has a peak, while Cases 1 and 3 do not. Furthermore, numerical calculation shows that when the loss is increased, the different peaks become broader, the peak amplitudes decrease, and the spacing between peaks near the resonance frequency becomes smaller. This is because the loss near the resonance frequency is high, causing the imaginary part of the Bessel function's argument to move further away from zero. Clearly, in different cases, the same frequency has different radiation spectral densities. Therefore, this characteristic can help obtain the maximum radiation spectral density at specific frequencies or the maximum radiation energy within the operating frequency band.

Next, the difference in radiation spectral density between the waveguide case and unbounded medium one is explored. In the case of fixed loss, the numerical results are shown in Fig. 3.24. From Fig. 3.24a, it can be seen that the radiation spectral density has different peaks at different frequencies in the waveguide filled with various anisotropic DNMs, while no peaks appear in unbounded anisotropic DNMs. Furthermore, the number of peaks increases with ω_{pz}. Numerical calculations indicate that as the loss increases, the peaks become wider, and the radiation spectral density approaches that of the unbounded case. The difference in spectral density between the waveguide and unbounded medium becomes smaller. This characteristic suggests that, at certain frequencies, the waveguide case is more likely to obtain larger radiation energy compared to the unbounded one.

Finally, the impact of waveguide radius, electron velocity, the z-direction loss, and the equivalent plasma frequency on the total radiation energy per unit length of charge is discussed. The influence of waveguide radius on the total radiation energy per unit length of charge is shown in Fig. 3.25. From Fig. 3.25a, it can be seen that at lower loss, for all cases (different anisotropic DNMs), the total radiation energy per unit

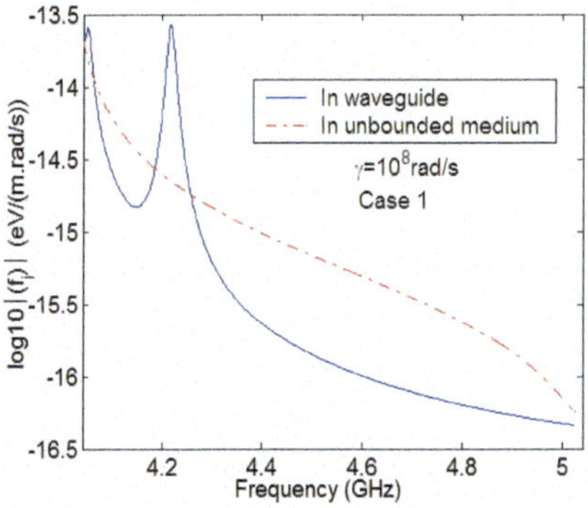

(a) Case 1

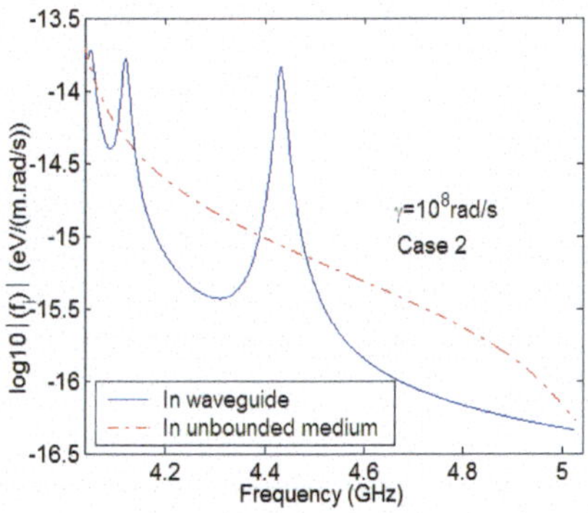

(b) Case 2

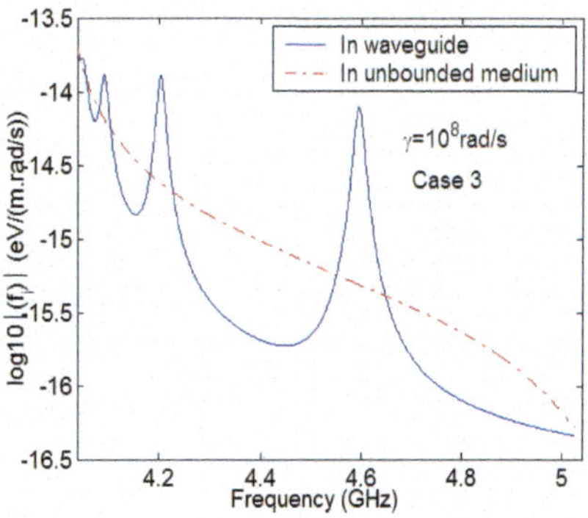

(c) Case 3

Fig. 3.24 Comparison between the spectral densities in the waveguide and unbounded medium at a given loss

length of charge increases with the increase in radius, reaching a peak. Figure 3.25b reveals that at higher loss, for Case 1, the total radiation energy per unit length of charge increases rapidly and then slowly decreases with the increase in waveguide radius. For Cases 2 and 3, the total radiation energy per unit length of charge increases rapidly and then slowly decreases with the increase in waveguide radius. In summary, when the waveguide radius is comparable to the operating wavelength, its influence on the total radiation energy is relatively small. In general, at lower loss, the radiation spectrum exhibits peaks, while at higher loss, it becomes a smoother function of frequency.

Furthermore, the influence of electron velocity on the total radiation energy per unit length of charge at lower loss and higher loss is discussed, as shown in Fig. 3.26, respectively. It can be observed that for all cases, as the electron velocity increases, the total radiation energy per unit length of charge also increases. As shown in Fig. 3.22, with the increase in electron velocity, the operating frequency band satisfying the RCR condition can expand, leading to an increase in total radiation energy. Additionally, for lower loss, there are some peaks in the total radiation energy at specific

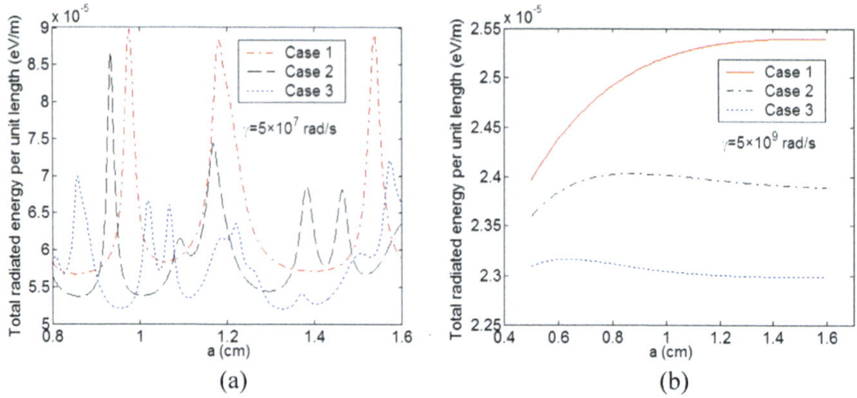

Fig. 3.25 Total radiation energy as a function of the waveguide radius for the smaller (**a**) and larger (**b**) loss

values of ω_{pz}, as shown in Fig. 3.27a. However, for higher loss, the total radiation energy decreases as a function of ω_{pz}, as shown in Fig. 3.27b. The difference in the variation of total radiation energy with respect to ω_{pz} under different loss conditions arises from the fact that the radiation spectrum density is not a smooth function at lower loss, whereas it becomes smooth at higher loss.

Two numerical integration methods are used in MATLAB to obtain the results under lower loss. One is the recursive adaptive Simpson integration algorithm, which is more effective when low precision or non-smooth integrands are present. The other is the recursive adaptive Lobatto integration algorithm, which is more effective at higher precision and when the integrand is smooth. Therefore, for lower loss, the former method is chosen for numerical results. However, at higher loss, both methods can yield the same numerical results.

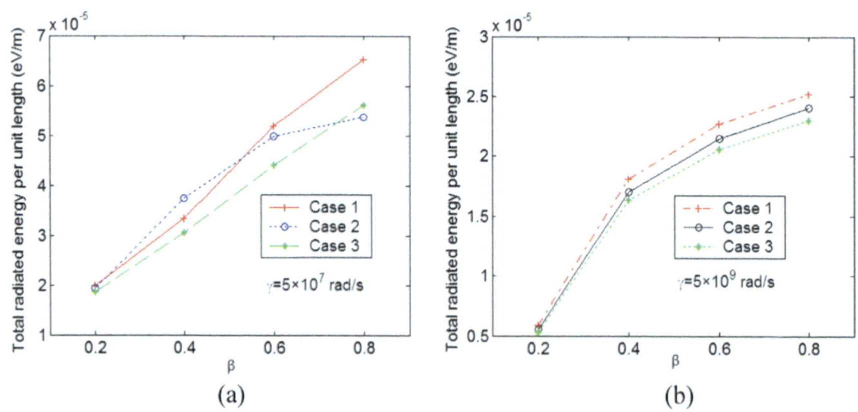

Fig. 3.26 Total radiation energy as a function of the particle velocity for the smaller (**a**) and larger (**b**) loss

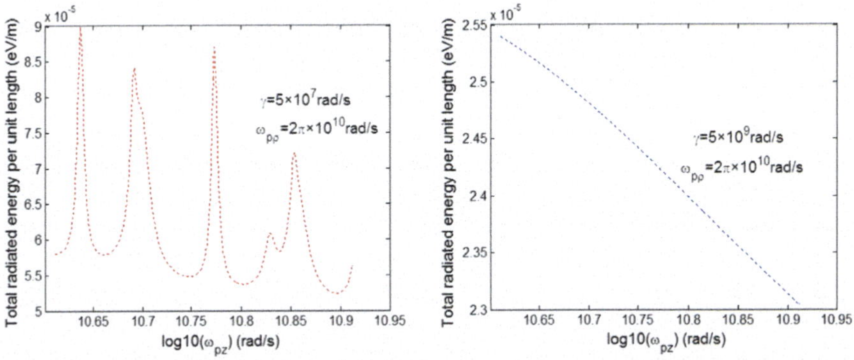

(a) (b)

Fig. 3.27 Relationship of the total radiation energy on ω_{pz} for the smaller (**a**) and larger (**b**) loss

A detailed discussion of the impact of electromagnetic loss on the total radiation energy is meaningful. As shown in Fig. 3.28, with increasing loss, the total radiation energy decreases for all cases, and the difference between the three cases becomes smaller. This is because the increase in loss leads to an overlap of the radiation energy spectral densities. In other words, in highly lossy media, the particle's radiation energy is quickly attenuated, and the electromagnetic wave reflected by the ideal waveguide is very weak, unable to form strong resonance within the circular waveguide. Due to the limitations imposed by the ideal waveguide, CR is not generated. As mentioned earlier, while the total radiation energy is not sensitive to the properties of the DNMs, there are clear differences in the radiation spectral densities for different DNMs.

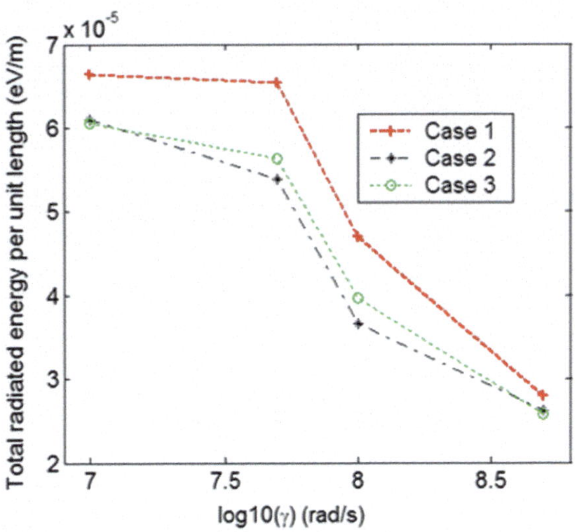

Fig. 3.28 Comparison of the total radiation energy for the three cases with different loss

3.3.3 Theoretical Analysis of Pencil Beam Model

In order to effectively enhance the energy of the RCR, a pencil electron beam is used. As shown in Fig. 3.29, for a pencil beam with a total charge q_T passing through the central channel (here vacuum region) of a circular waveguide locally filled with anisotropic DNMs, the field components are generated. Two sets of cylindrical coordinates are used. Here (ρ, θ, z) represents the position where the fields are calculated, and (ρ', θ', z') represents the position of the source. Anisotropic DNMs are characterized by tensors of permittivity and permeability, whose elements are represented by the formulas (3.1) and (3.2) in [26]. For simplicity, it is assumed that space charge effects can be neglected, and the charged particles are uniformly distributed. The pencil beam has a radial radius of ρ_0 and an axial length of $2z_0$, and it moves uniformly with a constant velocity $\boldsymbol{v}$ along the z-axis. The axial length of the pencil beam should satisfy the condition: $2z_0 < \lambda_{cp}$, where $\lambda_{cp} = v/f$ is the wavelength of the charged particles, and f is the operating frequency. The charge density ρ_P and current density $\boldsymbol{J}$ can be written in the following forms:

$$\rho_p = \frac{q_0 N}{2\pi\rho_0^2 z_0} T(t,\ z'), \tag{3.88}$$

$$\boldsymbol{J}(z', t) = \hat{z} v \frac{q_0 N}{2\pi\rho_0^2 z_0} T(t,\ z') \tag{3.89}$$

Here, $T(t, z') = 1,\ \ (z' \in (vt - z_0, vt + z_0))$, q_0 is the charge of a single electron, and N is the total electron number.

Performing a Fourier transform on the current density of the pencil beam, one can obtain its frequency-domain representation:

$$\boldsymbol{J}(z', \omega) = \frac{1}{2\pi}\int_{-\infty}^{\infty} dt \boldsymbol{J}(z', t) e^{j\omega t} = \hat{z}\frac{v q_0 N}{2\pi^2\rho_0^2 z_0} e^{j\omega z'/v}\frac{1}{\omega}\sin\frac{\omega z_0}{v}. \tag{3.90}$$

The cross-section of the circular waveguide shown in Fig. 3.29 is divided into three parts. The first part is pencil beam $(0 < \rho < \rho_0)$, the second part is vacuum $(\rho_0 < \rho < a)$, and the third part is filled with the anisotropic DNMs $(a < \rho < b)$.

For the first part, using the method of separation of variables and the Maxwell's equations as described in [26, 27], one can obtain the scalar wave equation:

$$\left[\frac{1}{\rho}\frac{\partial}{\partial\rho}\left(\rho\frac{\partial}{\partial\rho}\right) + k_\rho^2\right] A_z(\rho,\ z,\ \omega) = -\mu_0 \frac{v q_0 N}{2\pi^2\rho_0^2 z_0}\frac{1}{\omega}\sin\frac{\omega z_0}{v} e^{\frac{j\omega}{v}z'} = -p(z',\ \omega) \tag{3.91}$$

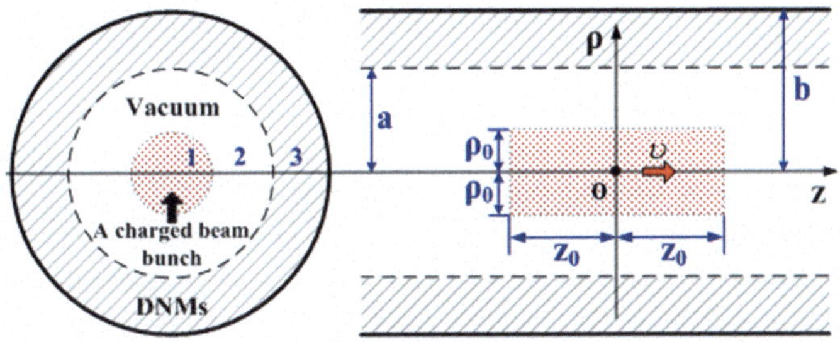

Fig. 3.29 Schematic diagram of a charged beam moving in a circular waveguide partially filled with anisotropic DNMs

where $k_\rho^2 = \omega^2\varepsilon_0\mu_0 - \omega^2/\upsilon^2$ represents the radial wave number, A_z is the z-component of the vector potential, ε_0 and μ_0 represent the permittivity and permeability in vacuum, respectively.

The analytical solution for the wave Eq. (3.91) is given by [50]:

$$A_z(\rho, z, \omega) = \int_V G(\rho, \rho', z, z', \omega)p(z', \omega)dV \tag{3.92}$$

where $G(\rho, \rho', z, z', \omega)$ is the Green's function solution for the non-uniform Helmholtz Eq. (3.91), and its solution satisfies the following equation [26, 27]:

$$\left[\frac{1}{\rho}\frac{\partial}{\partial\rho}\left(\rho\frac{\partial}{\partial\rho}\right) + k_\rho^2\right]G(\rho,\ \rho',\ z,\ z',\ \omega) = -\frac{1}{2\pi\rho}\delta(\rho - \rho')\delta(z - z') \tag{3.93}$$

Once the solution of (3.93) is obtained, the analytical expressions for the field components can be derived based on the electromagnetic potential method.

Similarly, by solving the homogeneous equation corresponding to (3.91), one can obtain the field components for the second and third parts. This section only focuses on the RCR in the third part [8]. Therefore, by applying boundary conditions at $\rho = \rho_0$, $\rho = a$, and $\rho = b$, the unknown coefficients for different parts can be determined completely. This allows for the characterization of RCR in the third part based on the time-averaged Poynting vector <*S*> and the derivation of the total radiation energy per unit length of charge. From the analytical expressions, it can be found that when the length of the pencil beam $2z_0$ is much smaller than the wavelength of RCR, the total radiation energy is proportional to the square of the electron number N. This result indicates that the total radiation energy of a short electron beam can be greatly enhanced, making RCR easier to observe.

3.3.4 Numerical Calculations of Pencil Beam Model

To validate the enhanced RCR, a typical example is provided here. Within the frequency range of 10.50–10.99 GHz, anisotropic metamaterials can exhibit double negative characteristics. The electron velocity is set to be $0.8c_0$. ρ_0 and z_0 are 0.1 mm and 2.5 mm, respectively. Since the fewer number of electron leads to the weaker space charge effect, N is chosen to be $N = 5.0 \times 10^7$ to neglect space charge effect in this case. The length of the electron beam in this example is smaller than the wavelength of the relevant radiated electromagnetic waves. The parameters a and b in Fig. 3.29 are set to be 1 mm and 10 mm, respectively. As a result, the RCR occurs within the narrow frequency band of 10.50–10.95 GHz.

In the third part, the reversed radiation characteristics of <*S*> indicate that due to the anisotropy of the metamaterial, the radiation directions at different positions are different, as shown in Fig. 3.30. The angle of the RCR can be changed by using different metamaterials or changing the electron velocity. After determining the sign of the refractive index n, according to Snell's law, the reflection and transmission of the Poynting vector at the discontinuities between the two media can be calculated.

The main factor influencing RCR is the peaks in the spectral density function f_i, as shown in Fig. 3.31. The impact of these peaks will be greater as the medium's loss decreases. In other words, as loss increases, the number of peaks in the radiation band decreases, and the trend of the radiation spectral density becomes smoother. Therefore, it can be concluded that when the synchronous radiation condition is met, there is maximum coherent radiation corresponding to the frequency at which the peak is located. Furthermore, the numerical results indicate that, like the case

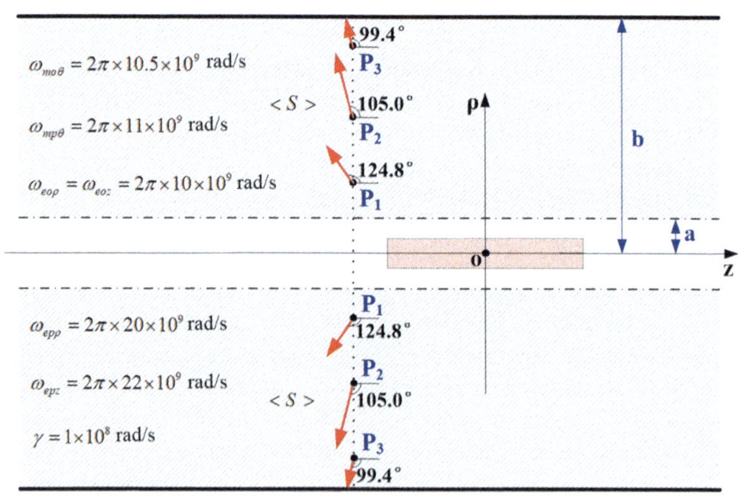

Fig. 3.30 Power flux densities at different positions that clearly show the reversed behavior of CR in DNMs

in unbounded DNMs, the radiation spectral density from the electron beam for the waveguide case is not sensitive to the anisotropy of DNMs [24].

The influence of electron velocity and medium loss on the total radiation energy per unit length of charge is shown in Fig. 3.32. The total radiation energy is an increasing function of electron velocity since the increase in electron velocity widens the operating frequency band for RCR. Similarly, the total radiation energy decreases with increasing loss, as shown in the radiation spectral density.

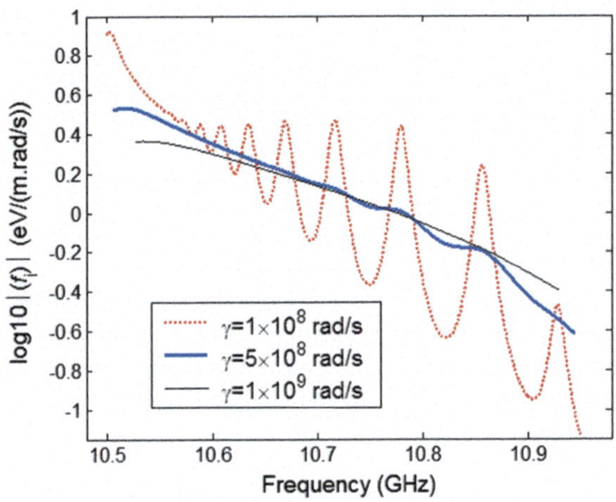

Fig. 3.31 Radiation energy spectral density over the radiation frequency band

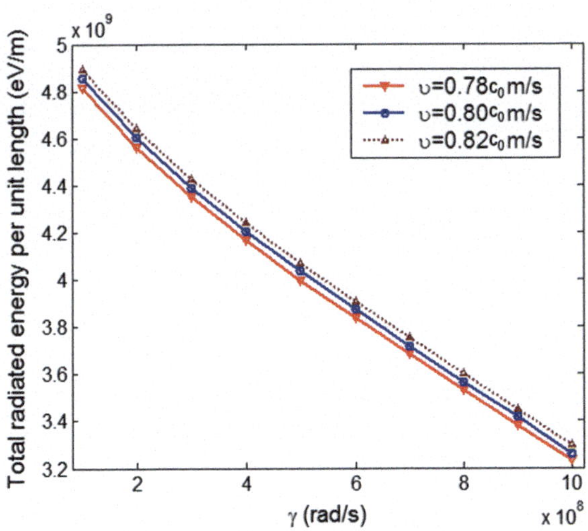

Fig. 3.32 Total radiation energy as a function of the loss for different charged particle velocities

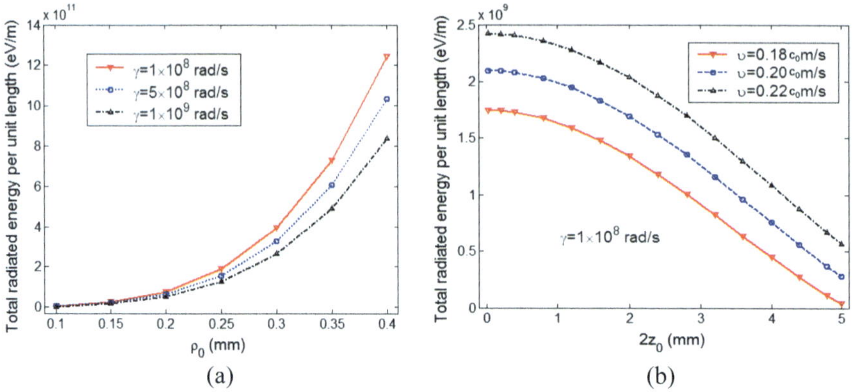

Fig. 3.33 **a** Total radiation energy as a function of the radius of the pencil beam with constant charge density; **b** total radiation energy as a function of the length of the pencil beam

Finally, the impact of the electron beam radius and its length on the total radiation energy is examined. When the charge density of the electron beam is kept constant, the numerical results are shown in Fig. 3.33. The total radiation energy per unit length of charge increases with the increase of the pencil beam radius. This is because when the radius of the pencil beam increases, the number of electrons N also increases. Therefore, the total radiation energy per unit length of charge can be greatly enhanced as it is directly proportional to N^2. However, the total radiation energy decreases with the increase in the electron beam length $2z_0$, indicating that the radiation from the electron beam becomes less coherent as the electron beam length increases.

The analysis results presented above indicate that the RCR from a short electron beam is stronger compared to the single-particle model [8]. For example, when N is 5×10^7 and γ is 5×10^8 rad/s, the RCR from a pencil beam with 2.5 mm length and 0.1 mm radius is approximately 2.3×10^{15} times stronger than that from a single particle, which is close to N^2. Therefore, RCR can be easily detected using a spectrum analyzer and a similar equipment. Furthermore, to further increase the total radiation energy, DNMs can be operated at higher frequencies [8].

References

1. Nahin PJ. Oliver Heaviside: The Life, Work, and Times of an Electrical Genius of the Victorian Age. Second Edition. Baltimore: Johns Hopkins University Press. 2002.
2. L'Annunziata MF. Radioactivity: Introduction and History, From the Quantum to Quarks. Second Edition. Amsterdam: Elsevier. 2016.
3. Sengupta P. Classical Electrodynamics. New Delhi: New Age International Publishers. 2000.
4. Marguet S. The Physics of Nuclear Reactors. Berlin: Springer. 2018.
5. Cherenkov PA. Visible luminescence of pure liquids under the influence of γ-radiation. Doklady Akademii Nauk SSSR. 1934; 2:8-451.

6. Tamm IY, Frank IM. Coherent radiation of fast electrons in a medium. Doklady Akademii Nauk SSSR. 1937; 14:3-107.
7. Bolotovskii BM. Theory of Cerenkov (III). Soviet Physics Uspekhi. 1962; 4:5-781.
8. Zrelov VP. Cherenkov Radiation in High-Energy Physics. Jerusalem: Israel Program for Scientific Translations. 1970.
9. Jelley JV. Cerenkov radiation and its applications. British Journal of Applied Physics. 1955; 6:7-227.
10. Chamberlain O, Segrè E, Wiegand C, et al. Observation of antiprotons. Physical Review. 1955; 100:3-947.
11. Aubert JJ, Becker U, Biggs PJ, et al. Experimental observation of a heavy particle J. Physical Review Letters. 1974; 33:23-1404.
12. Tsimring SE. Electron Beams and Microwave Vacuum Electronics. Hoboken: John Wiley & Sons. 2006.
13. Hruszowiec M, Nowak K, Szlachetko B, et al. The microwave sources for EPR spectroscopy. Journal of Telecommunications and Information Technology. 2017; 2-18.
14. Nusinovich GS, Sinitsyn OV, Rodgers JC, et al. Phase locking in backward-wave oscillators with strong end reflections. Physics of Plasmas. 2007; 14:5-053109.
15. Duan ZY, Gong YB, Wang WX, et al. Effect of attenuation on backward-wave oscillation start oscillation condition. IEEE Transactions on Plasma Science. 2004; 32:6-2184.
16. Duan ZY, Gong YB, Wei YY, et al. Impact of attenuator models on computed traveling wave tube performances. Physics of Plasmas. 2007; 14:9-093103.
17. Carusotto I, Artoni M, La Rocca GC, et al. Slow group velocity and Cherenkov radiation. Physical Review Letters. 2001; 87:6-064801.
18. Luo CY, Ibanescu M, Johnson SG, et al. Cerenkov radiation in photonic crystals. Science. 2003; 299:5605-368.
19. Afanasiev GN, Kartavenko VG, Zrelov VP. Fine structure of the Vavilov-Cherenkov radiation. Physical Review E. 2003; 68:6-066501.
20. Liu SG, Zhang YX, Yan Y, et al. Cherenkov radiation by an electron bunch moving in Hermitian medium. Journal of Applied Physics. 2007; 102:4-044901.
21. Chen HS, Chen M. Flipping photons backward: Reversed Cherenkov radiation. Materials Today. 2011; 14:1-34.
22. Smith DR, Padilla WJ, Vier DC, et al. Composite medium with simultaneously negative permeability and permittivity. Physical Review Letters. 2000; 84:18-4184.
23. Lu J, Grzegorczyk TM, Zhang Y, et al. Čerenkov radiation in materials with negative permittivity and permeability. Optics Express. 2003; 11:7-723.
24. Duan ZY, Wu BI, Lu J, et al. Reversed Cherenkov radiation in unbounded anisotropic double-negative metamaterials. Journal of Physics D: Applied Physics. 2009; 42:18-185102.
25. Antipov S, Spentzouris L, Liu W, et al. Wakefield generation in metamaterial-loaded waveguides. Journal of Applied Physics. 2007; 102:3-034906.
26. Duan ZY, Wu BI, Lu J, et al. Reversed Cherenkov radiation in a waveguide filled with anisotropic double-negative metamaterials. Journal of Applied Physics. 2008; 104:6-063303.
27. Duan ZY, Wu BI, Lu J, et al. Cherenkov radiation in anisotropic double-negative metamaterials. Optics Express. 2008; 16:22-18479.
28. Duan ZY, Guo C, Chen M. Enhanced reversed Cherenkov radiation in a waveguide with double-negative metamaterials. Optics Express. 2011; 19:15-13825.
29. Duan ZY, Guo C, Zhou J, et al. Novel electromagnetic radiation in a semi-infinite space filled with a double-negative metamaterial. Physics of Plasmas. 2012; 19:1-013112.
30. Duan ZY, Guo C, Guo X, et al. Double negative-metamaterial based terahertz radiation excited by a sheet beam bunch. Physics of Plasmas. 2013; 20:9-093301.
31. Averkov YO, Yakovenko VM. Cherenkov radiation by an electron bunch that moves in a vacuum above a left-handed material. Physical Review B. 2005; 72:20-205110.
32. Galyamin SN, Tyukhtin AV, Kanareykin A, et al. Reversed Cherenkov-transition radiation by a charge crossing a left-handed medium boundary. Physical Review Letters. 2009; 103:19-194802.

33. Galyamin SN, Tyukhtin AV. Electromagnetic field of a moving charge in the presence of a left-handed medium. Physical Review B. 2010; 81:23-235134.
34. Kats AV, Savel'ev S, Yampol'skii VA, et al. Left-handed interfaces for electromagnetic surface waves. Physical Review Letters. 2007; 98:7–073901.
35. Duan ZY, Wu BI, Xi S, et al. Research progress in reversed Cherenkov radiation in double-negative metamaterials. Progress in Electromagnetics Research. 2009; 90:75.
36. Bliokh YP, Savel'ev S, Nori F. Electron-beam instability in left-handed media. Physical Review Letters. 2008; 100:24-244803.
37. So JK, Won JH, Sattorov MA, et al. Cerenkov radiation in metallic metamaterials. Applied Physics Letters. 2010; 97:15-151107.
38. Zhou J, Duan ZY, Zhang YX, et al. Numerical investigation of Cherenkov radiations emitted by an electron beam bunch in isotropic double-negative metamaterials. Nuclear Instruments and Methods in Physics Research Section A: Accelerators, Spectrometers, Detectors and Associated Equipment. 2011; 654:1-475.
39. Kong JA. Electromagnetic Wave Theory. Second Edition. Hoboken: Wiley-Blackwell. 1990.
40. Lu J. Novel electromagnetic radiation in left-handed materials. Cambridge: Massachusetts Institute of Technology. 2006.
41. Shadrivov IV, Sukhorukov AA, Kivshar YS. Guided modes in negative-refractive-index waveguides. Physical Review E. 2003; 67:5-057602.
42. Ziolkowski RW, Kipple AD. Causality and double-negative metamaterials. Physical Review E. 2003; 68:2-026615.
43. Nguyen KT, Pasour JA, Antonsen TM, et al. Intense sheet electron beam transport in a uniform solenoidal magnetic field. IEEE Transactions on Electron Devices. 2009; 56:5-744.
44. Booske JH, Basten MA, Kumbasar AH, et al. Periodic magnetic focusing of sheet electron beams. Physics of Plasmas. 1994; 1:5-1714.
45. Booske JH, Kumbasar AH, Basten MA. Periodic focusing and ponderomotive stabilization of sheet electron beams. Physical Review Letters. 1993; 71:24-3979.
46. Smorenburg PW, Root Op't WPEM, Luiten OJ. Direct generation of terahertz surface plasmon polaritons on a wire using electron bunches. Physical Review B. 2008; 78:11-115415.
47. Marqués R, Martín F, Sorolla M. Metamaterials with Negative Parameters: Theory, Design, and Microwave Applications. Hoboken: John Wiley & Sons. 2011.
48. Moser HO, Jian LK, Chen HS, et al. All-metal self-supported THz metamaterial-the meta-foil. Optics Express. 2009; 17:26-23914.
49. Smith SJ, Purcell EM. Visible light from localized surface charges moving across a grating. Physical Review. 1953; 92:4-1069.
50. Shiozawa T. Classical Relativistic Electrodynamics: Theory of Light Emission and Application to Free Electron Lasers. Berlin: Springer. 2004.

Chapter 4
Experimental Studies on Reversed Cherenkov Radiation

As stated in Chap. 1, RCR is a novel electromagnetic property that appears in LHMs [1, 2]. Since the LHM was first realized by D. R. Smith [3], this emerging field has undergone rapid development, with RCR receiving extensive research both domestically and internationally [4–7]. This new type of electromagnetic radiation has significant applications in vacuum electronics, particle physics, optics, and other fields [7, 8].

Various research institutions around the world have conducted experimental studies of RCR in LHMs. One of the earliest studies dates back to 2002, when G. V. Eleftheriades's group proposed a type of LHM composed of coplanar waveguide microstrip transmission lines. This LHM exhibits a negative effective refractive index around 15 GHz, with the period of a unit cell of ~ 0.2λ (λ is the wavelength in free space), as shown in Fig. 4.1. In the experiment, a physical phenomenon similar to RCR was observed [9].

In 2007, J. A. Kong's group proposed a novel type of LHM for verifying RCR [10]. For the TM mode, this unit cell exhibits negative effective permittivities in the x and y directions and negative effective permeability in the z direction. They utilized an antenna array to simulate the monochromatic electromagnetic wave. This study demonstrates that this LHM could be used for experimental verification of RCR. Based on the work of B. I. Wu, S. Xi successfully fabricated an LHM in 2009 [11], which exhibits a negative effective refractive index in the frequency range of 8.1–9.5 GHz. They used a phased antenna array to simulate the effect of moving charged particles, thus generating full-spectrum radiation. And they reported the observation of RCR in LHMs within the negative refractive index frequency range. In the review paper [12], S. Zhang noted that the work reported by S. Xi could be considered as a special case of an experiment on negative refractive index, suggesting that a more direct verification method of RCR should involve observing the electromagnetic radiation characteristics through the interaction of real charged particles and LHMs.

In 2008, S. Antipov proposed a rectangular waveguide filled with DNMs, as shown in Fig. 4.2a, and conducted experimental study using a pulsed electron beam [13].

Z. Duan, *Metamaterial-Based Electromagnetic Radiations and Applications*,
https://doi.org/10.1007/978-981-97-8108-9_4

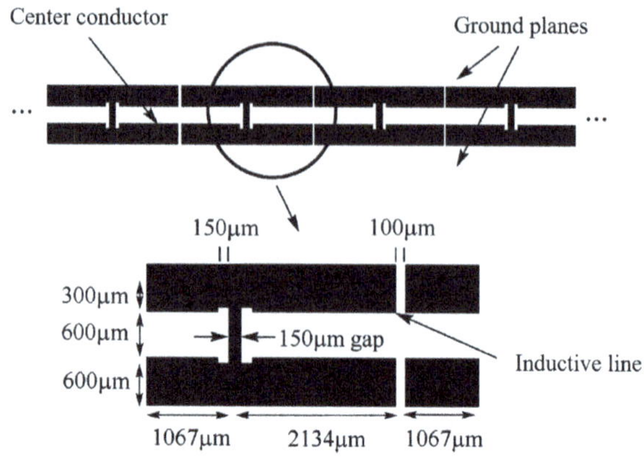

Fig. 4.1 Schematic diagram of the LHM composed of coplanar waveguide transmission lines

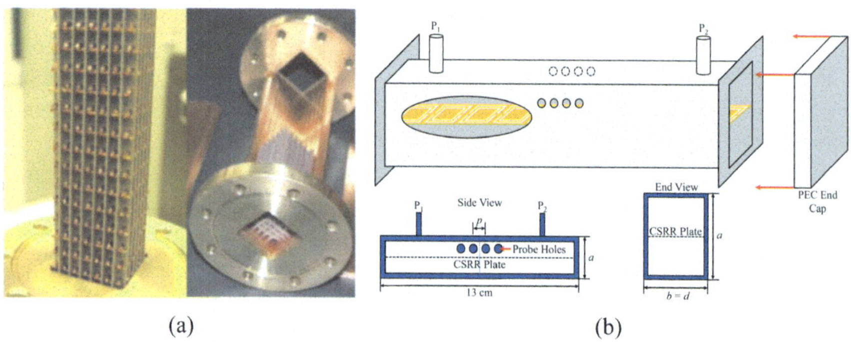

Fig. 4.2 **a** LHM sample. **b** Schematic diagram of the experimental setup

They observed radiation wave signals within the negative refractive index band of the DNMs, providing some evidence of CR generated in negative refractive index materials. However, they were unable to determine the direction of the radiated electromagnetic wave in their experiments. In the same year, D. Y. Shchegolkov reported a method for observing RCR in circular waveguides filled with DNMs [14], but no experiment has been reported in public. In 2015, N. A. Estep proposed an LHM suitable for vacuum environments, as shown in Fig. 4.2b. They only conducted transmission characteristic experiments [15], obtaining the dispersion curve of the structure using S parameters, and demonstrated that within the negative refractive index frequency range of the LHM, the phase velocity and group velocity of the propagated electromagnetic waves are in opposite directions. However, they still did not conduct experiments on the direct generation of RCR using real charged particles.

These research findings have significantly advanced the experimental study of RCR. The main reasons for the difficulty in experimentally verifying RCR in metamaterials using real charged particles stem from two factors: ① the high-speed movement of charged particles requires a high vacuum environment, and suitable all-metal metamaterials for high vacuum environments are not proposed; ② efficient signal output coupler is not proposed to extract the RCR signal generated by charged particles in LHMs [16].

Z. Duan's group was aware of the two issues around 2012. Based on their extensive theoretical work (for details, see Chap. 3), his group embarked on highly successful efforts [17–22]. In close collaboration with R. J. Temkin's group and M. Chen, they creatively proposed an entirely new type of planar all-metal metamaterial unit cell [23], which was periodically loaded into an empty square metal waveguide to construct an LHM. Additionally, they developed a coaxial output coupler suitable for this all-metal LHM and successfully detected the RCR signal excited by real charged particles in the LHM [24]. The following sections provide a detailed introduction to the experimental research on RCR.

4.1 Metamaterial Slow-Wave Structure

Conducting experimental research on RCR in an LHM directly using real charged particles requires observing the interaction between the charged particles and the LHM. A prerequisite for this experimental research is the innovative development of an LHM suitable for high vacuum environments with double negative characteristics. However, traditional LHMs, composed of metallic wires and SRRs, require fixing the metallic wires and SRRs onto a dielectric substrate. The negative effective permittivity is achieved through the array of metallic wires, and the negative effective permeability is achieved through the array of SRRs, thereby realizing double negative characteristics [3]. However, dielectric substrates are prone to outgassing, making them unsuitable for high vacuum environments. In addition, the dielectric substrate inevitably introduces electromagnetic loss. Therefore, traditional LHMs are not applicable for VEDs.

4.1.1 Effective Electromagnetic Parameters of Metamaterials

An all-metal planar CeSRR is proposed using the Babinet principle, as shown in Fig. 4.3a [23]. The CeSRR possesses axial symmetry and is easily fabricated in the S band. By periodically arranging CeSRRs along the z-axis with a period of p, a CeSRR array is formed. When the constructed CeSRR array is placed in the middle of the y direction of an empty metal rectangular waveguide and parallel to the x–z plane, a novel all-metal MSWS is constructed, as shown in Fig. 4.3b, with the structure parameters listed in Table 4.1. The so-called SWS refers to an electromagnetic structure that can propagate electromagnetic waves with a phase velocity slower than the velocity of light in free space. As can be seen from Fig. 4.3b, this MSWS has two natural channels for sheet electron beam (here simply called sheet beam). When the sheet beam moves along the z-direction above the surface of the

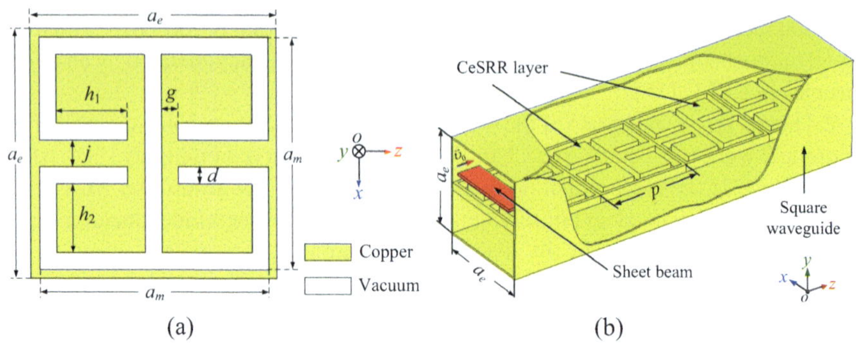

Fig. 4.3 **a** All-metal CeSRR. **b** Schematic diagram of MSWS

Table 4.1 Structure parameters of the CeSRR

Parameter	Value (mm)	Parameter	Value (mm)
a_e, p	14.5	a_m	13.5
h_1	4	h_2	4.25
g	1	d	1
j	1.5	t	1.2

CeSRR array, it can interact with the longitudinal electric field above the surface of the CeSRR array, allowing for efficient energy exchange.

The resonance frequency of this CeSRR is approximately 3 GHz, with the corresponding wavelength (λ) in free space being about 10 cm. From the data listed in Table 4.1, it is clear that the size of the CeSRR is significantly smaller than the wavelength in free space, approximately $\lambda/7$, indicating that the CeSRR exhibits distinct subwavelength characteristics. Based on EMT [25, 26] and the S-parameters retrieval method using the Kramers–Kronig relations introduced in Chap. 2 [27], the effective electromagnetic parameters of the CeSRR are obtained. The relationships of ε_{zz} and μ_{xx} with frequency for the CeSRR are shown in Fig. 4.4a. It can be seen that near the resonance frequency of 3 GHz, the values of ε_{zz} and μ_{xx} for polarization in the z direction are approximately equal to 1. Similarly, the variations of ε_{xx} and μ_{zz} with frequency for polarization in the x direction are shown in Fig. 4.4b. It is evident that near the resonance frequency of 3 GHz, ε_{xx} is less than 0, and μ_{zz} is approximately equal to 1.

From the above analysis, it is clear that for the CeSRR array, there is a negative effective permittivity but not a negative effective permeability. This is because the symmetry of the CeSRR results in the mutual cancellation of the magnetic response generated by the currents on the ring [23], meaning that the CeSRR only exhibits an electric response without a magnetic response [28]. The study in [23] theoretically predicts that the mode of electromagnetic waves propagated in this metamaterial is a quasi-TM mode. In fact, it is a hybrid mode. For the TM mode, an empty metal rectangular waveguide operating below the cut-off frequency can be considered as

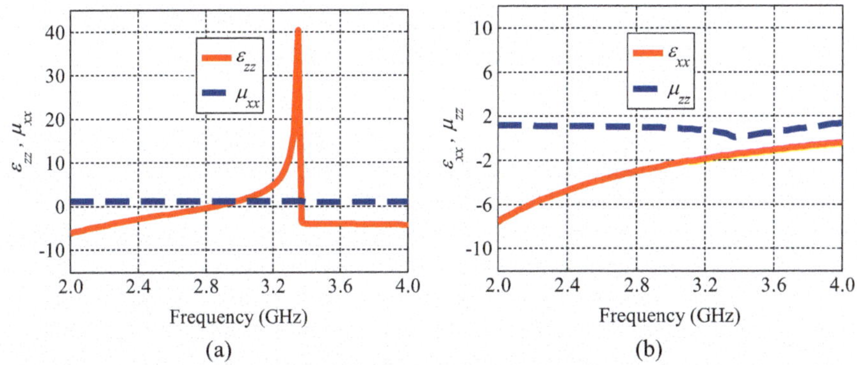

Fig. 4.4 **a** The relationships of ε_{zz} and μ_{xx} with frequency. **b** The relationships of ε_{xx} and μ_{zz} with frequency

a one-dimensional plasma [29–31], allowing the calculation of the variation of the relative effective permeability (the black dashed line in Fig. 4.5).

For comparison, Fig. 4.5 also presents the curve of the relative effective permittivity of the CeSRR with frequency. As can be seen from Fig. 4.5, within the frequency range of 2.81–3.05 GHz, the metamaterial exhibits both negative ε_{xx} and negative μ_{eff}. The negative ε_{xx} originates from the electric response of the CeSRR, while the negative μ_{eff} is derived from the empty rectangular waveguide operating below the cut-off frequency of the TM mode. Therefore, this MSWS exhibits double negative or left-handed characteristics, meaning it can be considered an LHM.

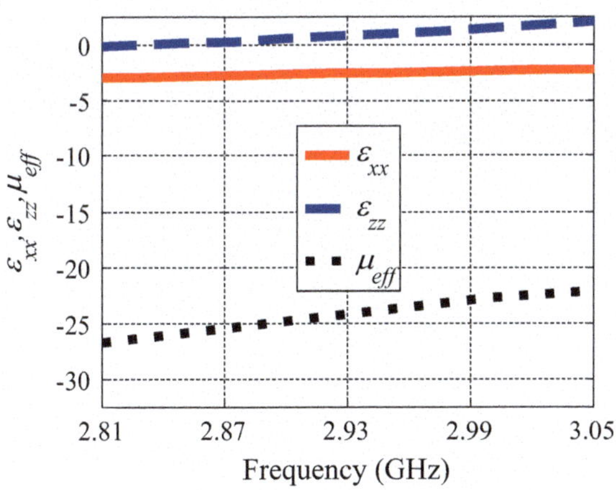

Fig. 4.5 ε_{xx}, ε_{zz}, and μ_{eff} versus frequency

4.1.2 Electromagnetic Properties of Metamaterial Slow-Wave Structure

Based on the aforementioned MSWS, the electromagnetic characteristics are studied using the HFSS simulation software. The HFSS simulation model, as illustrated in Fig. 4.6, consists of the CeSRR and a vacuum cube with edge length a_e. Considering that the x and y directions of the metamaterial point to the waveguide walls of a square metal waveguide, the surfaces in the x and y directions are set as perfect electric boundaries. To simulate the periodic nature of the metamaterial in the z direction, master–slave boundary conditions are applied to the surfaces in the z direction, with a phase shift φ set between the master and slave boundaries.

Dispersion characteristics, which represent the relationship between the phase velocity and frequency of electromagnetic waves propagating through the SWS, are critical for characterizing SWS. To clearly reflect the relationship between phase velocity and group velocity, the Brillouin diagram, i.e., the frequency-phase shift curve, is used to describe the dispersion characteristics. Using the above HFSS simulation model and performing the parameter sweep on the phase shift, the dispersion curve is obtained for phase shifts ranging from 0 to 2π, as shown by the solid line in Fig. 4.7. The diagram indicates that the zeroth-order spatial harmonic of this MSWS exhibits backward-wave characteristics, meaning that the direction of phase velocity is opposite to that of the group velocity. This signifies that the direction of phase velocity of the electromagnetic waves excited by the electron beam is opposite to the direction of motion of the electron beam, consistent with the characteristics of RCR. The operating point is chosen near the resonance frequency of the CeSRR, around 3 GHz, corresponding to a voltage about 160 kV, indicated by the triangular dashed line in Fig. 4.7. Note that the square dashed line corresponds to the 99.7 kV electron beam line.

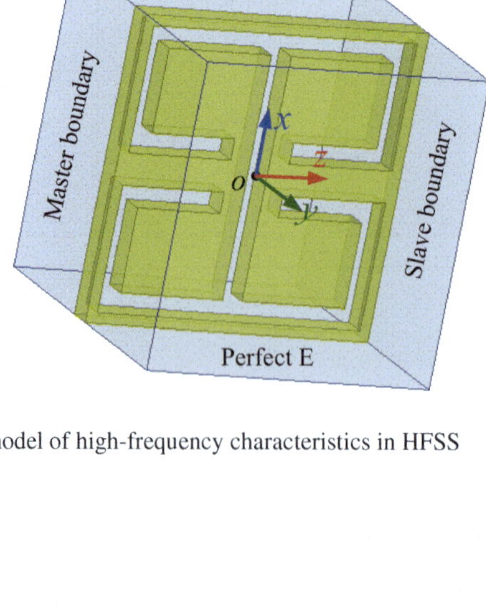

Fig. 4.6 Simulation model of high-frequency characteristics in HFSS

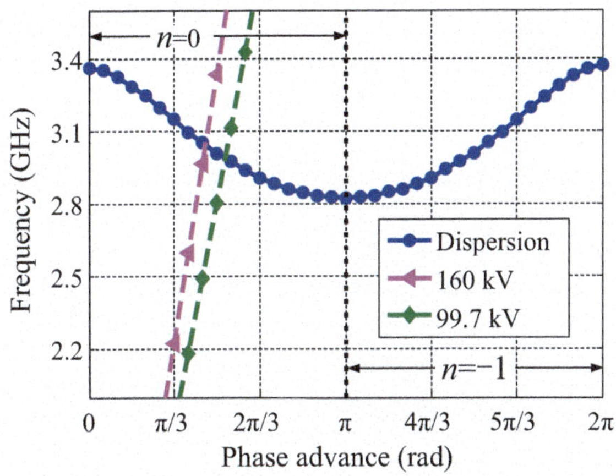

Fig. 4.7 Dispersion characteristics of MSWS

Interaction impedance is another important parameter characterizing the SWS, representing the interaction degree between the electron beam and the electromagnetic field within the SWS. According to the definition of interaction impedance [32, 33], for the zeroth-order spatial harmonic of the fundamental mode, the expression for the interaction impedance K_c is given as:

$$K_c = \frac{|E_{z0}|^2}{2\beta_0^2 P_w}, \tag{4.1}$$

where $|E_{z0}|$ is the longitudinal electric field amplitude of the zeroth-order spatial harmonic of the fundamental mode, $\beta_0 = \varphi/L$ is the phase constant of the zeroth-order spatial harmonic of the fundamental mode, and P_w is the total power flow propagated in the SWS.

According to the Floquet theorem, electromagnetic waves in a periodic system can be decomposed into infinite spatial harmonics. The longitudinal electric field E_{z0} of the zeroth-order spatial harmonic of the fundamental mode is expressed in (4.2):

$$E_{z0} = E_z(x, y, z)e^{-j\beta_0 z}. \tag{4.2}$$

Averaging E_{z0} over a period p yields the result:

$$\begin{aligned} E_{z0} &= \frac{1}{p}\int_0^p E_z(x, y, z)e^{-j\beta_0 z}dz \\ &= \frac{1}{p}\int_0^p \left[\mathrm{Re}(E_z) + j\mathrm{Im}(E_z)\right]e^{-j\beta_0 z}dz. \end{aligned} \tag{4.3}$$

Squaring the modulus of E_{z0} from (4.3) gives the following expression:

$$|E_{z0}|^2 = \left\{\frac{1}{p}\int_0^p [\mathrm{Re}(E_z)\cos(\beta_0 z) + \mathrm{Im}(E_z)\sin(\beta_0 z)]dz\right\}^2 + \left\{\frac{1}{p}\int_0^p [\mathrm{Re}(E_z)\sin(\beta_0 z) - \mathrm{Im}(E_z)\cos(\beta_0 z)]dz\right\}^2. \quad (4.4)$$

Additionally, the total power flow propagated in the SWS is given below:

$$P_w = \iint_S \mathrm{Re}(S_z)dS, \quad (4.5)$$

where $\mathrm{Re}(S_z)$ is the real part of the z-component of the Poynting vector $\boldsymbol{S}$.

By calculating the electric field and power flow in the SWS through simulation and combining it with the dispersion curve, the interaction impedance can be calculated using (4.1) with MATLAB. To better reflect the interaction degree between the electron beam and the electromagnetic field in the SWS, the average interaction impedance over the cross-section of the electron beam is typically used. Unless specifically stated, the interaction impedance referred to in this monograph refers to the absolute value of the average interaction impedance.

For the MSWS with parameters listed in Table 4.1, taking a rectangular cross-section sheet beam with dimensions of 12 mm × 2 mm as an example, the interaction impedance at nine black dot positions on the cross-section (shown in the inset of Fig. 4.8) is calculated, and their arithmetic mean is taken to obtain the average interaction impedance, as shown in Fig. 4.8. It can be seen that the average interaction impedance of the MSWS exceeds 750 Ω, which is significantly higher than those of traditional S-band helix SWSs (100–200 Ω) [32, 34] and coupled-cavity SWSs (300–400 Ω) [35]. This feature benefits from the strong resonant characteristics of the CeSRRs, which generate a strong local longitudinal electric field component above the surface of the CeSRRs [23]. From the above analysis, it can be found that as a brand-new SWS, the MSWS has several significant advantages: ① an all-metal structure, offering high power capacity; ② miniaturization, due to the subwavelength characteristics of the CeSRR and the fact that the empty square metal waveguide operates below the cut-off frequency [36, 37]; ③ high interaction impedance, resulting from the strong resonant characteristics of the CeSRRs leading to the enhancement effect of the local electric field; ④ fabrication simplicity in the microwave frequency band.

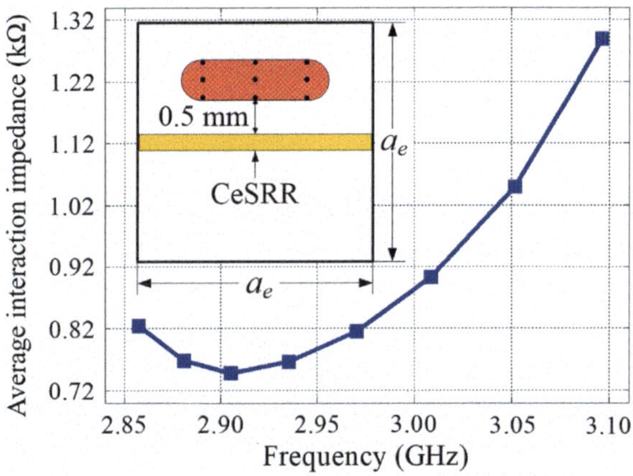

Fig. 4.8 Average interaction impedance of the MSWS versus frequency

4.2 Formation and Transportation of Sheet Beam

A sheet-shaped electron beam, commonly referred to as a sheet beam, is characterized by its rectangular or elliptical cross-section and a large aspect ratio. Compared to traditional pencil electron beam, sheet beam offers several advantages, such as high current and large interaction areas. VEDs with a sheet beam as electron source demonstrate significant advantages in power, efficiency, and gain. The introduction of the sheet beam paves a new way for the development of VEDs. Sheet beams have been successfully applied to VEDs operating at high-frequency bands, such as the W-band and G-band [38].

4.2.1 Sheet Beam Electron Gun

To ensure the advantages of sheet beam VEDs, the generation of the sheet beam is crucial. There are mainly two methods for generating sheet beam: direct and indirect methods [38]. The direct method involves generating sheet beam directly using an electron gun without additional components. This method uses either a circular cathode or a cathode with an approximate rectangular or elliptical cross-section to directly emit a planar symmetric sheet beam. By designing the aperture shape of the focusing electrode as rectangular or elliptical, the electrostatic field between the cathode and the focusing electrode can compress the sheet-shaped electron stream, resulting in a sheet beam with high current density [39]. The indirect method means that the beam transformation components at the end of a conventional pencil beam electron gun is added. The magnets are commonly used in beam transformation components to compress the pencil beam into a high-ellipticity sheet beam. Two typical compression methods are magnetic quadrupole compression [40] and elliptical solenoid compression [41].

In the experimental verification work of RCR, a sheet beam is generated using a rectangular cross-section graphene cathode [42], as detailed in Sect. 4.4.4 of this Chapter.

4.2.2 Transportation of Sheet Beam in Metamaterial Slow-Wave Structure

Due to the instability of sheet beam transport under uniform axial magnetic field [43], the related research has not received sufficient attention. It was not until the 1990s that sheet beam began to gradually get attention, mainly due to the following reasons: ① The periodic cusped magnet (PCM) was found to effectively focus sheet beam [44–46]; ② It was discovered that increasing the height filling ratio of sheet beam could effectively suppress their transport instability [47, 48]; ③ Sheet beam has a larger interaction area, which is helpful for improving device performance [38]. However, sheet beam exhibits Diocotron instability during its transport process [49, 50], making it difficult to maintain stable transport over a long distance, thereby limiting its technical advantages. Common magnetic focusing methods for sheet beam include uniform magnetic field [51] and periodic magnetic field. Periodic magnetic field mainly includes periodic Wiggler magnetic field [52] and PCM magnetic field [44], with the latter including various modified structures [38].

Under the combined effect of a uniform axial magnetic field and the horizontal component of the space charge field, the shape of the sheet beam will gradually change during transport [38], as shown in Fig. 4.9. The degree of deformation can be described by using the vertical displacement Δx near the edges of the sheet beam [53]:

$$\Delta x = \frac{|E_{y\max}|}{B_z} \frac{L}{v_z}, \tag{4.6}$$

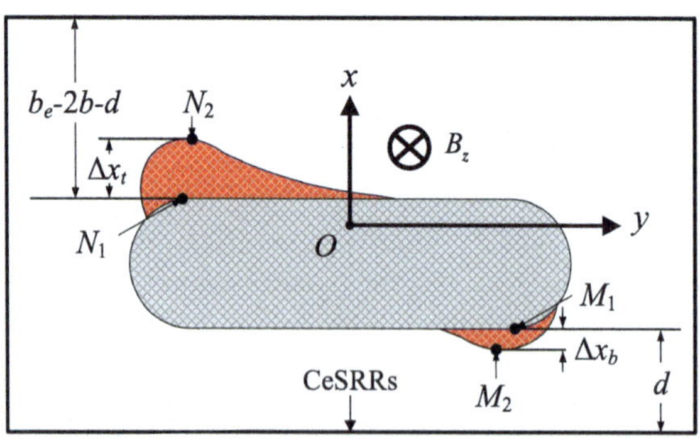

Fig. 4.9 Diagram of instability of sheet beam focused by uniform axial magnetic field

where B_z is the axial magnetic flux density, L is the transport distance of the sheet beam, v_z is the axial velocity of the sheet beam, and $|E_{ymax}|$ is the modulus of the horizontal component of the space charge field at the edge of the sheet beam.

During the actual transport of the sheet beam, the vertical displacement is generally small. Therefore, it can be assumed that the horizontal component of the space charge field remains approximately constant during the transition from M_1 to M_2 (or N_1 to N_2). Substituting the space charge field at points M_1 and N_1 into (4.6) yields the vertical displacements Δx_t and Δx_b on the upper and lower surfaces of the sheet beam edge, respectively:

$$\Delta x_t = \frac{|E_{yt}(N_1)|}{B_z}\frac{L}{v_z}, \tag{4.7}$$

$$\Delta x_b = \frac{|E_{yb}(M_1)|}{B_z}\frac{L}{v_z}. \tag{4.8}$$

For a transport distance of $L = 145$ mm, the relationship between vertical displacement and axial magnetic flux density, as well as d, can be calculated using (4.7) and (4.8). Figure 4.10 shows the computational results for $d = 0.3$ mm, $d = 0.5$ mm, and $d = 0.7$ mm, respectively. These figures indicate that at a fixed distance d, an increase in axial magnetic flux density leads to a decrease in vertical displacement, which means that the transport instability of the sheet beam is somewhat suppressed. However, in actual VEDs, increasing the magnetic field means increasing the current in the solenoid magnetic focusing configuration, which increases energy consumption and thus reduces the overall efficiency of the device. Therefore, an increase in the magnetic field should be considered comprehensively. Additionally, the black or blue curves in Fig. 4.10 show that at the same axial magnetic flux density, the vertical displacement of the sheet beam decreases with a reduction in d. The reason is that reducing d can decrease the modulus of the horizontal component of the sheet beam's space charge field. This result indicates that reducing d can also suppress the transport instability of the sheet beam to a certain extent. Moreover, according to [23], the longitudinal electric field in the MSWS becomes stronger as it approaches the surface of the CeSRR array, which is more useful for beam-wave interaction. Therefore, in devices with MSWS, reducing d is not only beneficial for the transport of the sheet beam but also for improving the efficiency of the device. However, d should not be too small due to manufacturing precision and assembly errors in actual devices. If d is too small, the sheet beam will be likely to be directly intercepted by the CeSRR array, significantly deteriorating the sheet beam. Taking all factors into consideration, d is set to 0.5 mm here.

To validate the aforementioned calculations, a simulation model was established in CST, as shown in Fig. 4.11a, to simulate the transport process of the sheet beam under a uniform axial magnetic flux density. In this model, the y–z plane of the coordinate system represents the upper surface of the CeSRR array. A rounded rectangular cathode with a width of 12 mm and a height of 2 mm is used to generate the sheet beam, as depicted in Fig. 4.11b. The CeSRR array, coupled with the empty square

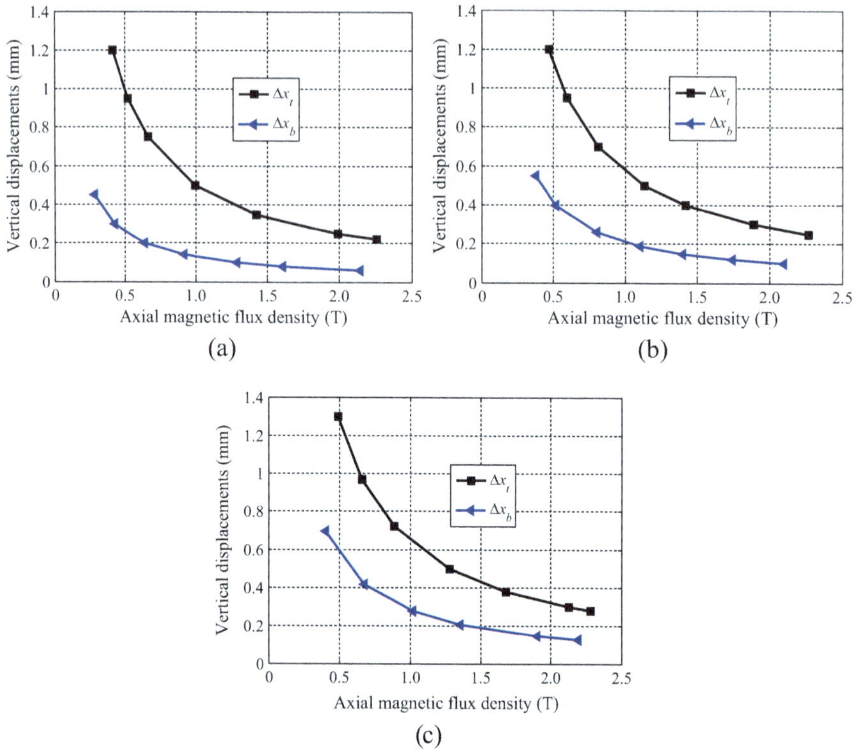

Fig. 4.10 The change of the vertical displacement of the sheet beam edge with the axial magnetic flux intensity. **a** $d = 0.3$ mm. **b** $d = 0.5$ mm. **c** $d = 0.7$ mm

waveguide, forms an MSWS with a length of $L = 145$ mm, as stated before. As mentioned earlier, the gap between the sheet beam and the CeSRR array d is 0.5 mm. The voltage and current of the sheet beam are set to be 160 kV and 100 A, respectively.

Using these settings, the simulation yields the electron trajectory of the sheet beam within the MSWS, as shown in Fig. 4.12. The trajectory reveals that the sheet beam remains unmodulated during the transport, with only minor changes in the velocity of the sheet beam. It is consistent with the theoretical analysis, where the axial velocity of the sheet beam is treated as constant.

To more clearly observe the Diocotron instability of the sheet beam during its transport, it is necessary to monitor the cross-section shapes of the sheet beam at the start ($z = 0$) and end ($z = 145$ mm) points of the MSWS. When the uniform axial magnetic flux density is 1 T, the cross-section shapes of the sheet beam for $d = 0.3$ mm, $d = 0.5$ mm, and $d = 0.7$ mm are simulated, as shown in Fig. 4.13. Simulated results indicate that the sheet beam undergoes different degrees of twisting deformation at 145 mm for all three cases. To better compare these with theoretical calculation results, the vertical displacements obtained from simulations and theory are listed in Table 4.2. The data reveal that the simulated results further validate

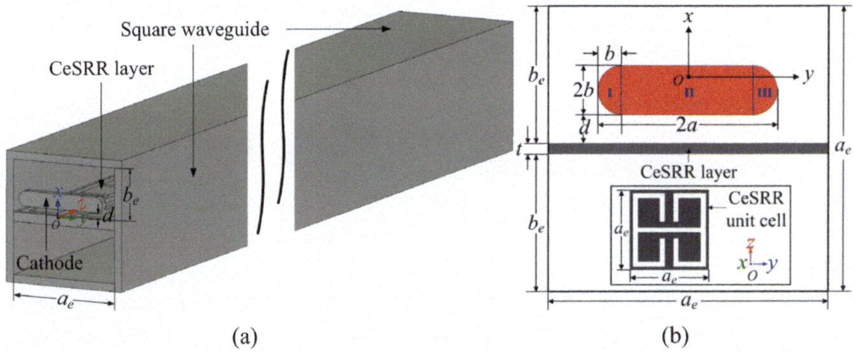

Fig. 4.11 **a** Simulation model of sheet beam transport. **b** Cross-section of sheet beam within the MSWS

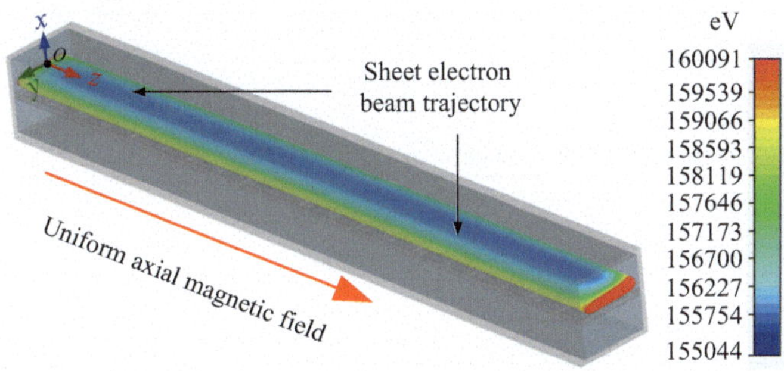

Fig. 4.12 Sheet beam trajectory

the theoretical analysis: increasing the axial magnetic flux density and decreasing d both reduce the vertical displacement of the sheet beam, thereby suppressing its transport instability to a certain extent. However, it is worth noting that the simulated results do not perfectly match the theoretical results, due to the approximations in the theoretical analysis and computational errors in the simulations.

To further investigate the influence of the CeSRR array in the MSWS on the transport of the sheet beam, the d is increased to 2.325 mm, and the axial magnetic flux density is kept constant at 1 T. The simulated cross-section shape of the sheet beam is shown in Fig. 4.14a, indicating that the degree of twisting of the sheet beam increases with d compared to Fig. 4.13. On the other hand, keeping $B_z = 1$ T and $d = 0.5$ mm constant, and removing the CeSRR array, yields the cross-section shape of the sheet beam at $z = 145$ mm, as depicted by the red area in Fig. 4.14b. These results show that removing the CeSRR array can deteriorate the transport instability of the sheet beam. Furthermore, in this case, significantly reducing the axial magnetic flux density from 1 to 0.2 T results in the cross-section shape of the sheet beam at $z =$

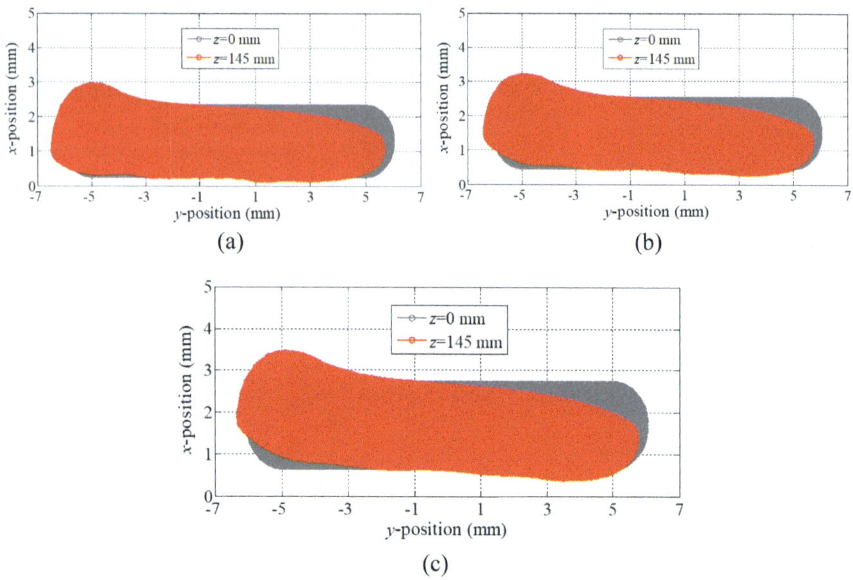

Fig. 4.13 Cross-section shapes of sheet beam. **a** $d = 0.3$ mm. **b** $d = 0.5$ mm. **c** $d = 0.7$ mm

Table 4.2 Comparison of theoretical and simulated results

Parameter/Unit	Value			
B_z/T	1	1	1	2
d/mm	0.3	0.5	0.7	0.7
Simulation Δx_t/mm	0.60	0.67	0.70	0.33
Theory Δx_t/mm	0.50	0.58	0.64	0.32
Simulation Δx_b/mm	0.11	0.18	0.25	0.15
Theory Δx_b/mm	0.13	0.21	0.29	0.14

145 mm, as shown by the blue area in Fig. 4.14b. It is evident from the figure that the transport instability of the sheet beam increases dramatically.

To further verify the above conclusions, theoretical calculations and simulation were conducted under the conditions of $L = 145$ mm, $B_z = 1$ T, and $d = 0.5$ mm with different sheet beam parameters. Two sets of different sheet beam parameters are presented here: In Example A, the parameters are $U = 200$ kV, $I = 150$ A, $a = 6.5$ mm, $b = 1.3$ mm. In Example B, the parameters are $U = 140$ kV, $I = 80$ A, $a = 5.5$ mm, $b = 0.8$ mm.

The theoretical and simulated results are listed in Table 4.3. From the data, it is evident that the theoretical and simulated results for the two examples agree well, further confirming the aforementioned theoretical analysis.

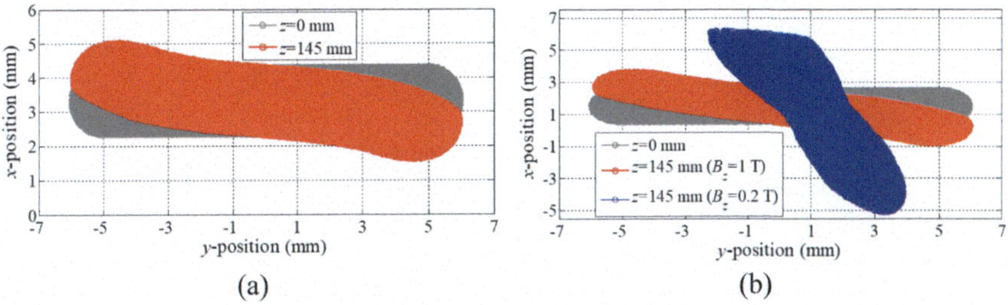

Fig. 4.14 Cross-section shape of the sheet beam. **a** $d = 2.325$ mm. **b** $d = 0.5$ mm

Table 4.3 Comparison of theoretical and simulated results in examples A and B

Parameter/Unit	Example A	Example B
Φ_0/V	– 251	– 205
Simulation Δx_t /mm	0.83	0.60
Theory Δx_t /mm	0.85	0.52
Simulation Δx_b /mm	0.19	0.21
Theory Δx_b /mm	0.22	0.22

4.3 Simulation Analysis of Reversed Cherenkov Radiation

Based on the analysis in Sects. 4.1 and 4.2, this section will analyze the interaction between real charged particles (here sheet beam) and LHMs, thereby elucidating the physical mechanism behind the generation of RCR.

4.3.1 Beam-Wave Interaction

With the assistance of CST simulation software, the beam-wave interaction between the sheet beam and the LHM is studied. The simulation model is shown in Fig. 4.15. In the simulation setup, all sides in the *x*, *y*, and *z* directions, except for the wave ports, are set as perfect electric boundaries. To ensure that the simulation conditions closely match experimental conditions, the following treatments are applied in the simulation: ① The grids in the region of the sheet beam and LHM are refined; ② Metal loss is considered to simulate the impact of surface roughness and assembly errors of the experimental components; ③ The total axial magnetic flux density is introduced into the beam-wave interaction model. Additionally, the voltage and the current of the sheet beam are 160 kV and 1.55 kA, respectively.

When the simulation model is built, the beam-wave interaction simulation is carried out. The normalized signal voltages over time on ports 1 and 2 are obtained, as shown in Fig. 4.16a. Further processing of the signal yields the peak output power over time, as shown in Fig. 4.16b. It is clear that the peak output power at port

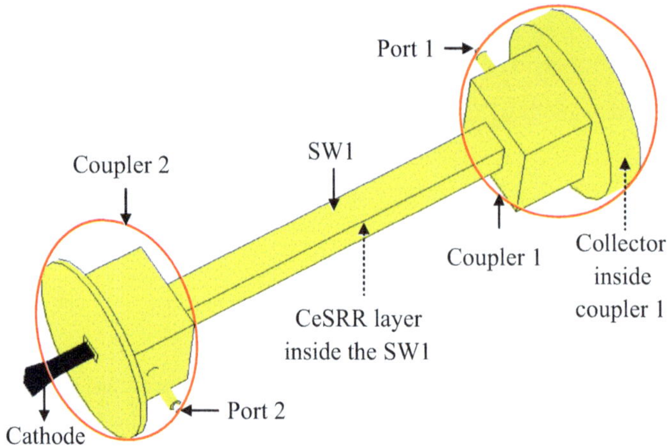

Fig. 4.15 Beam-wave interaction model in CST

2 can reach up to the megawatt level, whereas the peak output power at port 1 is significantly lower. Fast Fourier Transform is applied to the time-domain signals to obtain the frequency spectra of the signals at the two ports, as shown in Fig. 4.16c.

From Fig. 4.16, it can be found that the frequency of the signal at port 2 is approximately 2.836 GHz, and the signal frequency at port 1 is about 2.829 GHz, with a frequency deviation of only 7 MHz. Therefore, within an acceptable range of simulation calculation errors, the signal frequencies at the two ports can be considered the same. The generation of this signal is possible when the velocity of the sheet beam is slightly greater than the phase velocity of the electromagnetic wave. If the voltage of the sheet beam is changed so that this condition is not met, the signal will not be generated. This fact indicates that the electromagnetic radiation is CR.

In order to determine the propagation direction of the CR signal, the normalized power flow distribution along the z-axis in the LHM is analyzed. Figure 4.17a shows the distribution of the normalized power flow along the z-axis ([$x = 0$, $y = 2.1$ mm, $z = -50$ mm], [$x = 0$, $y = 2.1$ mm, $z = 350$ mm]) at 20 ns. It can be seen that the amplitude of the normalized power flow of the signal increases in the direction opposite to the movement of the sheet beam, indicating that the propagation direction of the generated CR signal is opposite to the movement direction of the sheet beam. These results reveal that the CR in the LHM exhibits reversed characteristics.

Furthermore, the CR angle is also studied through simulation. The CR angles at different positions in the LHM are obtained, as shown in Fig. 4.17b. It is evident that the CR angles vary with positions and are greater than 150°. These phenomena indicate that the LHM is anisotropic, where the directions of the phase velocity and group velocity of the electromagnetic wave are not completely opposite. These simulated results show the fundamental physical characteristics of RCR in anisotropic LHMs.

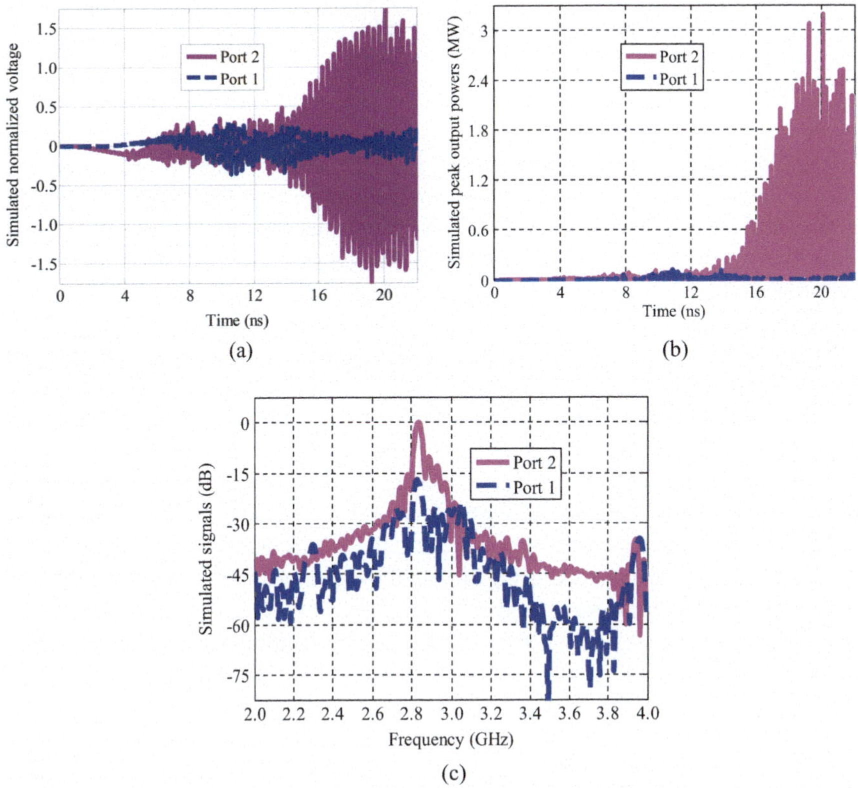

Fig. 4.16 Simulated results. **a** Normalized signal voltages versus time. **b** The relationship of peak output power with time. **c** The spectra of output signals

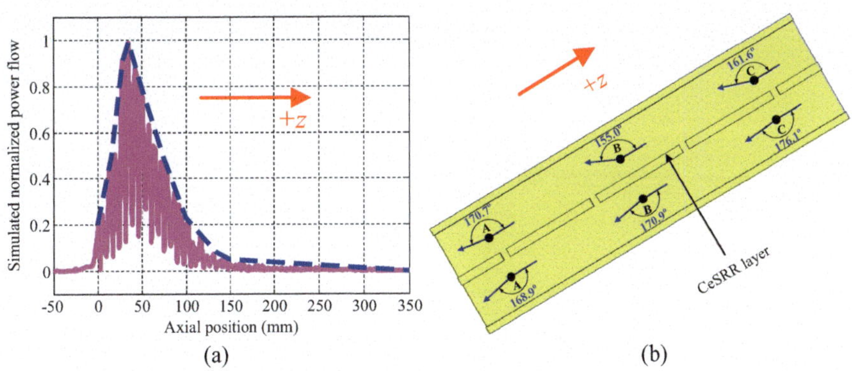

Fig. 4.17 **a** Normalized power flow distribution along the z-axis. **b** Simulated CR angles at different positions

4.3.2 Mode Analysis in Metamaterials

In Chap. 2, it is known that the negative effective permeability of an empty rectangular waveguide operating below the cut-off frequency requires a critical precondition: the propagation mode of the electromagnetic wave in the metamaterials should be TM mode or quasi-TM mode. If the propagation mode is quasi-TE mode, the empty rectangular waveguide can be considered a metamaterial with a negative effective permittivity, which is not conducive to effective beam-wave interaction due to the weak longitudinal electric field. Therefore, mode analysis in metamaterials is of great importance.

The quasi-TM mode in this metamaterial is theoretically analyzed in [23]. To further illustrate its validity, the electromagnetic field distribution in the metamaterial is simulated using CST Microwave Studio. The distribution of the electric and magnetic fields on the $x = 0$ mm plane is shown in Fig. 4.18a and c, respectively. It is evident that the electric field has a strong z-component, while the z-component of the magnetic field is almost zero. This fact confirms that it is the quasi-TM mode propagated in the metamaterial. Furthermore, the electromagnetic field distribution in assumed LHMs is analyzed using simulation. Here, an assumed LHM model is established in CST Microwave Studio, with the permittivity and permeability of the LHM taken from the results in Fig. 4.5. The distribution of simulated electric and magnetic fields on the $x = 0$ mm plane in the LHM are shown in Fig. 4.18b and d, respectively.

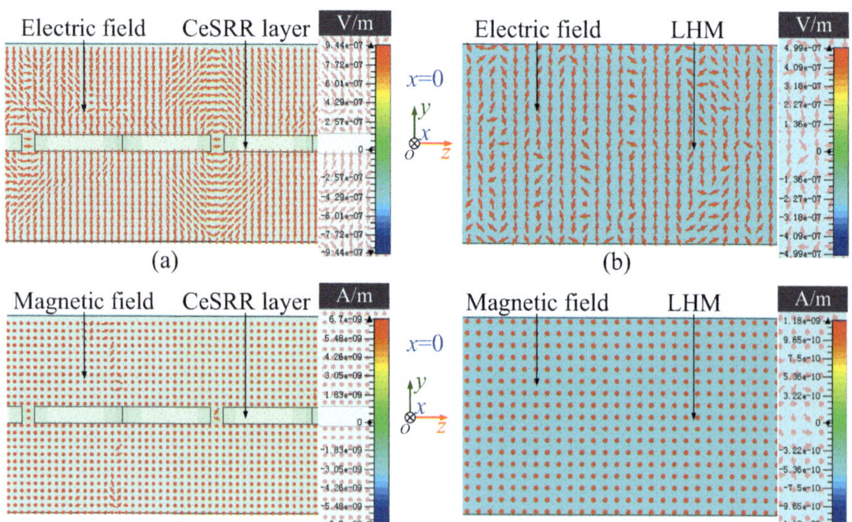

Fig. 4.18 The electromagnetic field distribution in real metamaterial and assumed LHM based on EMT. **a** Real metamaterial. **b** Assumed LHM. **c** Real metamaterial. **d** Assumed LHM

From Fig. 4.18, it is evident that the electromagnetic field distribution in the assumed LHM based on EMT is similar to that in the real metamaterial. Therefore, the mode of the electromagnetic wave is quasi-TM. These simulated results further confirm that this metamaterial can be considered an LHM.

4.3.3 Analysis and Discussion of Simulated Results

From Fig. 4.16 along with the dispersion curve, it's clear that the frequency of the generated signal is lower than the frequency corresponding to the 160 kV electron beam line, indicating that the actual beam-wave interaction voltage is lower than 160 kV. This can be attributed to two main reasons: ① Effective beam-wave interaction requires a synchronization condition, meaning that the velocity of the sheet beam should be slightly larger than the phase velocity of the electromagnetic wave [33], which will lead to a lower signal frequency; ② The voltage of the sheet beam can decrease due to its own space charge field effect from the cathode emission surface to the entrance of the metamaterial. This phenomenon is known as the voltage drop effect [54], causing the actual voltage of the sheet beam to be lower than 160 kV and leading to a decrease in signal frequency. Consequently, the actual voltage of the sheet beam at the entrance of the metamaterial is analyzed. The distribution of the electron trajectory of the sheet beam at the entrance of the metamaterial is shown in Fig. 4.19a. After the sheet beam leaves the cathode emission surface, its kinetic energy will decrease to a certain extent.

To quantitatively analyze the actual kinetic energy of the sheet beam involved in the beam-wave interaction process, the electron velocity information of the sheet beam at the entrance of the metamaterial is extracted. The proportion of electron numbers at different velocities is shown in Fig. 4.19b. Note that the horizontal axis

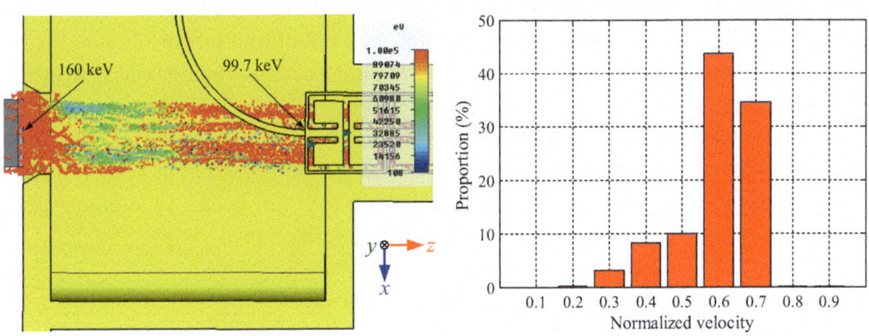

(a) (b)

Fig. 4.19 Schematic diagram of sheet beam voltage drop effect. **a** Trajectory of sheet beam. **b** Normalized velocity distribution of electrons at the entrance of the metamaterial represents the electron velocity normalized to the speed of light in vacuum. The average velocity $\langle \upsilon_0 \rangle$ of the sheet beam can be calculated using the following formula:

$$\langle \upsilon_0 \rangle = \frac{1}{N} \sum_{n=1}^{N} \upsilon_n \hat{z}, \tag{4.9}$$

where N is the total number of electrons of the sheet beam at the entrance of the metamaterial, and υ_n means the velocity of the nth electron. Then, the average kinetic energy of the sheet beam at the entrance of the metamaterial is obtained as 99.7 keV, which indicates that the actual beam-wave interaction voltage of the sheet beam should be 99.7 kV. The analysis presented above confirms that the actual beam-wave interaction voltage of the sheet beam decreases from 160 to 99.7 kV, and the frequency of the generated RCR signal is lower than the frequency corresponding to the intersection point of the 160 kV electron beam line and the dispersion curve.

Furthermore, the power capacity of the coaxial line is analyzed for field breakdown and thermal breakdown to validate that the coaxial line at the output port shown in Fig. 4.15 can withstand the generated output power.

The relationship between the power P and the field amplitude E on the coaxial line is [33]:

$$P = E^2 \frac{\pi \sqrt{\varepsilon_r} r_{cli}^2}{\sqrt{\mu_0 / \varepsilon_0}} \ln\left(\frac{r_{clo}}{r_{cli}}\right). \tag{4.10}$$

The parameters of the coaxial line at the output port are listed here: the inner radius of the coaxial line is $r_{cli} = 0.6$ mm; the outer radius of the coaxial line is $r_{clo} = 2.7$ mm; the supporting medium in the coaxial line is PTFE, with a relative permittivity ε_r of 2.1. Additionally, according to [55], the range of the breakdown field strength for PTFE is 25–40 kV/mm. Taking the breakdown field strength as 25 kV/mm as an example, the breakdown power of the coaxial line is ~ 4 MW. It can be noted that this breakdown power is greater than the output power obtained from simulation. Therefore, the field breakdown will not occur in the coaxial line at the output port.

Moreover, since the sheet beam used in the experiment is a single pulse with a nanosecond-level pulse width, the heat generated by the peak output power is only at the millijoule level, and the thermal breakdown will not occur. In summary, the coaxial line at the output port can withstand the generated peak output power.

4.4 Experimental Verification of Reversed Cherenkov Radiation

On the basis of the interaction between the sheet beam and the MSWS introduced in Sect. 4.3, the sheet beam electron gun, MSWS, output couplers, solenoid magnetic focusing configuration, and collector were fabricated and assembled. The experiment was then carried out.

4.4.1 Experimental Setup and Platform

An experimental setup was established to validate the presence of RCR using real charged particles. As shown in Fig. 4.20a, the experimental setup includes a cathode, solenoid magnetic focusing configuration, MSWS, two output couplers, and a collector. The physical principle of the experiment is described here: the sheet beam emitted from the cathode is focused under the uniform axial magnetic field generated by the solenoid magnetic focusing configuration and successfully transports through the electron beam channel of the MSWS. During this process, the sheet beam interacts with the axial electric field. When the synchronization condition for generating RCR is met, the coherent electromagnetic radiation is excited. This radiation is then outputted through output coupler 2 (whose port is named port 2) or output coupler 1 (whose port is named port 1). Finally, the electrons that have completed the beam-wave interaction are collected by the collector.

The cathode is connected to the inner conductor of a Tesla transformer to generate the required sheet beam. The emitting surface of the cathode has a rectangular cross-section with measured dimensions of 12 mm $\times$ 2 mm. The solenoid, which generates a uniform axial magnetic field for focusing the sheet beam, consists of a cylindrical supporting structure with flanges and coils wounding around its outer surface. One end of the solenoid is connected to the Tesla transformer, and the other end is connected to a baffle. The MSWS, composed of a 20-period CeSRR array and an empty square waveguide SW1, serves as the site for beam-wave interaction. The output coupler near the end of cathode and the output coupler near the end of collector can be used to output the generated electromagnetic radiation, with their local zoomed photos shown in Fig. 4.20b and c, respectively. The collector is a rectangular block with dimensions of 55 mm $\times$ 55 mm $\times$ 20 mm.

The cathode component and collector were fabricated, as shown in Fig. 4.21. The cathode component includes a cathode rod, cathode pad, and cathode. The cathode rod is made of stainless steel, which is used to connect the inner conductor of the Tesla transformer; the cathode pad is made of aluminum, and the cathode is made of graphene. The collector is designed as a flat plate made of graphene, with two through holes for fixing the collector to the output coupler 1.

The experimental platform is shown in Fig. 4.22. It consists of the Tesla transformer, vacuum system, connecting cylinder, experimental setup, baffle, etc. The

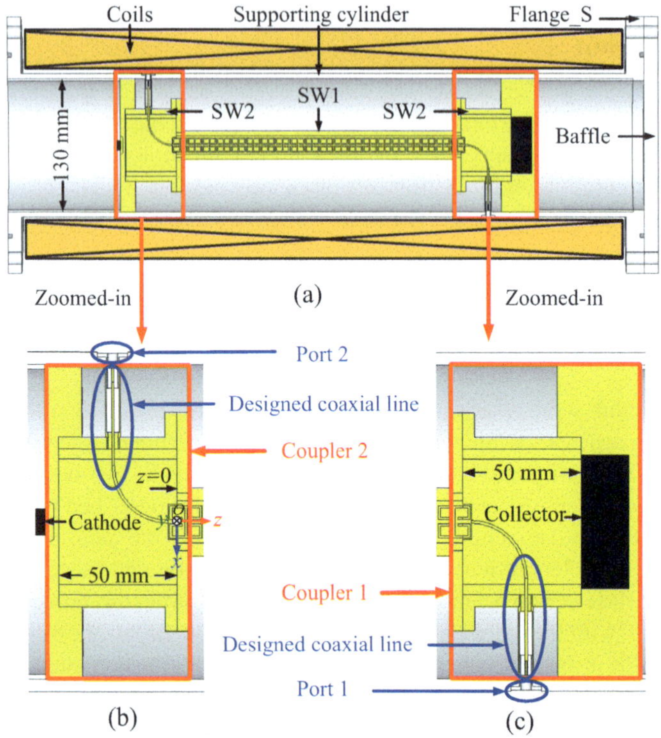

Fig. 4.20 Experimental setup. **a** Overall diagram. **b** Coupler 2. **c** Coupler 1

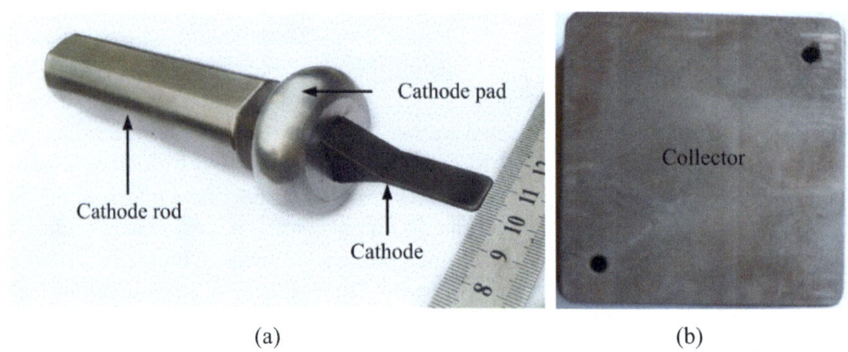

Fig. 4.21 Fabricated components. **a** Cathode component. **b** Collector

Fig. 4.22 Experimental platform

connecting cylinder serves as a transition structure connecting the Tesla transformer and the experimental setup, ensuring the compatibility of the entire experimental system. Additionally, there are grooves between the Tesla transformer and the vacuum system, between the vacuum system and the connecting components, between the connecting components and the experimental setup, and between the experimental setup and the baffle. Rubber sealing rings are placed in these grooves to ensure the high vacuum environment. The operating voltage of the Tesla transformer is approximately 140–180 kV. The vacuum system includes a direct-coupled pump and a molecular pump to establish a dynamic vacuum environment, with a vacuum degree reaching 3.8×10^{-4} Pa.

4.4.2 *Transmission Characteristics Analysis of Output Couplers*

For the experimental study of RCR, a suitable output coupler is needed to couple the generated microwave signals. For this purpose, a novel coaxial output coupler is proposed [16]. To investigate the transmission and reflection characteristics of electromagnetic waves in the metamaterial with this output coupler, a simulation model is established in CST Microwave Studio, as shown in Fig. 4.23a, with its cross-section view on the $y = 0$ plane shown in Fig. 4.23b. The simulation model includes the MSWS, output coupler 1, and output coupler 2.

Both output couplers include a waveguide component and a coaxial component. The waveguide component consists of a quarter CeSRR and a short waveguide with a cross-section size of 60 mm × 60 mm and an axial length of 50 mm. The coaxial component is an element of a coaxial line, with its inner conductor being a 90° arc-shaped probe. The outer radius of the probe is 0.6 mm, the inner radius of its outer conductor is 2.7 mm, and the filling material is PTFE with a relative permittivity of 2.1 and a loss tangent of 0.0002. The two output couplers have a small difference:

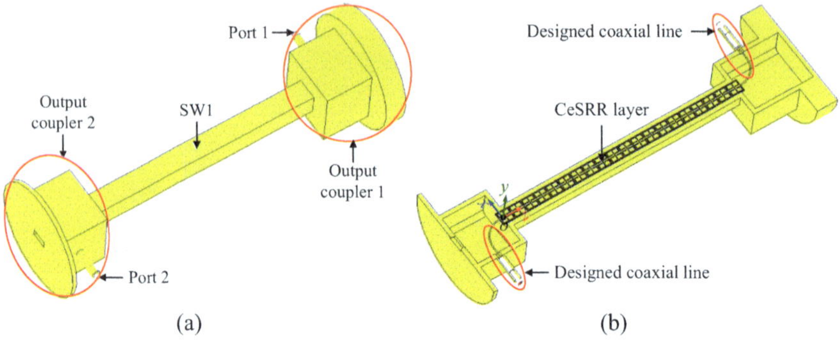

Fig. 4.23 **a** Simulation model of transmission characteristics. **b** Cross-section view on the $y = 0$ plane

the front end of output coupler 2 has a square hole as a prepared channel for the sheet beam, while the end of output coupler 1 has a prepared deep groove for fixing the collector.

The transmission characteristics between ports 1 and 2 are simulated using the above simulation model. The simulated results are shown in Fig. 4.24. These results indicate that the MSWS has a transmission passband in the 2.83–3.05 GHz range, with good transmission and low reflection characteristics. The amplitude of S_{12} is about − 4 dB and the amplitude of S_{22} is about − 10 dB at 2.85 GHz. Therefore, when the RCR signal is excited and outputted from one port, there will be a minimal reflection signal output at the other port. The above study provides an important basis for identifying the RCR signal, namely that the larger signal from the output port is the RCR signal.

Based on the simulated results, the experiment on transmission and reflection characteristics was carried out. The fabricated components are shown in Fig. 4.25a. For feasible manufacture, output coupler 1 is divided into baffle 1, SW1, flange_

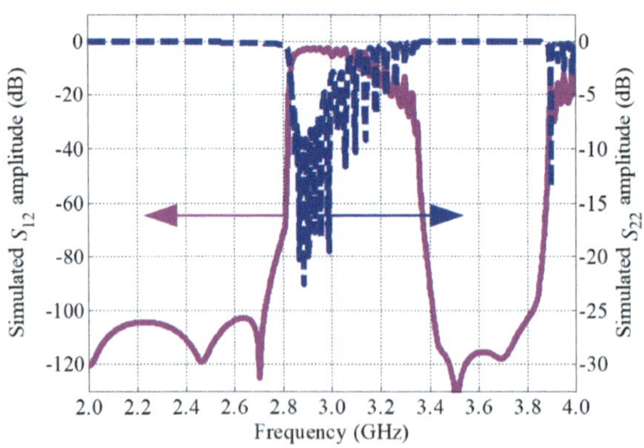

Fig. 4.24 Simulated S-parameters

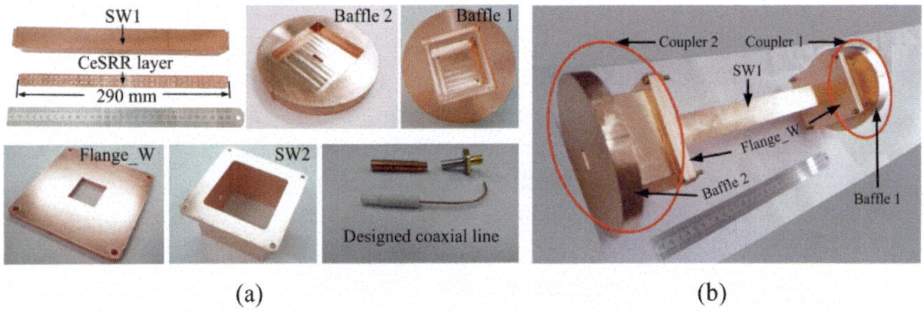

Fig. 4.25 **a** Fabricated components. **b** Device under test

W, and a coaxial line; output coupler 2 is divided into baffle 2, SW2, flange_W, and a coaxial line. The materials of the CeSRR array, SW1, SW2, baffles 1, baffle 2, and the flange_W are all made of oxygen-free copper. It should be noted that the CeSRR array is the core component of the experimental setup, and changes in its dimensions have a significant impact on the experimental results, especially on the radiation frequency. Therefore, high-precision manufacturing methods should be used to ensure that the dimensions of the fabricated CeSRR array match those in the simulation model as closely as possible.

The experimental components were fabricated using the methods described above and precisely assembled. First, a flange was embedded at each end of the SW1 and fixed by brazing. Next, SW2 was connected and fixed to the flange using screws. And then, the CeSRR array was fixed, and the inner conductor of the designed coaxial line was connected to the CeSRR array using soldering techniques. At the same time, the outer conductor and filling material of the designed coaxial line were assembled. Finally, the SW2 was embedded into baffles 2 and 1 and bonded using AB glue, which is suitable for vacuum environment. The assembled device under test is shown in Fig. 4.25b.

When the device under test was assembled, an experimental platform was established, as shown in Fig. 4.26. The transmission characteristics were tested using the Agilent N5230A PNA-L vector network analyzer.

The curves of S-parameter amplitudes with frequency were obtained by analyzing the experimental data, as shown in Fig. 4.27. It can be found that although there are minor differences, the simulation and experimental results agree well. The differences mainly arise from computational errors in the simulation, such as mesh generation, experimental processing errors, assembly errors, and unavoidable testing errors during the experiment. Both simulation and experimental results indicate that: ① The device possesses a transmission passband from 2.83 to 3.05 GHz, demonstrating the left-handed characteristics of the metamaterial; ② The device has good transmission and low reflection characteristics within the passband, which is beneficial for distinguishing RCR signals from their reflection signals. This work lays the foundation for further research on RCR.

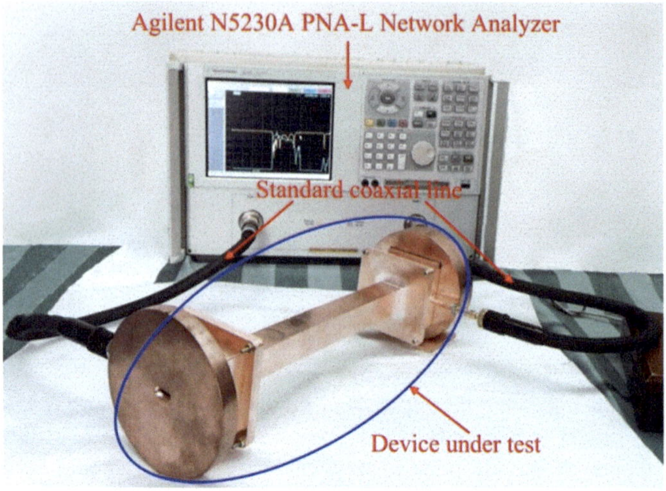

Fig. 4.26 Experimental platform of transmission characteristics

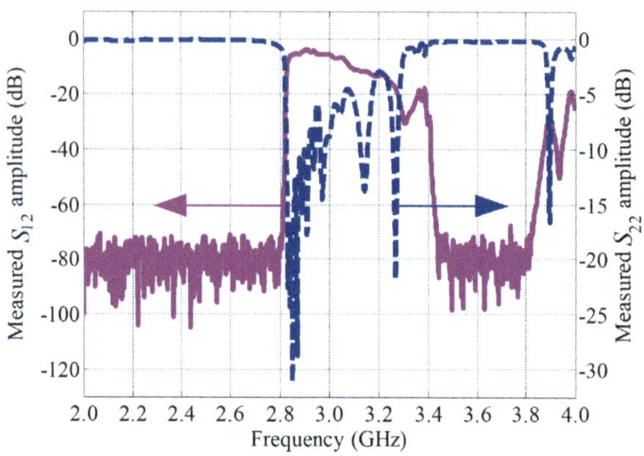

Fig. 4.27 Measured S-parameters

Based on the previous simulation and experimental results, the transmission phase shift with frequency for the MSWS can be obtained, as shown in Fig. 4.28.

Now, we take the simulated results in Fig. 4.28a as an example, to show the process of obtaining the dispersion curve from the transmission phase shift. First, another representation of the phase shift can be obtained using CST Microwave Studio, as shown in Fig. 4.29a, where the phase shift is continuous and not limited between $-\pi$ and π. Second, by dividing the transmission phase shift in Fig. 4.29a by the number of periods of the CeSRR array, the transmission phase shift per period with frequency is obtained, as shown in Fig. 4.29b, where the phase shift varies between $-\pi$ and 0. To compare with the dispersion curve obtained from HFSS (solid bule line), the curve

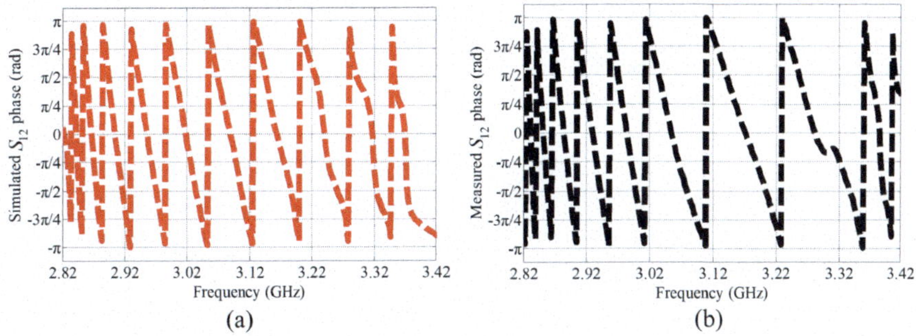

Fig. 4.28 Transmission phase shift versus frequency. **a** Simulation. **b** Experiment

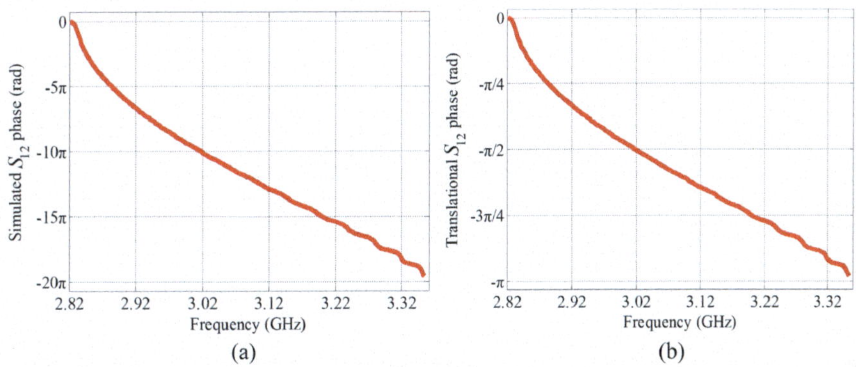

Fig. 4.29 The curves of transmission phase shift with frequency. **a** Another form for transmission phase shift. **b** Transformed transmission phase shift

in Fig. 4.29b is vertically shifted by π to bring the transmission phase shift within the range of 0 to π. Then, by swapping the horizontal and vertical coordinates, the dispersion curve (dotted red line) is obtained, as shown in Fig. 4.30. Similarly, using the experimental transmission phase shift, the dispersion curve (dashed black line) in Fig. 4.30 is obtained. From the three dispersion curves, it's evident that the dispersion curve derived from the transmission phase shift agrees well with the dispersion curve obtained from HFSS simulation, demonstrating the left-handed or backward-wave characteristics within the range of 2.83–3.05 GHz.

4.4.3 Design of the Solenoid Magnetic Focusing Configuration

In the experimental verification of RCR, it is necessary to study the interaction mechanism between the sheet beam and the electromagnetic waves in the metamaterial.

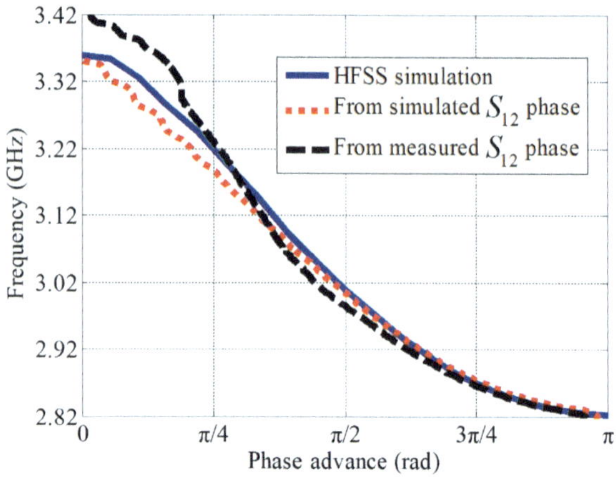

Fig. 4.30 Dispersion curves obtained from different methods

As noted in Sect. 4.2.2, the transport of the sheet beam in a uniform magnetic field will show Diocotron instability. Therefore, an appropriate focusing method needs to be designed to ensure that the sheet beam can transport stably through the MSWS and interact effectively with the electromagnetic waves. The theoretical analysis previously discussed indicates that the axial uniform magnetic field is feasible to focus the sheet beam. Additionally, the sheet beam produced by the Tesla transformer has the feature of high current, making it suitable for the axial uniform magnetic field. Hence, a solenoid magnetic focusing configuration is used in this experiment. It consists of a solenoid and a solenoid pulse power supply, and the latter provides the excitation current for the former.

According to the features of the experimental setup shown in Fig. 4.25b, two holes need to be drilled on the solenoid supporting cylinder to connect the output couplers 1 and 2, respectively. To ensure the stable transport of the sheet beam, the axial magnetic field in the interaction area needs to be as uniform as possible, with a magnetic flux density of 1 T. Based on these requirements, a solenoid coil wound with rectangular cross-section flat copper wire with a width of $w_1 = 6.4$ mm and a thickness of $t_1 = 3.4$ mm is designed. During the winding process, the copper wire needs to bypass the two output ports, which leads the solenoid coil being axially non-uniform. Therefore, in the theoretical calculations, the solenoid coil is considered as three segments, as shown in Fig. 4.31. Although the three segments are divided for calculations, the excitation current in the three segments of the solenoid coil is consistent.

After careful optimization, the structure parameters of the three segments of the solenoid coil are given here: the first segment of the solenoid coil has 8 layers, with 18 turns per layer from the 1st to the 6th layer, and 8 turns per layer for the 7th and 8th layers. The second segment of the solenoid coil has 6 layers, with 51 turns per layer from the 1st to the 5th layer. On the 6th layer, 8 turns and 7 turns are respectively wound near the output port close to the cathode and the output port close to the collector. The third segment of the coil has 5 layers, with 18 turns per layer.

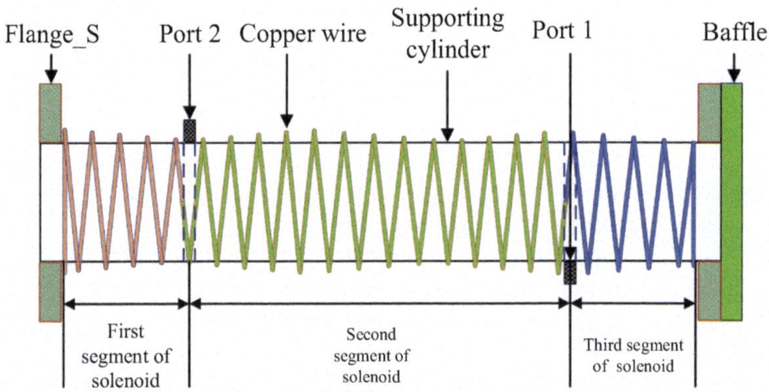

Fig. 4.31 Schematic diagram of a solenoid magnetic focusing configuration

The first segment of the solenoid coil has significantly more layers than the other two segments to address the issue of the magnetic field drop effect at the edges of the solenoid, ensuring the uniformity of the axial magnetic field. The two ends of the second segment of the solenoid coil have an additional layer to compensate for the magnetic field drop effect caused by the absence of coils at the output ports. The third segment of the solenoid coil has the fewest layers, allowing the axial magnetic field to decrease rapidly near the collector, facilitating the collection of electrons by the collector after beam-wave interaction.

The relationship between the axial magnetic flux density B_{zj} near the central axis inside the jth ($j = 1, 2, 3$) segment of the solenoid coil and the axial coordinate z_j can be expressed as:

$$B_{zj} = \frac{\mu_0 n_1 n_2 I_L}{2}\left[\left(\frac{L_{sj}}{2} + z_j - z_{0j}\right)\ln\frac{r_{oj} + \sqrt{r_{oj}^2 + \left(\frac{L_{sj}}{2} + z_j - z_{0j}\right)^2}}{r_{ij} + \sqrt{r_{ij}^2 + \left(\frac{L_{sj}}{2} + z_j - z_{0j}\right)^2}} + \left(\frac{L_{sj}}{2} - z_j + z_{0j}\right)\ln\frac{r_{oj} + \sqrt{r_{oj}^2 + \left(\frac{L_{sj}}{2} - z_j + z_{0j}\right)^2}}{r_{ij} + \sqrt{r_{ij}^2 + \left(\frac{L_{sj}}{2} - z_j + z_{0j}\right)^2}}\right], \tag{4.11}$$

where I_L represents the excitation current of the solenoid coil, $n_1 = 1/w_1$ and $n_2 = 1/t_1$ denote the number of coil turns per unit length and the number of coil layers per unit thickness, respectively, z_{0j} represents the axial coordinate of the midpoint of the jth segment of the solenoid coil, r_{ij} is the inner radius of the jth segment, r_{oj} is the outer radius of the jth segment, and L_{sj} means the length of the jth segment of the solenoid coil. The total magnetic flux density near the central axis inside the solenoid coil is the sum of the contributions from the three segments, that is, $B_z = B_{z1} + B_{z2} + B_{z3}$.

When the excitation current is set to 1.09 kA, the axial magnetic flux density as a function of the position along the axis is calculated using (4.11), as shown in Fig. 4.32. It is clear that the generated axial magnetic field is approximately uniform in the beam-wave interaction area, and the axial magnetic flux density is about 1.06 T, which can meet the experimental requirements.

Based on theoretical analysis, a solenoid coil was fabricated, as shown in Fig. 4.33a. Measurements of the axial magnetic flux density along the central axis inside the solenoid were conducted using a digital Tesla meter SG-42, as shown in Fig. 4.33b.

First, the probe of the Tesla meter was fixed on a ruler, with the end of the probe aligned with the zero scale of the ruler, defining the $z = 0$ position as the entrance of the MSWS. Next, the probe was fixed along the internal central axis of

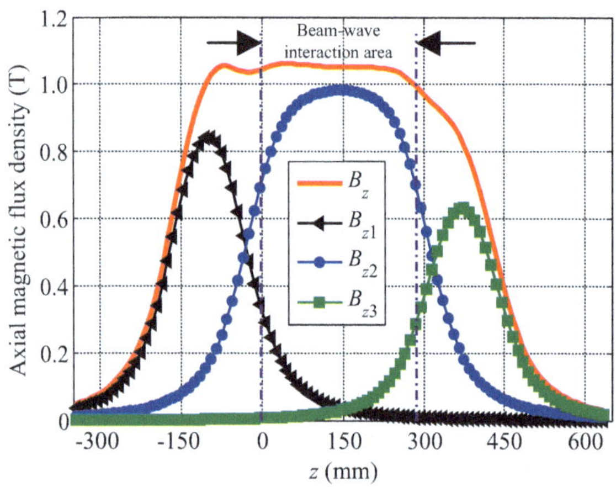

Fig. 4.32 Theoretical calculation results of the axial magnetic flux density

Fig. 4.33 **a** Sample of solenoid. **b** Experimental platform

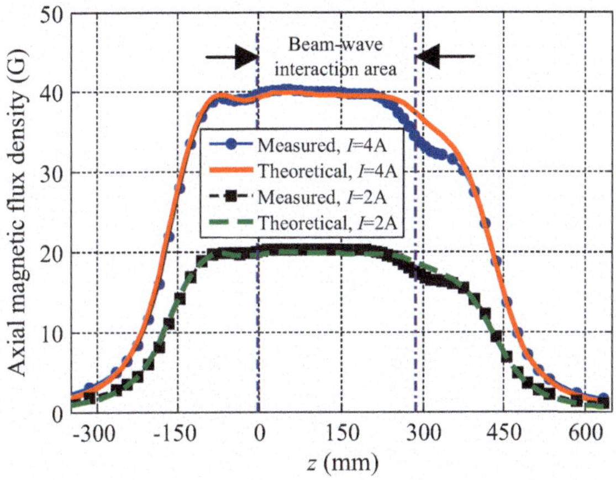

Fig. 4.34 Axial magnetic flux density versus the axial position on the central axis of the solenoid

the solenoid, and its axial position was varied between − 350 mm and 650 mm. Lastly, the constant current source was adjusted, and the magnetic flux density was measured every 10 mm when the excitation current was 2 A or 4 A. Thus, the axial magnetic flux density along the central axis was obtained, as shown in Fig. 4.34.

The theoretical values for the excitation currents of 2 A and 4 A cases were calculated using (4.11). It can be found that the theoretical and experimental results are in good agreement.

In the solenoid experiment, a low constant current source was used to provide the excitation current for the solenoid. Since the actual excitation current is time-varying, it is difficult to measure the axial magnetic flux density at a specific moment. However, due to the narrow pulse width of the sheet beam, which is only 20 ns, the axial magnetic field should remain uniform during this time. This requires that the designed excitation current pulse width should be wide, on the order of milliseconds, so that the excitation current can be considered constant over the 20 ns duration. Therefore, a constant current source is used instead of an alternating current source to test the axial magnetic flux density. In addition, according to (4.11), the magnetic flux density is directly proportional to the excitation current. Therefore, the use of a small current instead of a large current in testing is feasible in theory.

To provide the excitation current for the solenoid, a solenoid pulse power supply was used, as shown in Fig. 4.35. L denotes the inductance of the solenoid coil, S is the thyristor switch, R is the total resistance, C and U_C are the capacitance of the capacitor and its peak voltage (with a breakdown voltage of 5 kV), respectively, R_1 is the current limiting resistor, D is the diode, and T is the adjustable transformer.

A brief introduction to the physical principle of the solenoid pulse power supply is provided here: First, the thyristor switch S is opened, and the capacitor is charged using a standard 220 V/50 Hz power supply. When the capacitor is fully charged, the thyristor switch S is closed. Using the unidirectional conductivity of the diode D, the capacitor quickly discharges into the solenoid coil circuit, thereby generating the necessary excitation current for the solenoid.

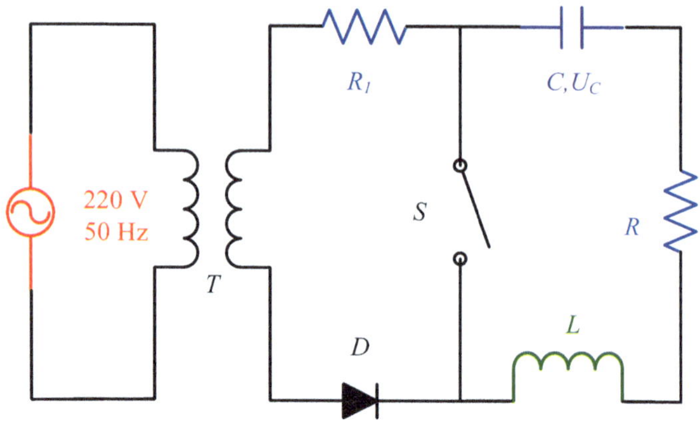

Fig. 4.35 Schematic diagram of the physical principle of the solenoid pulse power supply

The capacitor voltage $u_C(t)$ over time can be expressed as:

$$u_C(t) = U_C e^{-\alpha t} \cos(\omega_d t). \tag{4.12}$$

The time-varying excitation current $i_L(t)$ in the solenoid can be found by multiplying the capacitor voltage $u_C(t)$ by the capacitance C and then taking the derivative with respect to time t [56, 57]:

$$i_L(t) = -CU_C[\alpha e^{-\alpha t} \cos(\omega_d t) + \omega_d \alpha e^{-\alpha t} \sin(\omega_d t)], \tag{4.13}$$

where $\alpha = R/2L$ is the attenuation coefficient with $R = 0.65\ \Omega$ and $L = 8.05$ mH, and $C = 900\ \mu$F. The peak value of $u_C(t)$, denoted as U_C, can be adjusted using the transformer T. $\omega_d = (\omega_0 + \alpha)(\omega_0 - \alpha)/4$ is the damped resonant angular frequency, where $\omega_0 = 1/2LC$ is the resonant angular frequency. Note that the peak value of i_L(t) is denoted as I_L.

To protect the capacitor from being punctured and burned out (not exceeding 5 kV) and to ensure that the solenoid can generate sufficient excitation current (about 1.09 kA) to maintain the required axial uniform magnetic field for the experiment, the peak voltage U_C is set to 3.8 kV. The relationship between the capacitor discharge voltage and the excitation current over time is obtained using (4.12) and (4.13), as shown by the solid and dashed lines in Fig. 4.36, respectively. Experimental measurements of the capacitor's discharge voltage were conducted, resulting in the discharge voltage over time, as shown by the dotted line in Fig. 4.36. It is clear that the measured discharge voltage curve agrees well with the calculated discharge voltage curve.

It can be observed from Fig. 4.36 that the excitation current in the solenoid is time-varying, which leads to a time-varying axial magnetic field. Since the sheet beam produced by the Tesla transformer is a single pulse with a pulse width of around 20 ns, which is significantly shorter than the millisecond-scale period of the

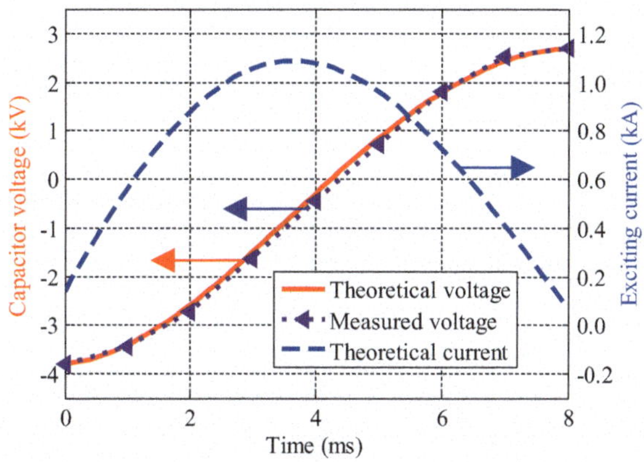

Fig. 4.36 Capacitor voltages and exciting currents versus time

axial magnetic field, the axial magnetic field in the solenoid can be considered to be uniform during the pulse period of the sheet beam. In the subsequent experiments, the trigger timing of the single-pulse sheet beam is adjusted to be around 3.8 ms to ensure that the required focusing magnetic field is produced when the peak voltage of the capacitor is 3.8 kV.

4.4.4 Experimental Measurement and Analysis

Before the experiment verification of RCR, the beam voltage, beam current, and cross-section shape of the sheet beam were tested. Initially, the sheet beam voltage was reduced using resistance division and attenuator methods, and then the voltage waveform was measured using an oscilloscope. The time-domain waveform of the sheet beam voltage is shown in Fig. 4.37a. It can be observed that the voltage curve is a damped oscillation, with the operating voltage near its first peak, where the voltage is about − 160 kV. The voltage in the curve is negative because the outer conductor of the Tesla transformer is grounded, and the voltage of the sheet beam generated by the inner conductor is negative high voltage. For the sake of description, this chapter uses positive values to describe the sheet beam voltage, i.e., 160 kV. Next, a Rogowski coil was used to detect the induced magnetic field produced by the varying current on the inner conductor of the Tesla transformer. The oscilloscope measured the electrical signal excited by the induced magnetic field, which was then used to infer the current of the sheet beam passing through the inner conductor of the Tesla transformer, approximately 1.55 kA [58]. Finally, a piece of white paper was fixed onto baffle 2 of output coupler 2. The high-speed electrons left burn marks on the paper, producing the cross-section shape of the sheet beam, as shown in Fig. 4.37b. It can be seen that the cross-section size of the sheet beam is approximately 11 mm × 1.7 mm, which matches the cathode's cross-section size of 12 mm × 2 mm.

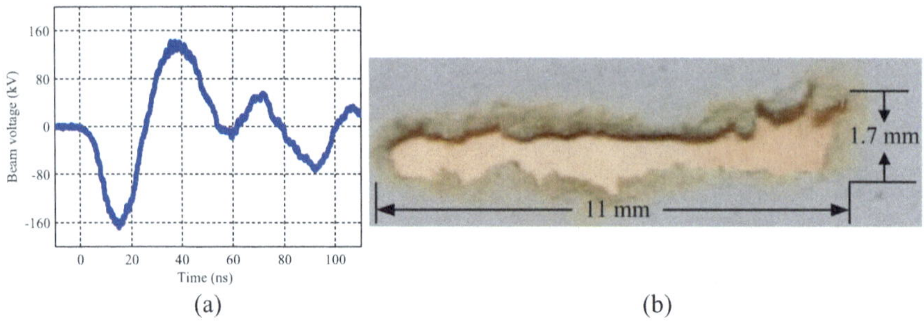

Fig. 4.37 Measured results of sheet beam. **a** Time-domain waveform of the sheet beam voltage. **b** Cross-section shape of the sheet beam

Subsequently, the experimental platform is established, as shown in Fig. 4.38. The instruments used in the experiment include two 50 dB attenuators, one 10 dB attenuator, two detectors, and one oscilloscope.

As previously mentioned, the sheet beam is generated by the Tesla transformer and the graphene cathode. The sheet beam passes through the electron beam channel of the MSWS under the focusing of a uniform axial magnetic field, thereby exciting a coherent electromagnetic wave signal. In the experiment, the sheet beam was produced at the peak of the solenoid's maximum excitation current by adjusting the trigger. Here, two sets of experiments were conducted, as stated below. The schematic diagram of the experiments is shown in Fig. 4.39.

In the first experiment, the experimental setup involved connecting a 50 dB attenuator, a 10 dB attenuator, a detector, and an oscilloscope in sequence to the back of output port 2. These experimental instruments were connected using coaxial lines, with a total attenuation of about 3 dB. At the back of output port 1, a 50 dB attenuator, a detector, and an oscilloscope were connected in sequence, with these instruments

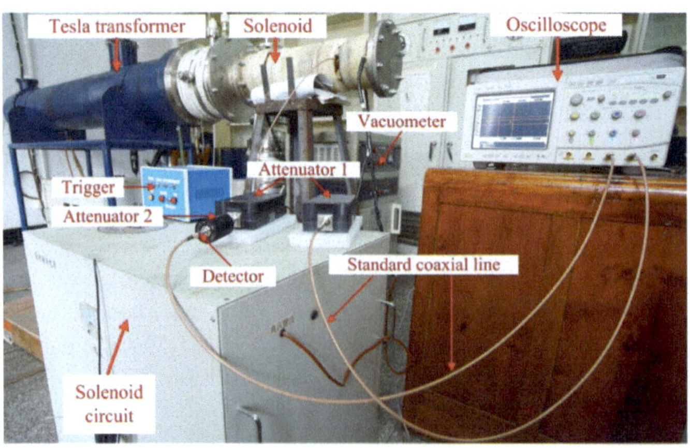

Fig. 4.38 Experimental platform of RCR

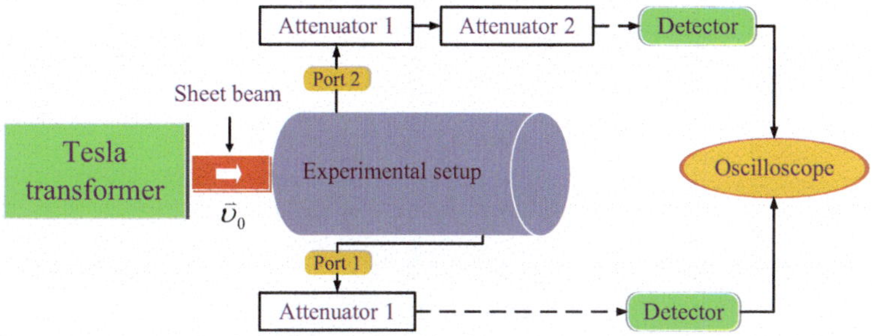

Fig. 4.39 Experimental schematic diagram of RCR

also connected using coaxial lines, and the total attenuation of the coaxial lines is about 3 dB. The detector converts the detected signal voltage envelope, allowing the output power of signals from the two ports to be calculated by comparing it with the detector's output power-to-voltage characteristic curve (shown in Fig. 4.40). Note that the detector measures the effective value of the signal voltage, corresponding to the average output power. Since the peak voltage is $\sqrt{2}$ times of the effective voltage, the peak output power is twice of the output power.

In the second experiment, the experimental setup involved connecting a 50 dB attenuator, a 10 dB attenuator, and an oscilloscope in sequence to the back of output port 2. These instruments were connected using coaxial lines, with a total attenuation of about 3 dB. At the back of output port 1, a 50 dB attenuator and an oscilloscope were connected in sequence using coaxial lines, with a total attenuation of about 3 dB. Since a detector was not used in this experiment, the captured signal was a time-varying voltage signal. Therefore, the spectra of the output signal were obtained from the time-varying information of the voltage signal through Fast Fourier Transform (FFT).

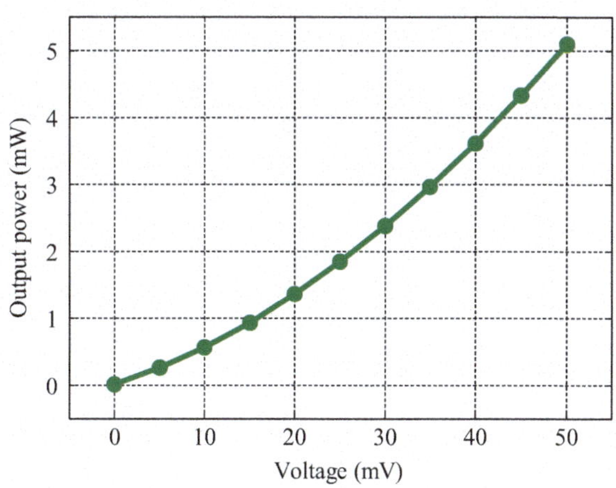

Fig. 4.40 The output power versus voltage

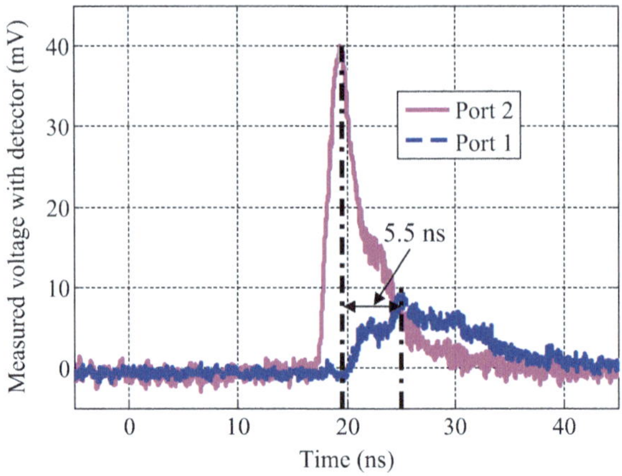

Fig. 4.41 The measured signal voltage over time with a detector

Initially, the first experiment was conducted. The signal envelope amplitude over time is directly provided by the oscilloscope, as shown in Fig. 4.41. It indicates that there is a clear delay of approximately 5.5 ns for the signal at port 1 compared to the signal at port 2, and the delay time equals the signal propagation time from port 2 to port 1. This conclusion further confirms that the signal at port 1 is the reflected signal generated by the output signal at port 2 in the experiment. Furthermore, from Fig. 4.40, the voltage amplitude of 40 mV at port 2 corresponds to an output power of 3.6 mW. Since this output power is the result after 63 dB of attenuation, the actual output power at port 2 is about 7.2 kW, and the corresponding peak output power is about 14.4 kW. Similarly, the actual peak output power at port 1 is calculated to be ~ 0.25 kW. Therefore, the peak output power at port 2 is significantly greater than that at port 1.

It should be noted that there is a significant discrepancy between the experimental and simulated results, with the measured peak output power at port 2 being far less than the simulated value. The reasons are given here: ① the transport rate of the sheet beam in the experiment was lower than that in the simulation due to the manufacturing and assembly errors; ② the beam current used in the experiment was less than the current set in the simulation. The actual current emitted from the cathode was much lower than the current through the inner conductor of the Tesla transformer in the sheet beam testing experiment, as well as lower than the current in the simulation. These results in a delay in the initiation of the backward-wave oscillation, preventing the system from reaching a steady state within the pulse duration and thus leading to a significantly lower peak output power of the signal observed in the experiment relative to the simulated results. However, the fact that the peak output power of the signal at port 2 is significantly greater than that at port 1 agrees with the simulation, confirming that the signal detected at port 2 is indeed the RCR signal.

Next, under the same conditions as the first experiment, the second experiment was conducted. An FFT on the attenuated time-domain signal measured by an oscilloscope was performed to obtain the signal spectra. The relative output powers of the signals at the two ports from the first set of experiments were used to obtain the normalized signal spectra, as shown in Fig. 4.42. It reveals that the frequencies of the signals at ports 2 and 1 are 2.850 and 2.847 GHz, respectively, which are almost identical, with only a 3 MHz measured error, and agree well with the simulated results. The experimental results reported above indicate that the interaction between the sheet beam and the MSWS indeed excites a coherent RCR signal.

Finally, with all other conditions remaining unchanged, the signal frequencies generated at different sheet beam voltages were obtained using both simulation and experimental methods. It can be seen from Fig. 4.43 that the signal frequencies correspond to different sheet beam voltages of 150, 160, and 170 kV. This fact indicates that as the sheet beam voltage increases, the corresponding signal frequency increases, which perfectly matches the dispersion characteristics of the metamaterial.

The simulation and experimental results presented above confirm the existence of RCR excited within a metamaterial. This novel electromagnetic characteristic of

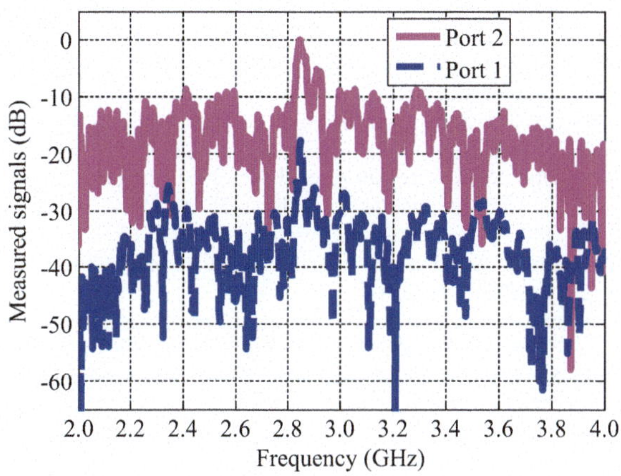

Fig. 4.42 Spectral analysis of the measured output signal

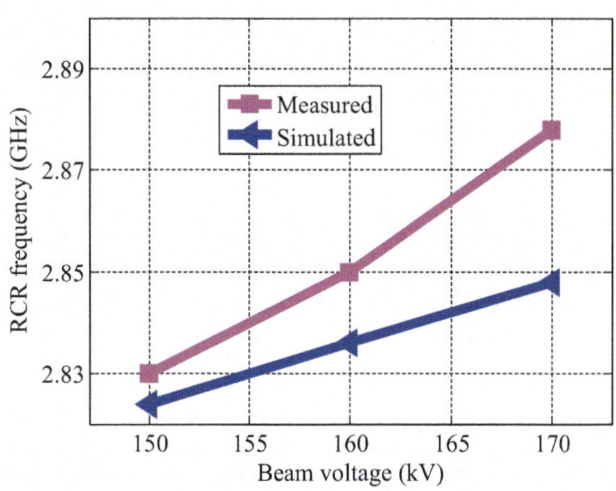

Fig. 4.43 The curves of RCR signal frequency as a function of beam voltage

RCR has received widespread attention. The experimental work introduced in this chapter strongly verifies the existence of RCR and lays a solid foundation for its application research.

In recent years, some research groups have reported advances in RCR. For instance, Y. Luo's group utilized topological structures to create a type of metamaterial and studied the RCR [59]. With the development of materials science, topology, and even mechanical engineering (such as 3D-printing technology), new fabrication methods are being proposed, leading to the realization of novel metamaterials. This, in turn, advances the research on RCR and its applications [60].

References

1. Kats AV, Savel'ev S, Yampol'skii VA, et al. Left-handed interfaces for electromagnetic surface waves. Physical Review Letters. 2007; 98:7–073901.
2. Bliokh YP, Savel'ev S, Nori F. Electron-beam instability in left-handed media. Physical Review Letters. 2008; 100:24–244803.
3. Shelby RA, Smith DR, Schultz S. Experimental verification of a negative index of refraction. Science. 2001; 292:5514–77.
4. Galyamin SN, Tyukhtin AV, Kanareykin A, et al. Reversed Cherenkov-transition radiation by a charge crossing a left-handed medium boundary. Physical Review Letters. 2009; 103:19–194802.
5. Vorobev VV, Tyukhtin AV. Nondivergent Cherenkov radiation in a wire metamaterial. Physical Review Letters. 2012; 108:18–184801.
6. Ginis V, Danckaert J, Veretennicoff I, et al. Controlling Cherenkov radiation with transformation-optical metamaterials. Physical Review Letters. 2014; 113:16–167402.
7. Duan ZY, Tang XF, Ma XW, et al. Novel metamaterial high-power microwave radiation sources. 2017; ZL 201510342200.X (in Chinese).
8. Lu J, Grzegorczyk TM, Zhang Y, et al. Čerenkov radiation in materials with negative permittivity and permeability. Optics Express. 2003; 11:7–723.
9. Grbic A, Eleftheriades GV. Experimental verification of backward-wave radiation from a negative refractive index metamaterial. Journal of Applied Physics. 2002; 92:10–5930.
10. Wu BI, Lu J, Kong JA, et al. Left-handed metamaterial design for Čerenkov radiation. Journal of Applied Physics. 2007; 102:11-114907.
11. Xi S, Chen HS, Jiang T, et al. Experimental verification of reversed Cherenkov radiation in left-handed metamaterial. Physical Review Letters. 2009; 103:19–194801.
12. Zhang S, Zhang X. Flipping a photonic shock wave. Physics. 2009; 2:91.
13. Antipov S, Spentzouris L, Gai W, et al. Observation of wakefield generation in left-handed band of metamaterial-loaded waveguide. Journal of Applied Physics. 2008; 104:1–014901.
14. Shchegolkov DY, Azad AK, O' Hara JF, et al. A proposed measurement of the reverse Cherenkov radiation effect in a metamaterial-loaded circular waveguide. International Conference on Infrared, Millimeter and Terahertz Waves. Pasadena, USA, 2008.
15. Estep NA, Askarpour AN, Alù A. Experimental demonstration of negative-index propagation in a rectangular waveguide loaded with complementary split-ring resonators. IEEE Antennas and Wireless Propagation Letters. 2015; 14–119.
16. Tang XF. Research on the mechanism of metamaterial sheet beam radiation sources. Chengdu: University of Electronic Science and Technology of China, 2017 (in Chinese).
17. Duan ZY, Wu BI, Lu J, et al. Reversed Cherenkov radiation in a waveguide filled with anisotropic double-negative metamaterials. Journal of Applied Physics. 2008; 104:6–063303.
18. Duan ZY, Wu BI, Lu J, et al. Cherenkov radiation in anisotropic double-negative metamaterials. Optics Express. 2008; 16:22–18479.

19. Duan ZY, Wu BI, Lu J, et al. Reversed Cherenkov radiation in unbounded anisotropic double-negative metamaterials. Journal of Physics D: Applied Physics. 2009; 42:18–185102.
20. Duan ZY, Guo C, Chen M. Enhanced reversed Cherenkov radiation in a waveguide with double-negative metamaterials. Optics Express. 2011; 19:15–13825.
21. Duan ZY, Wang YS, Mao XT, et al. Experimental demonstration of double-negative metamaterials partially filled in a circular waveguide. Progress in Electromagnetics Research. 2011; 121:215.
22. Duan ZY, Guo C, Zhou J, et al. Novel electromagnetic radiation in a semi-infinite space filled with a double-negative metamaterial. Physics of Plasmas. 2012; 19:1–013112.
23. Duan ZY, Hummelt JS, Shapiro MA, et al. Sub-wavelength waveguide loaded by a complementary electric metamaterial for vacuum electron devices. Physics of Plasmas. 2014; 21:10–103301.
24. Duan ZY, Tang XF, Wang ZL, et al. Observation of the reversed Cherenkov radiation. Nature Communications. 2017; 8–14901.
25. Smith DR, Schultz S, Markoš P, et al. Determination of effective permittivity and permeability of metamaterials from reflection and transmission coefficients. Physical Review B. 2002; 65:19–195104.
26. Smith DR. Analytic expressions for the constitutive parameters of magnetoelectric metamaterials. Physical Review E. 2010; 81:3–036605.
27. Szaboó Z, Park GH, Hedge R, et al. A unique extraction of metamaterial parameters based on Kramers–Kronig relationship. IEEE Transactions on Microwave Theory and Techniques. 2010; 58:10–2646.
28. Chen HT, O'Hara JF, Taylor AJ, et al. Complementary planar terahertz metamaterials. Optics Express. 2007; 15:3–1084.
29. Xu H, Wang ZY, Hao JM, et al. Effective-medium models and experiments for extraordinary transmission in metamaterial-loaded waveguides. Applied Physics Letters. 2008; 92:4–041122.
30. Marqués R, Martel J, Mesa F, et al. Left-handed-media simulation and transmission of EM waves in subwavelength split-ring-resonator-loaded metallic waveguides. Physical Review Letters. 2002; 89:183901.
31. Esteban J, Camacho-Peñalosa C, Page JE, et al. Simulation of negative permittivity and negative permeability by means of evanescent waveguide modes-theory and experiment. IEEE Transactions on Microwave Theory and Techniques. 2005; 53:4–1506.
32. Chernin D, Antonsen TM Jr, Levush B. Exact treatment of the dispersion and beam interaction impedance of a thin tape helix surrounded by a radially stratified dielectric. IEEE Transactions on Electron Devices. 1999; 46:7–1472.
33. Wang WX. Microwave engineering technology. Beijing: National Defense Industry Press, 2009 (in Chinese).
34. Editorial Committee of the Electron Tube Design Manual. O-Type Backward Wave Tube Design Manual. Beijing: National Defense Industry Press, 1985 (in Chinese).
35. Yu Y. Simulation study and design of coupled-cavity traveling-wave tube slow-wave structures. Nanjing: Southeast University, 2011 (in Chinese).
36. Shapiro MA, Trendafilov S, Urzhumov Y, et al. Active negative-index metamaterial powered by an electron beam. Physical Review B. 2012; 86:8–085132.
37. Duan ZY, Wu BI, Xi S, et al. Research progress in reversed Cherenkov radiation in double-negative metamaterials. Progress in Electromagnetics Research. 2009; 90:75.
38. Lyu ZF, Zhang CQ, Jiang SK, et al. Terahertz sheet beam vacuum electron devices. Journal of Infrared and Millimeter Waves. 2023; 42:1-26 (in Chinese).
39. Tang XF, Duan ZY, Wang ZL, et al. The design method of millimeter-wave sheet-beam electron gun. Journal of Infrared and Millimeter Waves. 2014; 33:6–619 (in Chinese).
40. Basten MA. Formation and transport of high-perveance electron beams for high-power, high-frequency microwave devices. Madison: University of Wisconsin-Madison, 1996.
41. Carlsten BE, Russell SJ, Earley LM, et al. Technology development for a mm-wave sheet-beam traveling-wave tube. IEEE Transactions on Plasma Science. 2005; 33:1–85.

42. Duan ZY, Tang XF, Diao P, et al. Elliptical sheet beam electron gun. 2015; ZL 201110455251.5 (in Chinese).
43. Knauer W. Diocotron instability in plasmas and gas discharges. Journal of Applied Physics. 1966; 37:2–602.
44. Booske JH, Kumbasar AH, Basten MA. Periodic focusing and ponderomotive stabilization of sheet electron beams. Physical Review Letters. 1993; 71:24–3979.
45. Booske JH, McVey BD, Antonsen TM Jr. Stability and confinement of nonrelativistic sheet electron beams with periodic cusped magnetic focusing. Journal of Applied Physics. 1993; 73:9–4140.
46. Booske JH, Basten MA, Kumbasar AH, et al. Periodic magnetic focusing of sheet electron beams. Physics of Plasmas. 1994; 1:5–1714.
47. Nguyen KT, Pasour JA, Antonsen TM Jr, et al. Intense sheet electron beam transport in a uniform solenoidal magnetic field. IEEE Transactions on Electron Devices. 2009; 56:5–744.
48. Pasour J, Nguyen K, Wright E, et al. Demonstration of a 100-kW solenoidally focused sheet electron beam for millimeter-wave amplifiers. IEEE Transactions on Electron Devices. 2011; 58:6–1792.
49. Kyhl RL, Webster HF. Breakup of hollow cylindrical electron beams. IRE Transactions on Electron Devices. 1956; 3:4–172.
50. Han MC, Jiang SK, Zhang XM, et al. Preliminary exploration of the Chinese translation of Diocotron in sheet beam transmission. Vacuum Electronics. 2022; 3:42 (in Chinese).
51. Antonsen TM Jr, Ott E. Velocity shear driven instabilities of an unneutralized electron beam. The Physics of Fluids. 1975; 18:9–197.
52. Scharlemann ET. Wiggle plane focusing in linear wigglers. Journal of Applied Physics. 1985; 58:6–2154.
53. Panda PC, Srivastava V, Vohra A. Analysis of sheet electron beam transport under uniform magnetic field. IEEE Transactions on Plasma Science. 2013; 41:3–461.
54. Pierce JR. Rectilinear electron flow in beams. Journal of Applied Physics. 1940; 11:8–548.
55. Ding MP. Research on Modification and Properties of Polytetrafluoroethylene. Xi'an: Northwestern Polytechnical University. 2006 (in Chinese).
56. Kroon DJ. Electromagnets. Boston: Boston Technical Publishers Inc. 1968.
57. Knoepfel HE. Pulsed High Magnetic Field. Amsterdam: North-Holland Publishing Co. 1970.
58. Zhang YB, Gong YB, Wang ZL, et al. Study of high-power Ka-band rectangular double-grating sheet beam BWO. IEEE Transactions on Electron Devices. 2016; 42:6–1502.
59. Hu H, Lin X, Zhang JJ, et al. Nonlocality induced Cherenkov threshold. Laser & Photonics Reviews. 2020; 14:10–2000149.
60. Zhang TY, Zhang XQY, Zhang ZC, et al. Tunable optical topological transition of Cherenkov radiation. Photonics Research. 2022; 10:7–1650.

Chapter 5
Oscillator and Amplifier Based on Reversed Cherenkov Radiation

In the field of vacuum electronics, there is a class of oscillators based on CR, commonly referred to as BWOs. BWOs can generate and amplify electromagnetic waves through the excitation of an electron beam when CR conditions are met [1]. In recent years, with the rapid development of terahertz science and technology, BWOs have received significant attention as a sort of high-power terahertz radiation source. However, their drawbacks, such as large size, heavy weight, and difficulty of integration capability, have restricted their applications in various areas. Taking advantage of the unique properties of metamaterials and the novel electromagnetic characteristics of RCR [2–5], a new type of device similar to the BWO, the RCR oscillator (RCRO), has been developed. Compared to traditional BWOs, the RCRO offers notable advantages, such as miniaturization and high efficiency. It has important application prospect in large-scale scientific facilities, biomedical imaging, microwave heating, microwave sterilization, and more [6].

Furthermore, there is another type of amplifier based on CR, known as TWTs. TWTs can be categorized into wideband TWTs such as helix TWT [7–11] and narrowband TWTs such as coupled-cavity TWT [12–14]. Helix TWTs have extensive applications in radar systems, electronic warfare, and communications as core electronic devices for microwave power amplification. With the growing demand in the satellite market, the demand for helix-based space TWTs in China is expected to rise substantially in the coming years. Coupled-cavity TWTs, due to their characteristics of high peak power, high average output power, high efficiency, and high duty cycle, are widely used in fire control, search, warning radar, and others [15, 16]. Similarly, a novel electronic device known as the RCR amplifier (RCRA) has gained significant attention from researchers due to its advantages of miniaturization, high efficiency, and high gain.

This chapter reports the innovative combination of metamaterials and vacuum electronics, harnessing the inherent advantages of metamaterials to pave a new path for the development of vacuum electronics. Here, it focuses on the miniaturized RCRO and RCRA.

Z. Duan, *Metamaterial-Based Electromagnetic Radiations and Applications*,
https://doi.org/10.1007/978-981-97-8108-9_5

5.1 Novel Metamaterial Slow-Wave Structures

There is an important class of VEDs known as O-type or linear-beam devices [15]. Within linear-beam devices, a crucial component is the SWS, which serves as the area for energy exchange between electromagnetic waves and the electron beam. When metamaterials are creatively introduced into the SWS, this novel type of SWS is referred to as an MSWS. Several research groups have conducted studies on MSWS [17–27]. This section primarily discusses circuit theory, high-frequency characteristics, transmission properties, and cold test experiments with metamaterials.

5.1.1 Circuit Theory

Z. Duan's group has creatively proposed an all-metal circular double-bridge CeSRR unit cell [28–31], as shown in Fig. 5.1a. An MSWS is constructed by periodically loading CeSRRs into a hollow circular waveguide, as depicted in Fig. 5.1b. Due to the structural complexity of the metamaterial unit cell, the analysis using field theory to obtain its electromagnetic properties is extremely challenging. Therefore, an equivalent circuit theory method is employed here for theoretical analysis.

CeSRRs, as subwavelength resonant unit cells, have specific resonant frequencies. Moreover, adjacent CeSRR unit cells form a resonant cavity with strong coupling characteristics, which also possess resonance frequencies. Based on the structural characteristics of the cavity walls and CeSRRs, an equivalent circuit model of the MSWS is constructed using a similar analysis method to the coupled-cavity SWS [15], as shown in Fig. 5.2. In this model, C_c is the capacitance of the cavity, L_s and C_s are the inductance and capacitance of the slot-line in a CeSRR unit cell respectively, and L_g and C_g are the inductance and capacitance of the metal bridge connecting the inner and outer loops of a CeSRR unit cell respectively. Additionally, L'_c and K' are the modified cavity inductance and coupling factor with CeSRR, respectively.

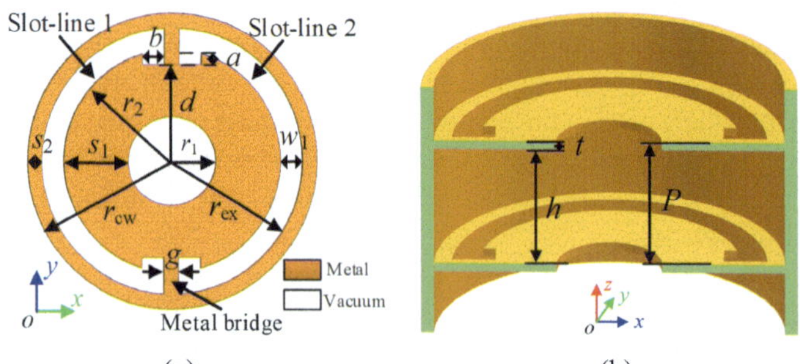

Fig. 5.1 **a** Circular double-bridge CeSRR unit cell. **b** MSWS

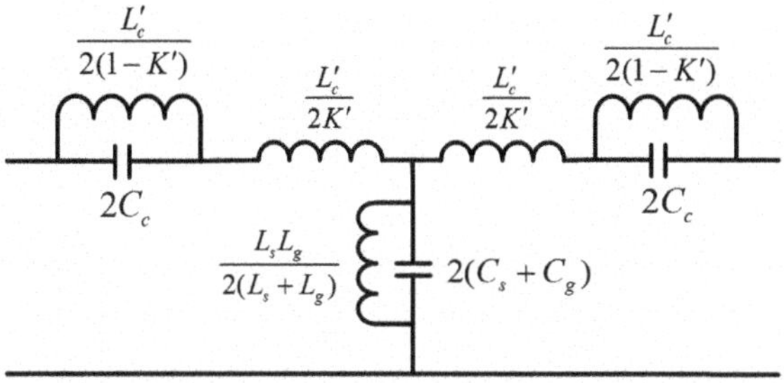

Fig. 5.2 Equivalent circuit model of the MSWS

The resonant frequency of the slot-line can be approximated as a short-circuited transmission line for calculation [32]. However, this method can only provide a rough estimate of the effective length of the slot-line, resulting in some inaccuracy in the calculated slot-line resonant frequency. Additionally, when the slot-line is non-uniform or irregular, it becomes challenging to determine the effective length of the slot-line. To address this issue, a new method for calculating the resonant frequency of the CeSRR unit cell is proposed by simulating the high-frequency current distribution in the slot-line.

In the conventional calculation method, regular slot-lines in a CeSRR can be directly used to determine their effective electrical length, which is used to calculate the inductance L_s and capacitance C_s of the slot-line, resulting in the resonant frequency. The capacitance and inductance are represented as:

$$C_s = \varepsilon_0 l_s (t + w_1)/w_1, \tag{5.1}$$

$$L_s = \frac{\mu_0 l_s w_1}{\pi^2 (t + w_1)}, \tag{5.2}$$

where ε_0 and μ_0 are the permittivity and permeability in vacuum, respectively. l_s represents the effective electrical length of the slot-line and is defined as:

$$l_s = \pi (r_2 + w_1/2) - g. \tag{5.3}$$

Based on (5.1), (5.2), and (5.3), the resonant frequency of the slot-line can be calculated as:

$$f = \frac{1}{2\pi \sqrt{L_s C_s}}. \tag{5.4}$$

For CeSRRs with irregular slot-lines, it is impossible to directly use (5.1) and (5.2) to determine their lumped parameters. Figure 5.3 illustrates the simulated high-frequency current distribution on single slot-line and double slot-line CeSRRs, indicating that the high-frequency current flows along the inner wall of the slot-line. The effective path for high-frequency current flow corresponds to the perimeter of the

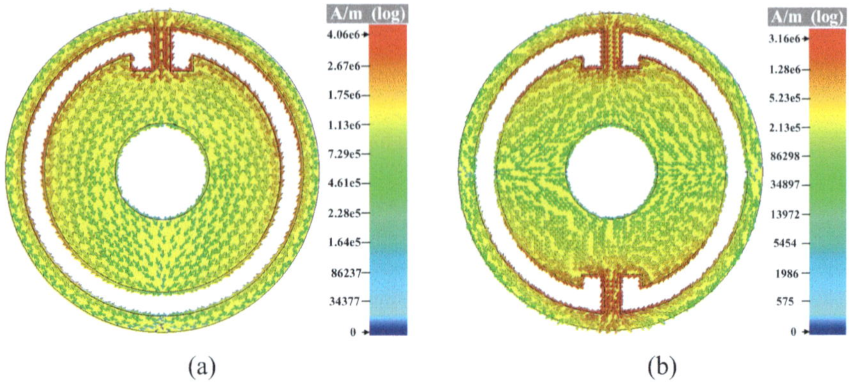

Fig. 5.3 High-frequency current distribution on CeSRRs. **a** Single slot-line CeSRR. **b** Double slot-line CeSRR

slot-line. Most importantly, regardless of whether the slot-line is uniform or regular, the length of the inner wall of the slot-line always represents the complete path for the high-frequency current flow.

According to transmission line theory, when the perimeter of the slot-line matches the path length of one period of high-frequency current, it can be equivalently represented as a short-circuited transmission line. Therefore, the resonant frequency of the CeSRR should be expressed as:

$$f_s = \frac{c_0}{C_{\text{all}}}, \tag{5.5}$$

where C_{all} is the perimeter of the inner wall of the slot-line. This method can be used to estimate the resonant frequency of the CeSRR based on the structural dimensions of the slot-line, regardless of whether the slot-line's shape is regular. It should be noted that (5.5) does not consider the thickness of the CeSRR. Hence, this calculation method may not be applicable if the thickness is numerically comparable to the length of the slot-line.

From the above analysis, it can be seen that the capacitance and inductance generated by the slot-line in the CeSRR determine its resonant frequency, and the capacitance and inductance values are derived from the physical dimensions of the structure itself. Analyzing the CeSRR using the equivalent circuit theory method is relatively straightforward. Therefore, based on the theoretical analysis, an electromagnetic simulation software is employed to conduct a comprehensive study of the circular double-bridge CeSRR. Its dimensional parameters are listed in Table 5.1.

The double-bridge CeSRR is modeled using HFSS simulation software, as shown in Fig. 5.4. Two interfaces perpendicular to the x-direction are set as wave ports. Two interfaces perpendicular to the z-direction are set as electric boundary conditions. Two interfaces perpendicular to the y-direction are set as magnetic boundary conditions. The reflection and transmission characteristics of electromagnetic waves propagating

Table 5.1 Dimensional parameters of the double-bridge CeSRR

Parameter	Value/mm
r_1	7.1
r_2	14.2
r_{ex}	18
r_{cw}	20
b	3
g	2
a	3
s_1	7.1
s_2	2
w_1	3.8
t	1
P	12

through the double-bridge CeSRR in free space are simulated. This provides the S-parameters of the double-bridge CeSRR, as shown in Fig. 5.5. Using the S-parameters retrieval method described in Chap. 2, the effective electromagnetic parameters of the metamaterial are retrieved [33, 34], as shown in Fig. 5.6a.

From Fig. 5.6a, it can be observed that the double-bridge CeSRRs exhibit a negative real part of effective permittivity in the frequency range of 2.2–2.6 GHz, while the real part of the effective permeability is positive. According to the theory proposed by Marqués et al. [35, 36], a waveguide operating below the cutoff frequency in the quasi-TM mode can be considered a one-dimensional plasma with a negative real part of the effective permeability. Therefore, according to the classical combination method and its extension described in Chap. 2, the hollow circular waveguide periodically loaded by the double-bridge CeSRRs can achieve a metamaterial with

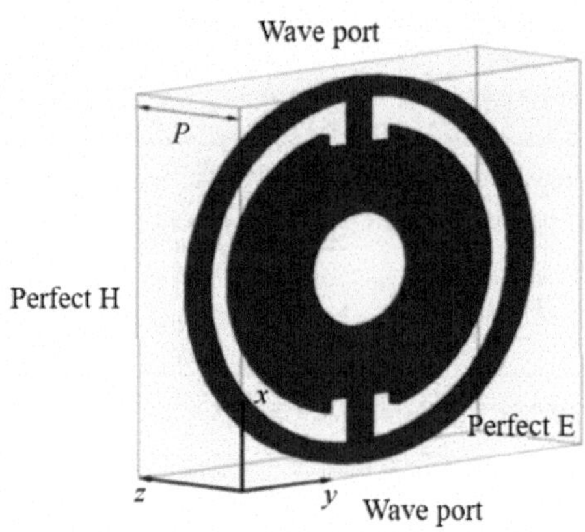

Fig. 5.4 Simulation model of the double-bridge CeSRR with $P = 12$ mm

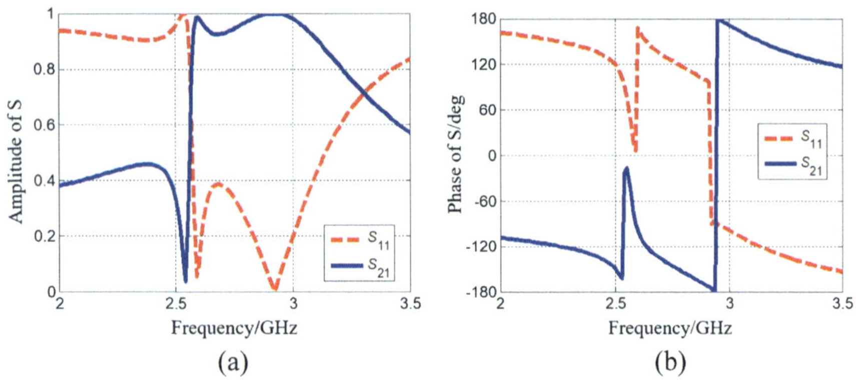

Fig. 5.5 S-parameters of the double-bridge CeSRR. **a** The amplitudes of S_{11} and S_{21}. **b** The phases of S_{11} and S_{21}

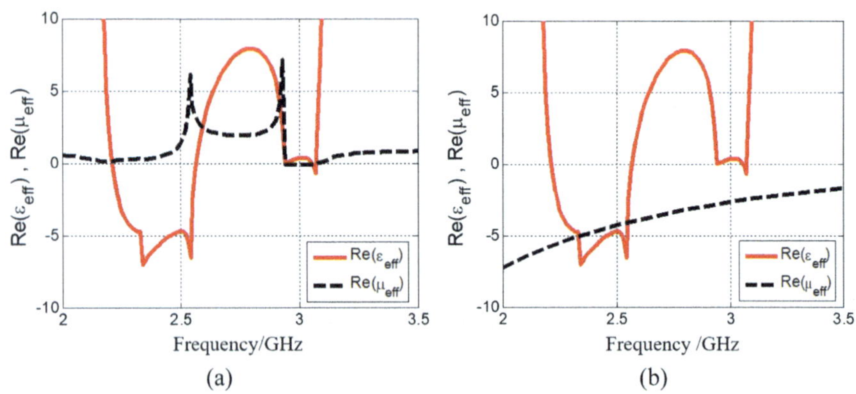

Fig. 5.6 **a** Effective electromagnetic parameters of the CeSRRs. **b** Effective electromagnetic parameters of the MSWS

double-negative characteristics, as shown in Fig. 5.1b, where the electromagnetic wave can propagate in this MSWS. The retrieved effective electromagnetic parameters of this MSWS are shown in Fig. 5.6b. It can be seen that the MSWS exhibits negative effective permittivity and negative effective permeability in the frequency range of 2.2–2.6 GHz.

5.1.2 High-Frequency Characteristics

The previous analysis covered the resonant frequency of the double-bridge CeSRR and the effective electromagnetic parameters of the MSWS which is formed by the hollow circular waveguide loaded with CeSRRs. Now, we will investigate the dispersion characteristics and interaction impedance of the MSWS.

Since the double-bridge CeSRR is a subwavelength all-metal disk-like structure with a certain thickness, its equivalent circuit parameters can be calculated by using the method of solving for coplanar waveguide circuit parameters. The circular double-bridge CeSRR shown in Fig. 5.1a can be treated as a nonuniform coplanar waveguide. The capacitance per unit length can be obtained as follows [32]:

$$C_{\mathrm{pul}} = 4\varepsilon_0\left(\frac{K(k')}{K(k)} + \frac{t}{4w_1}\right), \tag{5.6}$$

where $K(k)$ is the first kind of complete elliptic integral, and for this CeSRR, k' can be expressed as:

$$k' = \frac{w_1}{s_1 + s_2 + w_1}. \tag{5.7}$$

According to the transmission line theory, the expression for the unit length inductance can be obtained as:

$$L_{\mathrm{pul}} = \frac{1}{c_0^2 C_{\mathrm{pul}}}. \tag{5.8}$$

The angle of a single slot-line is expressed as:

$$\theta = \pi - \frac{g}{r_2 + w_1/2}. \tag{5.9}$$

Next, the capacitance and inductance of the double-bridge CeSRR are given as follows:

$$C_s = \frac{\theta r C_{\mathrm{pul}}}{4}, \tag{5.10}$$

$$L_s = \theta r_{\mathrm{av}} L_{\mathrm{pul}}, \tag{5.11}$$

where r and r_{av} are the effective radius and average radius of the slot-lines in the double-bridge CeSRR, respectively. For the metal bridge connecting the inner and outer loops of the double-bridge CeSRR, its inductance can be calculated using the formula for straight conductor inductance. Using the parallel plate capacitor formula [37], the capacitance can be approximated as:

$$L_g = 2(r_{\mathrm{cw}} - d)\left[\log\left(\frac{2(r_{\mathrm{cw}} - d)}{g + t}\right) + 0.5 + 0.2335\frac{g + t}{r_{\mathrm{cw}} - d}\right], \tag{5.12}$$

$$C_g = \varepsilon_0 \frac{t w_1}{g}. \tag{5.13}$$

Based on the distribution of high-frequency current on the double-bridge CeSRR, the inductance and capacitance of the slot-lines can be obtained as follows:

$$\begin{cases} L_s' = \theta r_{\text{av}} L_{\text{pul}} + L_g \\ C_s' = \dfrac{\theta r C_{\text{pul}}}{4} + C_g \end{cases}. \tag{5.14}$$

By substituting (5.14) into (5.4), the resonant frequency of the slot-lines can be determined. The above calculation method can also be directly applied to situations without grooves on both sides of the metal bridge (i.e., $d = r_2$). In fact, due to the influence of the grooves on both sides of the metal bridge connecting the inner and outer loops of the double-bridge CeSRR, an additional capacitance C_i is introduced. It can be obtained using the parallel plate capacitor formula:

$$C_i = 2\varepsilon_0 \left(\frac{at}{b} + \frac{bt}{a + d} \right). \tag{5.15}$$

Thus, for the double-bridge CeSRR, as shown in Fig. 5.1a, the capacitance of the slot-lines is:

$$C_s'' = C_s' + C_i. \tag{5.16}$$

By substituting (5.16) and (5.14) into (5.4), the resonant frequency of the double-bridge CeSRR with grooves on both sides of the metal bridge can be determined.

Next, the equivalent circuit parameters of the resonant cavity formed by two double-bridge CeSRRs are analyzed. K. Fujisawa et al. provided a formula for the inductance of a cavity with a height of h when they studied the equivalent circuit parameters of a resonant cavity of klystron [38]:

$$L_c = \frac{\mu_0 h}{2\pi} \ln \left(\frac{r_{\text{cw}}}{r_1} \right). \tag{5.17}$$

For the capacitance of the resonant cavity without drift tube, it can be calculated based on the specific dimensions of the resonant cavity from the perspective of electromagnetic storage energy. For the TM_{010} mode of a cylindrical resonant cavity, the electric field energy is given [1]:

$$U = \frac{\pi \varepsilon_0 E_0^2}{2} r_{\text{cw}}^2 h |J_1(2.405)|^2, \tag{5.18}$$

where $J_1(2.405)$ is the first-order Bessel function of the first kind with respect to its argument 2.405. For the TM_{010} mode of a cylindrical resonant cavity, according to the relationship between the electric field energy and resonant cavity capacitance [1], the capacitance of the cavity can be calculated as:

$$C_c = \frac{\pi\varepsilon_0 r_{cw}^2}{h}|J_1(2.405)|^2. \tag{5.19}$$

Taking into the coupling between the slot-lines account, a series inductance L_3 is formed and is expressed:

$$L_3 = \frac{\mu_0 t}{4\pi K}\ln\left(\frac{r_2 + w_1}{r_2}\right). \tag{5.20}$$

In this case, the inductance and coupling factor of the resonant cavity are rewritten as:

$$L_c' = L_c\left[1 - \frac{2K^2L_3}{L_c + 2KL_3}\right]^{-1}, \tag{5.21}$$

$$K' = \frac{K}{1 + 2K(1 - K)L_3/L_c}. \tag{5.22}$$

According to the definition of the coupling factor [39], the coupling factor between the CeSRRs is:

$$K = 1 - \frac{g}{r_2 + w_1/2}. \tag{5.23}$$

Using the Curnow formula [40], the dispersion equation for the MSWS can be derived as:

$$\cos\varphi = 1 + \frac{\left(1 - \omega^2/\omega_c^2\right)\left(1 - \omega^2/\omega_s^2\right)}{k_0(1 - K')\left(1 - \omega^2/\omega_k^2\right)}, \tag{5.24}$$

where φ represents phase shifts, and they are expressed as:

$$k_0 = \frac{K'L_s'}{L_c'},\ \omega_k = \sqrt{\frac{1 - K'}{L_c'C_c}},\ \omega_c = \sqrt{\frac{1}{L_c'C_c}},\ \omega_s = \sqrt{\frac{1}{L_s'C_s''}}, \tag{5.25}$$

where ω_k, ω_c, and ω_s represent the partial angular frequency, the cavity angular frequency, and the slot-line angular frequency, respectively.

From the dispersion curve, it is possible to determine whether this MSWS exhibits left-handed characteristics and the frequency range where left-handed characteristics occur. Based on the dispersion equation, the interaction impedance of the MSWS can be obtained [40]:

$$K_c = \frac{-2\omega L_c'^2\left(1 - (\omega/\omega_s)^2\right)}{L_s'\sin(\varphi)\left(1 - K' - (\omega/\omega_c)^2\right)^2}. \tag{5.26}$$

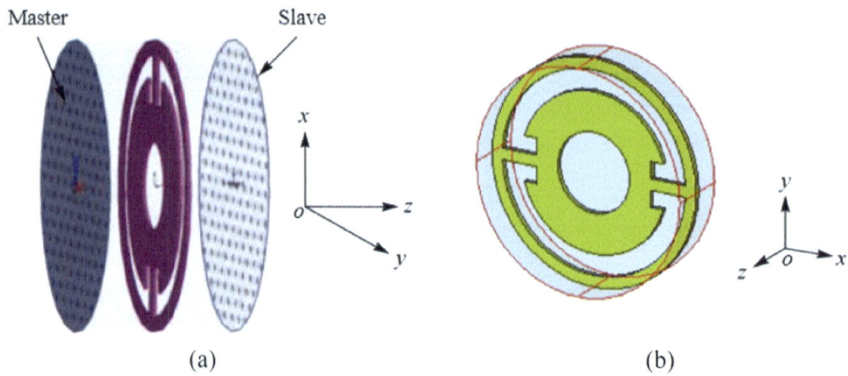

Fig. 5.7 Simulation model for high-frequency characteristics. **a** Simulation model in HFSS. **b** Simulation model in CST

A larger interaction impedance [1] indicates a higher degree of interaction between electromagnetic waves and the electron beam, leading to more efficient energy exchange.

Based on the theoretical analysis of the dispersion equation and interaction impedance for the MSWS, further investigations of the high-frequency characteristics are conducted by using simulation software such as HFSS or CST.

The simulation model for high-frequency characteristics is shown in Fig. 5.7. The periodic length of the MSWS is set as $P = 12$ mm, with both vertical interfaces to x and y directions set as electric boundary conditions. In the HFSS, the two interfaces perpendicular to the z-direction are set as master and slave boundary conditions. In the CST, both interfaces are set as periodic boundary conditions. The simulated results are shown in Fig. 5.8.

Figure 5.8a shows dispersion curves obtained through simulations with CST and HFSS. Simulated results in CST indicate that the MSWS exhibits left-handed characteristics in the frequency range of 2.16–2.64 GHz. Simulated results in HFSS show left-handed characteristics in the frequency range of 2.21–2.68 GHz. Furthermore, as mentioned in the previous section, the MSWS exhibits negative effective permittivity and negative effective permeability within the frequency range of 2.2–2.6 GHz, indicating good agreement between the dispersion passband of this structure and the frequency range of its negative effective electromagnetic parameters.

Figure 5.8b presents the variation curve of the interaction impedance of the MSWS with frequency. The HFSS simulated results indicate that the interaction impedance of the MSWS is greater than 110 Ω within the frequency range of 2.1–2.6 GHz. The CST simulated results show that the interaction impedance of the MSWS is greater than 130 Ω within the same frequency range. The normalized phase velocity-frequency relationship is derived using the dispersion curves obtained through HFSS and CST simulations, as illustrated in Fig. 5.8c. It can be observed that although there are some differences in the simulated results obtained from different software, the overall trends match well, demonstrating the validity of the theoretical analysis.

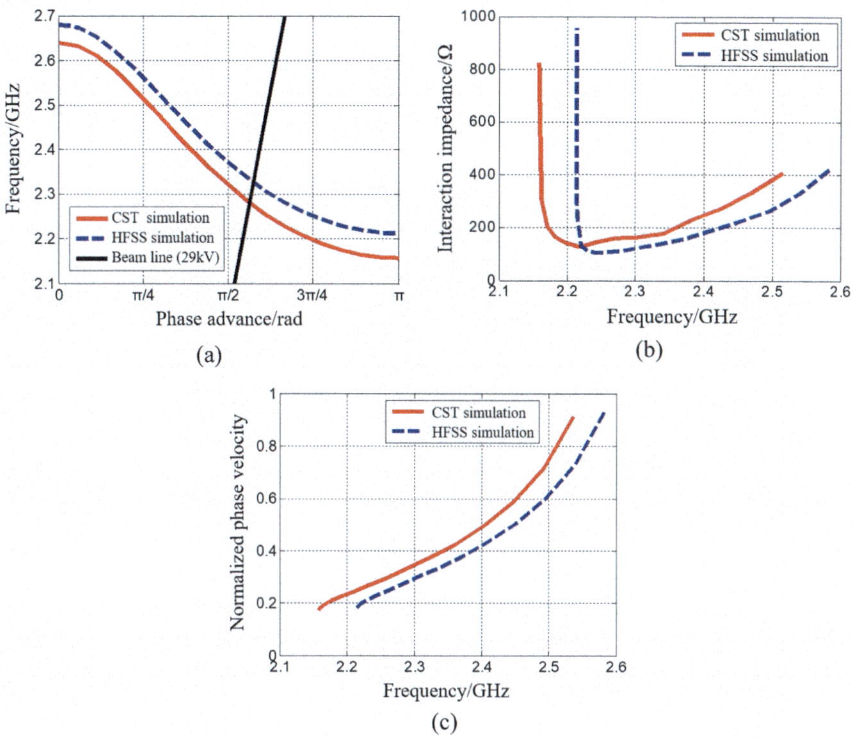

Fig. 5.8 Simulated results of high-frequency characteristics. **a** Dispersion curves. **b** Interaction impedances. **c** Normalized phase velocities

5.1.3 Transmission Characteristics

In this section, the passband of the MSWS is determined by studying its dispersion characteristics. Furthermore, an in-depth investigation into the transmission and loss characteristics of this MSWS within the passband is also conducted.

Transmission characteristics are a valuable indicator of the high-frequency loss and the matching properties between ports when electromagnetic waves pass through the MSWS. In Fig. 5.1b, eight double-bridge CeSRRs are periodically loaded into a hollow circular waveguide, thus an MSWS with ports 1 and 2 is constructed. The double-bridge CeSRR and hollow circular waveguide are both made of oxygen-free copper TU1. To efficiently couple electromagnetic waves into the MSWS, coaxial input coupler (port 1) and coaxial output coupler (port 2) are employed after the output power magnitude and the assembly method of the focusing magnetic field are considered. The MSWS with coaxial input and output couplers is shown in Fig. 5.9. And $r_5 = 0.525$ mm, $r_6 = 1.7$ mm, and $r_7 = 1.9$ mm.

A substantial portion of electromagnetic wave loss within the MSWS stems from its Ohmic loss, which is closely related to the conductivity of the materials. However,

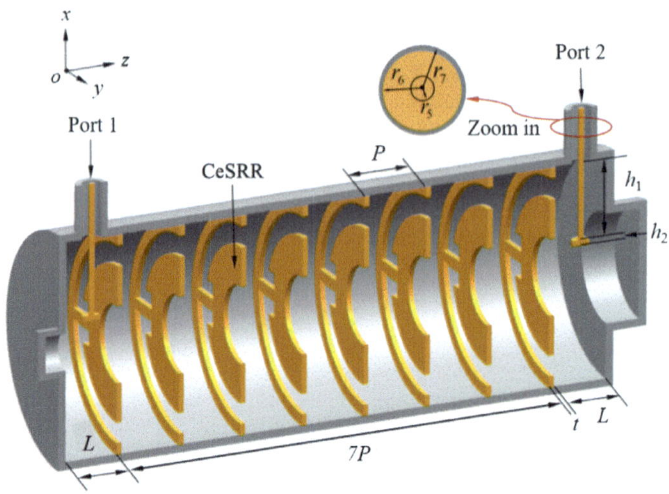

Fig. 5.9 The schematic diagram of the MSWS

the effective conductivity of the same material will vary at different frequencies. Hence, it is necessary to employ the effective conductivity of oxygen-free copper TU1 in place of a perfect conductivity for calculating the transmission characteristics of the MSWS, as shown in Fig. 5.9.

The relationship between the conductivity σ_s of metal under the smooth surface condition and the effective conductivity σ_{eff} under the rough surface condition is represented by the correction factor K_{rough} [41]:

$$\sigma_{\text{eff}} = \frac{\sigma_s}{K_{\text{rough}}^2}. \tag{5.27}$$

Currently, the H-B model [42] and the Groiss model [43] are the two most commonly used physical models for considering the relationship between the effective conductivity of metal materials and surface roughness. The formulas for calculating the correction factor K_{rough} in the two models are given as follows:

$$K_{\text{rough-HB}} = 1 + \frac{2}{\pi}\arctan\left[1.4\left(\frac{\Delta}{\delta}\right)^2\right], \tag{5.28}$$

$$K_{\text{rough-G}} = 1 + \exp\left[-\left(\frac{\delta}{2\Delta}\right)^{1.6}\right], \tag{5.29}$$

where Δ represents surface roughness and $\delta = 1/\sqrt{\pi f \mu \sigma}$ is the skin depth of the electromagnetic wave.

Due to the use of materials such as oxygen-free copper, fernico, molybdenum, and nickel-copper in the MSWS and the coaxial input and output couplers, the surface roughness of these metal materials can affect the transmission characteristics of electromagnetic waves. Therefore, the effective conductivity of oxygen-free copper (with a theoretical conductivity of approximately 5.8×10^7 S/m), fernico expansion sealing alloy 4J34 (with a theoretical conductivity of approximately 2.22×10^6 S/m), molybdenum (with a theoretical conductivity of approximately 1.82×10^7 S/m), and nickel-copper NCu40-2-1 (with a theoretical conductivity of approximately 1.86×10^6 S/m) are calculated using the two models, respectively, as shown in Fig. 5.10. Taking oxygen-free copper as an example, in the frequency range of 2.1–2.7 GHz, the effective conductivity calculated by the H-B model ranges from 4.03×10^7 to 4.32×10^7 S/m, while the effective conductivity calculated by the Groiss model ranges from 3.23×10^7 to 3.61×10^7 S/m.

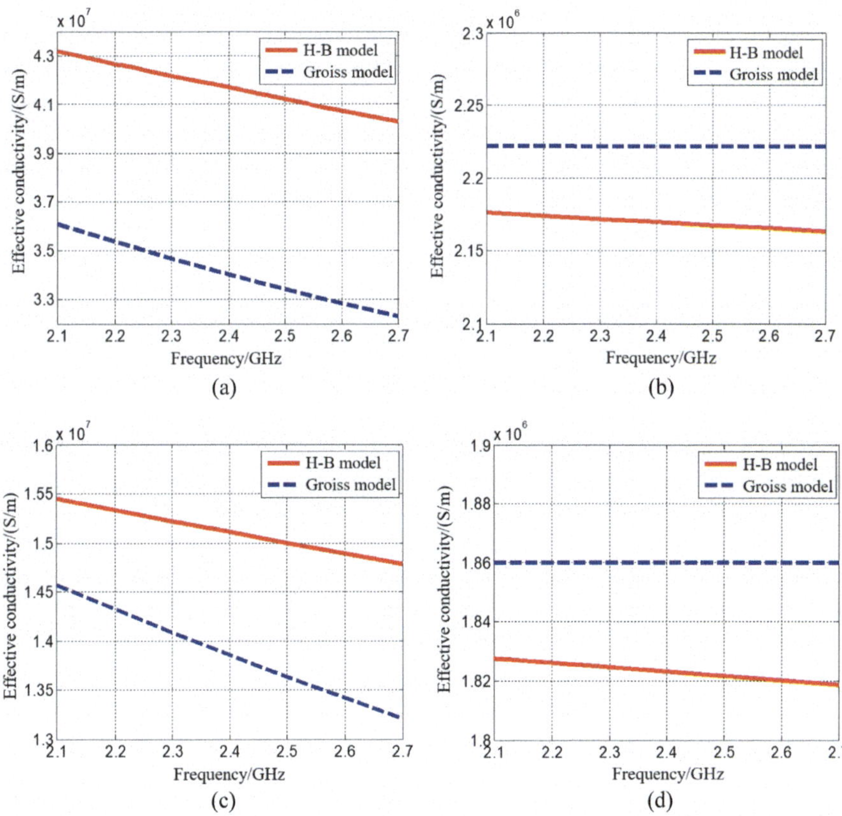

Fig. 5.10 Effective conductivity calculated by different models. **a** Oxygen-free copper (surface roughness $\Delta = 0.6$ μm). **b** Fernico ($\Delta = 0.8$ μm). **c** Molybdenum ($\Delta = 0.8$ μm). **d** Nickel-copper ($\Delta = 0.8$ μm)

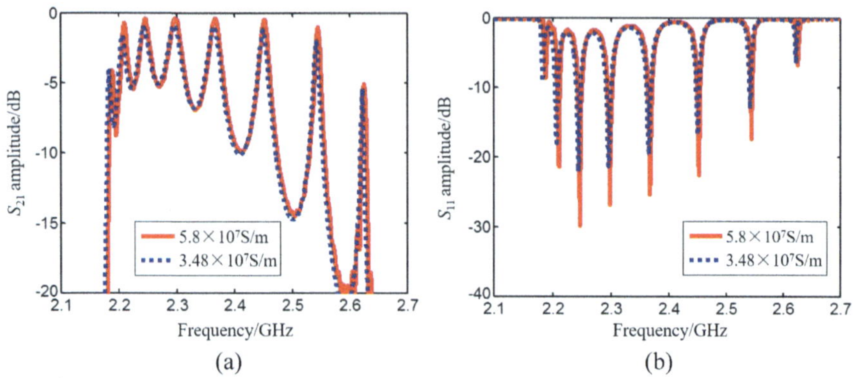

Fig. 5.11 Transmission characteristics of the MSWS obtained from CST. **a** S_{21}. **b** S_{11}

Next, the perfect conductivity is replaced with the predicted effective conductivity from the H-B and Groiss models, and the transmission characteristics of this MSWS are simulated using both CST and HFSS. It is found that in the microwave frequency band, the predicted effective conductivity of the Groiss model is closer to the cold test experimental results than that of the H-B model. Therefore, only the simulated results obtained using the Groiss model's effective conductivity are presented.

The CST simulated results are shown in Fig. 5.11. The solid line represents the simulated results using perfect conductivity, and the dashed line represents the simulated results using the effective conductivity from the Groiss model at 2.3 GHz.

In Fig. 5.11a, it can be seen that the transmission characteristics simulated using effective conductivity are slightly worse than the results with perfect conductivity, resulting in higher transmission loss. However, the overall trend remains consistent. When oxygen-free copper TU1 is assumed to have a perfect conductivity of 5.8×10^7 S/m, the maximum value of S_{21} at 2.3 GHz is approximately − 0.45 dB. When the effective conductivity of 3.48×10^7 S/m is used, the maximum value of S_{21} at 2.3 GHz is approximately − 0.89 dB.

To facilitate analysis, the transmission characteristics of this MSWS are also simulated by using HFSS. The surface roughness of the model is set to be 0.6 μm, and the simulation is rerun. The results are shown in Fig. 5.12. When the perfect conductivity of 5.8×10^7 S/m is used, the maximum value of S_{21} at 2.33 GHz is approximately − 1.4 dB. However, when the influence of surface roughness is considered, the maximum value of S_{21} at 2.33 GHz is approximately − 1.81 dB.

5.1.4 Cold Test Experiment

After the high-frequency characteristics of the MSWS are studied, the relevant components are designed and processed, as shown in Fig. 5.13a. To facilitate the assembly of the device, the hollow circular waveguide was divided into seven

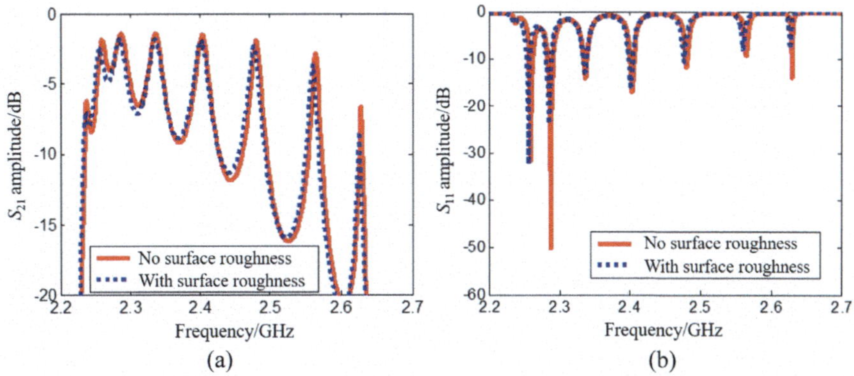

Fig. 5.12 Transmission characteristics of the MSWS obtained from HFSS. **a** S_{21}. **b** S_{11}

segments and processed separately. Next, they were fixed together with the double-bridge CeSRRs using positioning holes and pins. Each component was positioned by positioning pins to ensure excellent concentricity. Alumina ceramic was used as the window material.

Once all the components were securely fixed, silver solder was used for welding to form the MSWS, as shown in Fig. 5.13b. Based on this, an experimental platform was then set up, which included an Agilent E8362B vector network analyzer, the MSWS, standard coaxial lines, and coaxial input and output couplers, as shown in Fig. 5.13c.

First, the usage specifications of the Agilent E8362B vector network analyzer for measuring S-parameters was strictly kept, and the vector network analyzer was calibrated to ensure the accuracy of the experimental results. Next, the two standard coaxial lines were used to connect the MSWS to the vector network analyzer for measuring the S-parameters. Finally, experimental data of S-parameters were

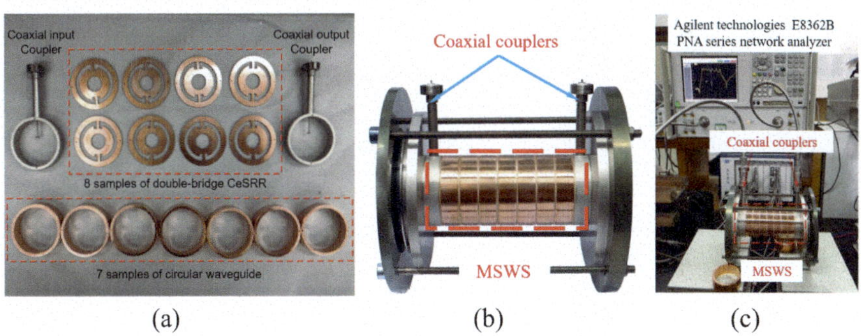

Fig. 5.13 Samples of the CeSRR, the MSWS, and the test platform. **a** Samples of the fabricated components. **b** The MSWS and coaxial input and output couplers. **c** A cold test experimental platform

retrieved from the vector network analyzer. It should be noted that the real and imaginary parts of the S-parameters are directly retrieved from the vector network analyzer. These data need to be post-processed and converted into the amplitude and phase information for comparison with the CST and HFSS simulated results.

Figure 5.14 shows the comparison of measured and CST simulated S-parameters. As listed in Table 5.2, the CST simulated results indicate that the amplitude of S_{21} is approximately − 0.89 dB at 2.3 GHz, and the test results show that the amplitude of S_{21} is approximately − 1.01 dB at 2.35 GHz. Figure 5.15 shows the transmission phase shift as a function of frequency for both CST simulation and cold test experiment, where the solid line represents CST simulated results and the dashed line represents the results of the cold test experiment.

Additionally, dispersion curves can be obtained by transforming the transmission phase shift, as detailed in Sect. 4.4.2. Figure 5.16 provides a comparison of dispersion curves obtained using different methods. The dispersion curve obtained by transforming the transmission phase shift from CST simulation indicates that the MSWS exhibits left-handed characteristics within the frequency range of 2.17–2.63 GHz. Similarly, the dispersion curve obtained by transforming the transmission phase shift from the cold test experiment shows that the MSWS exhibits left-handed characteristics within the frequency range of 2.18–2.62 GHz.

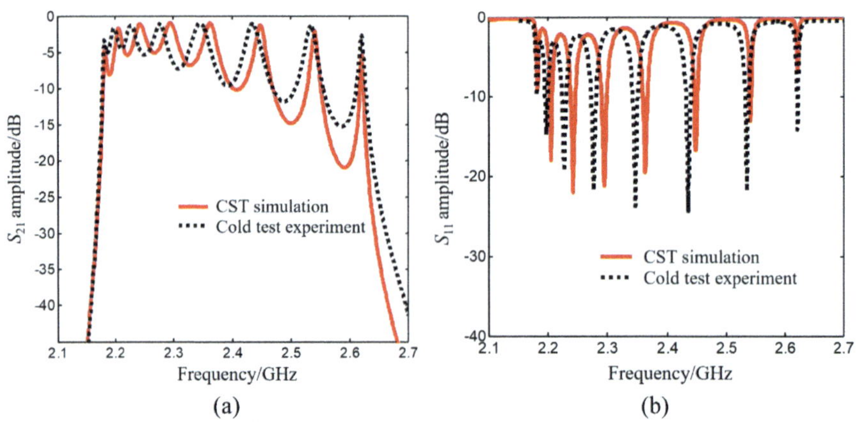

Fig. 5.14 Comparison between the results of CST simulation and cold test experiment. **a** S_{21}. **b** S_{11}

Table 5.2 S_{21} obtained from CST simulation and cold test experiment at different frequencies

CST simulated results	Frequency/GHz	2.18	2.21	2.24	2.3	2.36	2.45	2.54	2.62
	S_{21}/dB	− 4.14	− 1.6	− 1.04	− 0.89	− 0.94	−1.21	− 2	− 5.37
Cold test results	Frequency/GHz	2.18	2.2	2.23	2.28	2.35	2.43	2.53	2.62
	S_{21}/dB	− 3.34	− 1.7	− 1.28	− 1.14	− 1.01	− 1.1	− 1.33	− 2.69

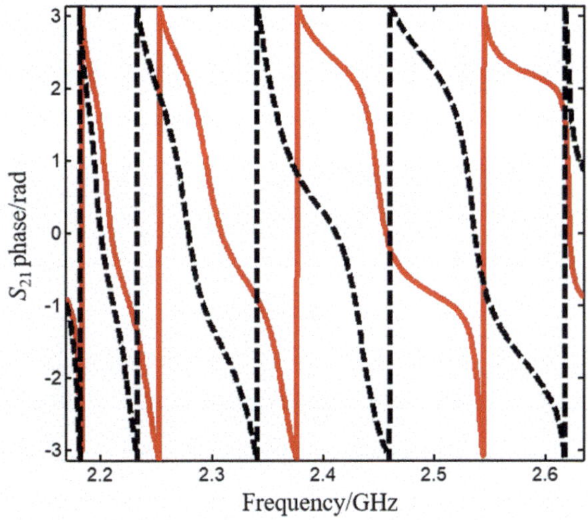

Fig. 5.15 Transmission phase shift versus frequency

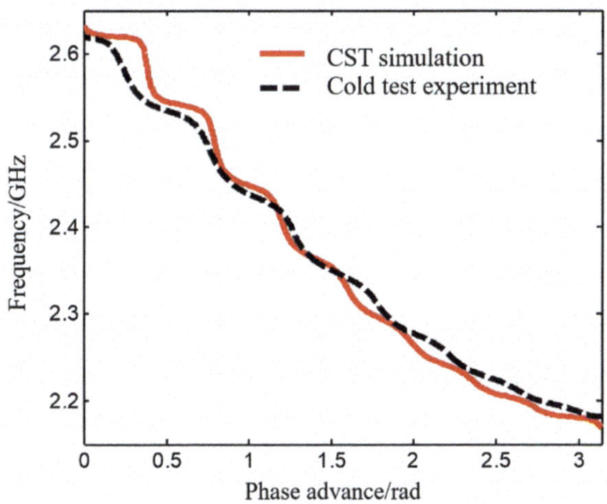

Fig. 5.16 Comparison of dispersion curves obtained by different methods

The same simulation analysis is conducted using HFSS. Figure 5.17 shows the measured S-parameters data and HFSS simulated results. As listed in Table 5.3, the HFSS simulated results indicate that the amplitude of S_{21} is approximately − 1.81 dB at 2.33 GHz, and the cold test results show that the amplitude of S_{21} is approximately − 1.01 dB at 2.35 GHz. The transmission phase shifts with frequency for both HFSS simulation and cold test experiment are shown in Fig. 5.18, where the solid line represents HFSS simulated results, and the dashed line represents the results of the cold test experiment.

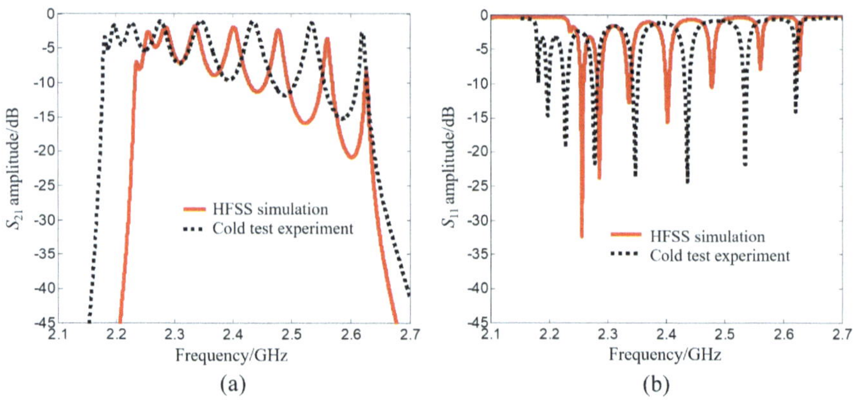

Fig. 5.17 Comparison between the results of HFSS simulation and cold test experiment. **a** S_{21}. **b** S_{11}

Table 5.3 S_{21} obtained from HFSS simulation and cold test experiment at different frequencies

HFSS simulated results	Frequency/GHz	2.24	2.26	2.29	2.33	2.4	2.48	2.56	2.63
	S_{21}/dB	− 6.98	− 2.53	− 1.84	− 1.81	− 1.89	− 2.35	− 3.64	− 8.44
Cold test results	Frequency/GHz	2.18	2.2	2.23	2.28	2.35	2.43	2.53	2.62
	S_{21}/dB	− 3.34	− 1.7	− 1.28	− 1.14	− 1.01	− 1.1	− 1.33	− 2.69

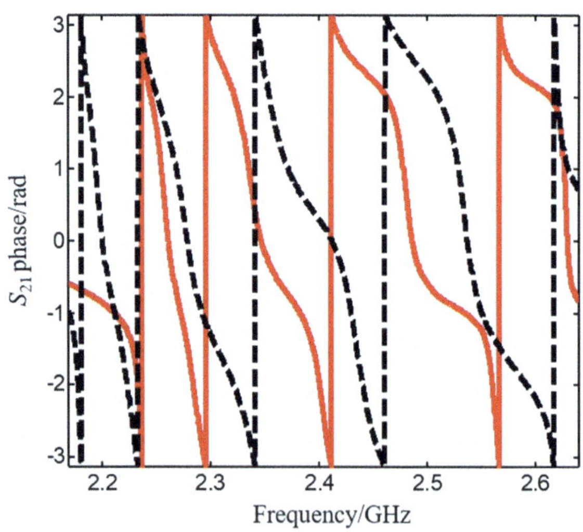

Fig. 5.18 Translational phase versus frequency

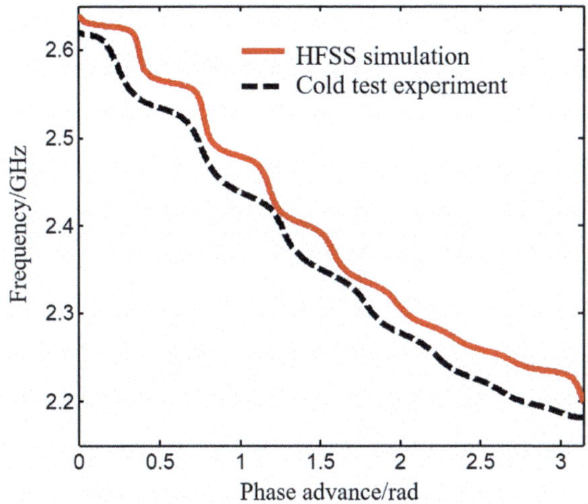

Fig. 5.19 Comparison of dispersion curves obtained by different methods

Dispersion curves can be obtained using the phase advance method, as shown in Fig. 5.19. It can be observed that the dispersion curve of the HFSS simulation indicates that the MSWS exhibits left-handed characteristics within the frequency range of 2.2–2.64 GHz. The dispersion curve of the cold test results shows that the MSWS exhibits left-handed characteristics within the frequency range of 2.18–2.62 GHz.

5.2 Electron-Optical System

In VEDs, the high-speed electron beam interacts with electromagnetic waves. Simultaneously, a magnetic focusing configuration is required to focus the electron beam due to the space charge effect. After beam-wave interaction, the residual electrons need to be efficiently collected for improving device efficiency. Therefore, VEDs have an electron-optical system, typically composed of an electron gun, a magnetic focusing configuration, and a collector, which are responsible for electron generation, focusing, and collection, respectively. The electron-optical system of RCR devices is introduced below.

5.2.1 Grid-Controlled Electron Gun

For the RCRO operating in the microwave band, a pencil beam usually meets its need. A grid-controlled electron gun is employed to effectively generate a pencil beam. It consists of a cathode, a shadow grid electrode, a control grid electrode, a focus electrode, and an anode, as shown in Fig. 5.20.

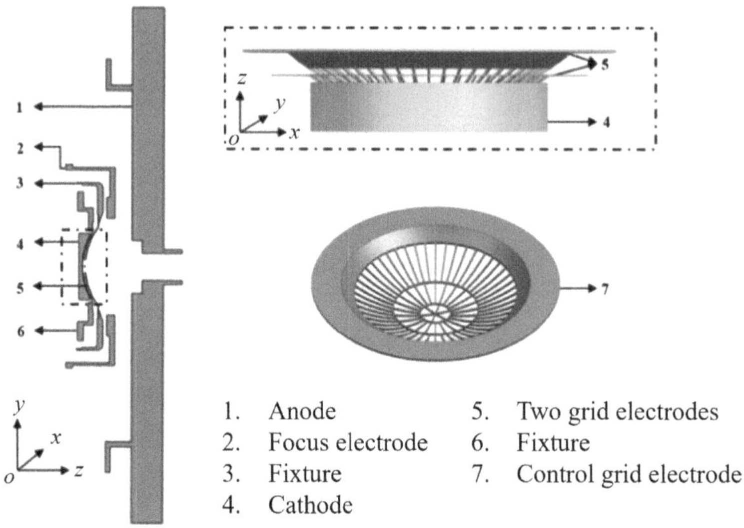

Fig. 5.20 Schematic diagram of the grid-controlled electron gun

A barium tungsten cathode based on the thermionic electron emission mechanism is used, and its surface is curved to increase the emission area. Under certain surface temperature conditions, the cathode surface emits lots of free electrons. There is a voltage difference between the cathode and anode, which creates an accelerating electric field. The free electrons emitted from the cathode surface move toward the anode under the influence of the accelerating electric field, forming an electron beam with a certain velocity. This electron beam enters the MSWS through the anode aperture and then interacts with the electromagnetic waves. The focus electrode has the same voltage as the cathode, and is used to focus the moving free electrons, eliminating the space charge forces within the electron beam, ensuring its stable shape, and preventing the divergence of the electron beam. The shadow grid electrode is adjacent to the cathode surface but does not contact it. It is electrically connected to the cathode through wires and is maintained at the same voltage to protect the control grid electrode. The control grid electrode has the same shape as the shadow grid electrode but is at a slightly higher voltage. The voltage of the control grid electrode can be independently adjusted to obtain the desired current when keeping the voltage difference between the cathode and anode constant. A three-level strip grid structure made of pure molybdenum is used for both the shadow grid electrode and control grid electrode, as shown in Fig. 5.20. It should be noted that in VEDs, for safe operation, the anode is typically connected to the body of the VED and is grounded, making the anode zero voltage. Thus, the cathode has a negative voltage.

The CST Particle Tracking solver is used to simulate this electron gun. In the simulation setup, the anode voltage is set to be zero. The voltages of the cathode, shadow grid electrode, and focus electrode will vary between − 24 and − 26.5 kV, and the control grid electrode operates at a voltage between 10 and 430 V (the voltage difference between the control grid electrode and cathode). A thermal emission model and a space-charge-limited current emission model are established to simulate the emission of electrons in the electron gun. Since the thermionic electron emission capability of the barium tungsten cathode does not saturate under the above simulation conditions, the main limiting factor for emission current in the grid-controlled electron gun is from the position and voltage of the virtual cathode caused by space charge effects. Therefore, the use of the space-charge-limited current emission model allows for a more accurate simulation of the electron emission capability.

In the simulation setup of the grid-controlled electron gun, open boundary conditions are applied, and hexahedral grids are used for meshing. To achieve more accurate simulated results, partial grid refinement is applied to the electron beam channel region. Considering the impact of thermal velocity on the emission of electrons in the electron gun, the emission temperature and diffusion angle are set in CST for simulation. The results show that when the cathode voltage is − 26 kV and the control grid electrode can operate at 130 V, the emitted electron beam current from the electron gun is 2 A, the radius of the beam waist is approximately 1.4 mm, and the distance between the beam waist position and the cathode surface is about 16.6 mm, as shown in Fig. 5.21a. Figure 5.21b indicates the electron distribution at the beam waist position. It can be seen that without considering the thermal velocity, the electron beam has a smaller radius and better laminar flow. However, when the thermal velocity is considered, the electron beam radius increases, and the laminar flow deteriorates.

After the design of the grid-controlled electron gun, the relevant components were manufactured, and the assembly of the grid-controlled electron gun was completed. To test the performance of this grid-controlled electron gun, a short tube of the electron gun was realized, as shown in Fig. 5.22. The short tube of the electron gun consists of a grid-controlled electron gun, a drift tube, and a collector. The inner conductor of the collector is made of oxygen-free copper and is insulated

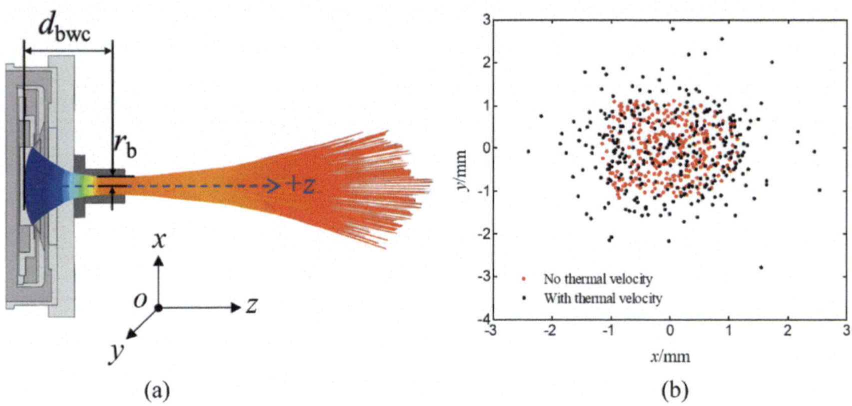

Fig. 5.21 Electron gun simulated results. **a** Schematic diagram of the electron beam emitted by the electron gun. **b** Electron distributions at the beam waist position with/without thermal velocity

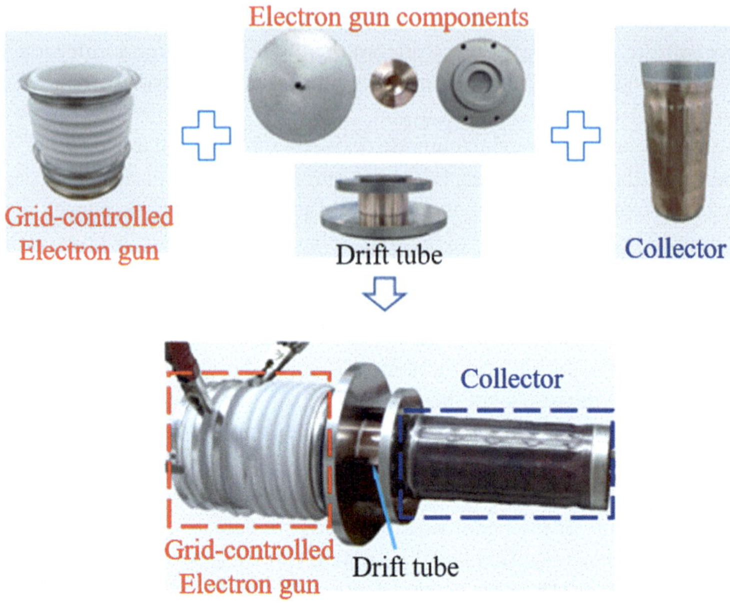

Fig. 5.22 Sample of the short tube of the electron gun

from the copper shell by beryllium oxide ceramics, ensuring both insulation and heat dissipation. The electron gun, drift tube, and collector were connected by using argon arc welding, and approximately 80 h of evacuation were then carried out. The final vacuum degree achieved was approximately 2.5×10^{-6} Pa, meeting the test requirements.

During the test of the short tube of the electron gun, the working temperature of the cathode surface was approximately 1100 °C, the control grid electrode operated at a voltage of 10–430 V, and the voltages on the cathode, shadow grid electrode, and focus electrode were located in the range of − 24 to − 26.5 kV. The parameters for the test of the short tube of the electron gun are listed in Table 5.4. Experimental results show that the adjustable range of emitted current for this grid-controlled electron gun is from 1.2 to 6.8 A, meeting the requirements of the RCRO.

5.2.2 Uniform Magnetic Focusing Configuration

An appropriate focusing configuration is needed to effectively focus the pencil beam generated by the grid-controlled electron gun. It can be found that a uniform permanent magnetic focusing configuration could provide a long and uniform magnetic field distribution with strong axial magnetic flux density and weak radial magnetic flux density. However, it also had drawbacks such as large volume and serious

Table 5.4 Parameters for the test of the short tube of the electron gun

Item	Symbol	Test value	Unit	Remark
Heater voltage of cathode	U_h	9.5~10.6	V	Adjustable
Heater current of cathode	I_h	2.5~2.7	A	Adjustable
Warm-up time of cathode	t_{kyr}	5	min	Adjustable
Cathode voltage	U_c	– 24 to – 26.5	kV	Adjustable
Electrode voltage for the control grid	U_{gp}	10~430	V	Adjustable
Duty cycle of power supply	D	≤ 1%		Adjustable
Beam current	I	1.2~6.8	A	

magnetic flux leakage [44]. When the axial dimensions of the device and the testing environment are considered, a uniform magnetic focusing configuration is ultimately chosen to focus the pencil beam. NdFeB N48SH, one of the strongest magnetic materials, is used to construct the magnet. Figure 5.23 shows a schematic model of the magnetic focusing configuration, consisting of four annular magnets, two pole pieces, and two magnetic shielding structures. Table 5.5 lists the structural dimensions of this magnetic focusing configuration.

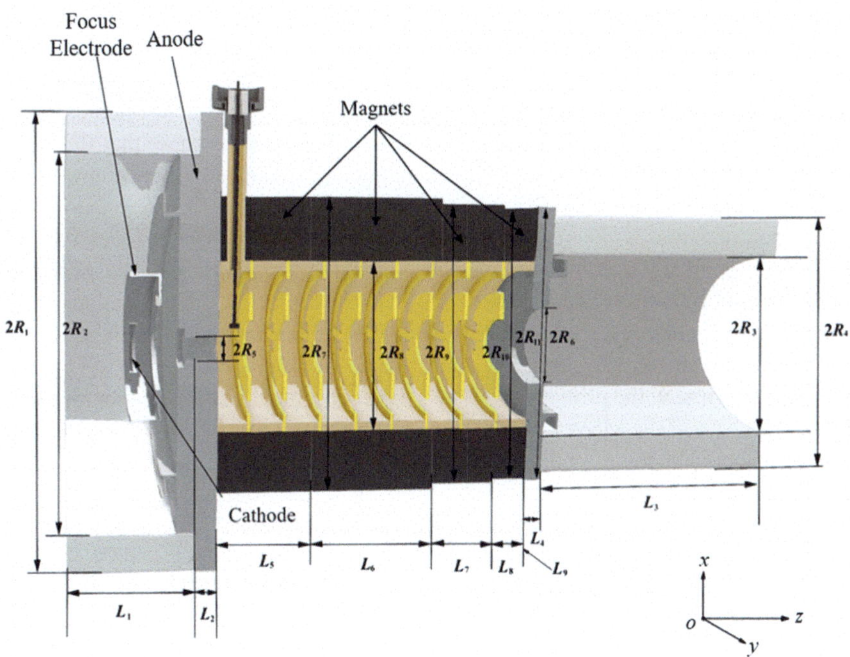

Fig. 5.23 Schematic diagram of the magnetic focusing configuration

Table 5.5 Structural dimensions of the magnetic focusing configuration

Parameter	Value/mm	Parameter	Value/mm
R_1	60	L_1	42.5
R_2	49.5	L_2	7
R_3	23.8	L_3	79.1
R_4	35	L_4	5.9
R_5	3.33	L_5	32
R_6	10	L_6	41
R_7	39	L_7	21
R_8	22.5	L_8	11
R_9	37.5	L_9	0.5
R_{10}	36.5	R_{11}	37

The magnetic field distribution of this uniform magnetic focusing configuration was simulated by using static magnetic field solver in CST. The measured remanence of the NdFeB N48SH material is 1.377 T, the recoil permeability is 1.046, the coercivity is 1041 kA/m, and the maximum magnetic energy product is 360.5 kJ/m^3. Open boundary conditions are employed in the simulation setup, and hexahedral grids are used for meshing. The simulated results are shown in Fig. 5.24. It can be seen that the axial magnetic flux density in the uniform region is approximately 1013 G, while the maximum radial magnetic flux density is only 1.5 G. The simulated magnetic field distribution information is imported into the grid-controlled electron gun model for the combined simulation. The simulated results show that the electron beam can transport well in the uniform magnetic focusing configuration without being intercepted by the MSWS, as shown in Fig. 5.25.

The components of the magnetic focusing configuration were processed and assembled to create a complete magnetic focusing configuration, as shown in Fig. 5.26a and b. A test platform was set up, as shown in Fig. 5.26c. It consists of a stepper motor and a Gauss meter. Subsequently, the measurement of the uniform magnetic focusing configuration was done.

In the experiment, the probe of the Gauss meter is controlled by a stepper motor to enter the interior of the MSWS, testing the axial magnetic flux density at intervals of 0.5 mm. The axial 0 position represented the position at the bottom of the cathode's arc-shaped emission surface. Figure 5.27 shows the tested axial magnetic flux density along the central axis of the MSWS and the CST simulated results. It can be observed that the tested axial magnetic flux density along the central axis of the MSWS in the uniform region is approximately 962 G, slightly lower than the CST simulated result of 1013 G.

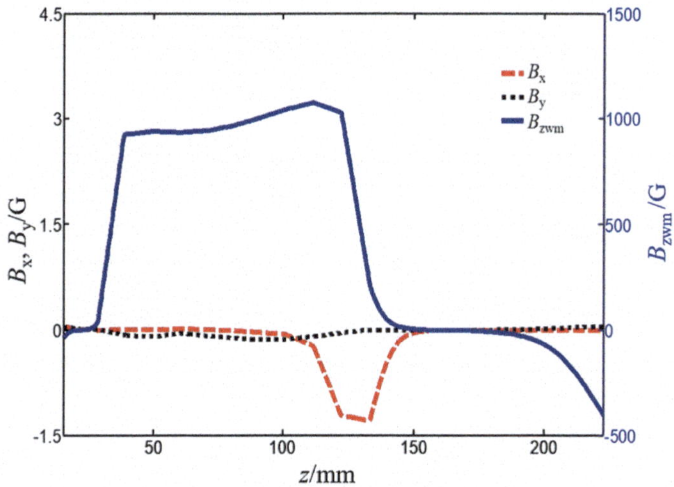

Fig. 5.24 Simulated magnetic flux densities of the magnetic focusing configuration

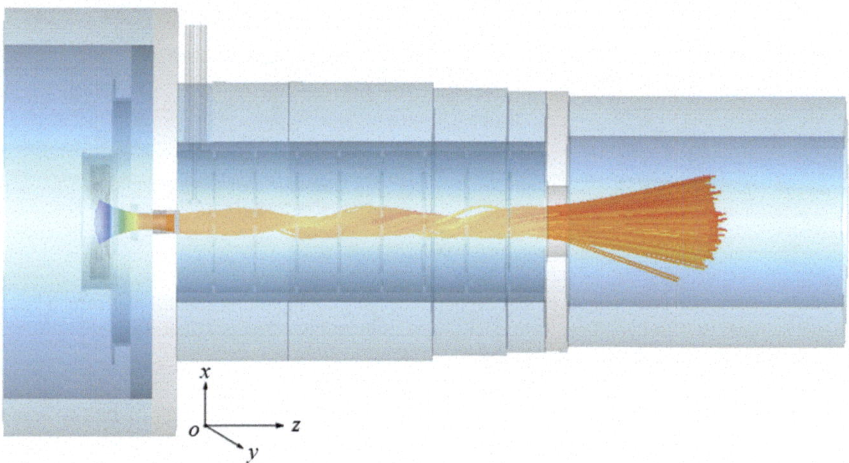

Fig. 5.25 Electron beam trajectory from the combined simulation of the electron gun and magnetic focusing configuration

5.3 Reversed Cherenkov Radiation Oscillator

The analysis of beam-wave interaction involves the simulation of the physical process where the electron beam generated by the electron gun, under the focusing of the magnetic focusing configuration, interacts with electromagnetic waves in the MSWS, thus leading to the energy exchange. Analysis of beam-wave interaction is crucial for predicting the VED's operational state and performance and plays a significant

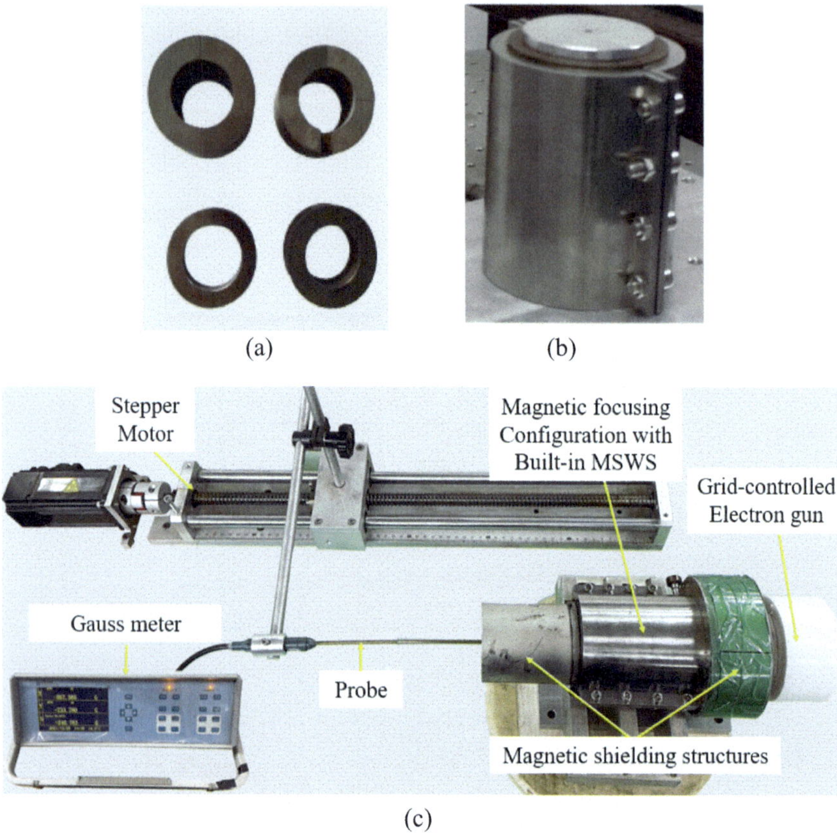

Fig. 5.26 Fabricated components of the magnetic focusing configuration and test platform. **a** Four magnet samples. **b** Assembled magnetic focusing configuration. **c** The test platform

role in the design and optimization of VEDs. Based on the analysis of beam-wave interaction, this section discusses the hot test experiment of the RCRO and evaluates its performance in a real environment.

5.3.1 Beam-Wave Interaction Analysis

After the MSWS, coaxial input and output couplers, and the electron-optical system are designed, the beam-wave interaction analysis is conducted by using the Particle-in-cell (PIC) solver in CST to predict the performance of the RCRO. In the simulation settings, all directions except the waveguide port are set as electric boundary conditions and hexahedral grids are used for meshing. To make the beam-wave interaction simulation more realistic, the information of the electron beam and the magnetic

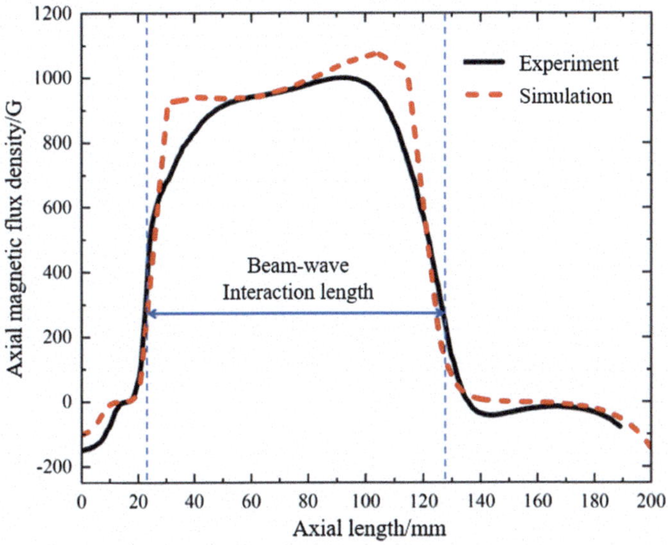

Fig. 5.27 Comparison of simulated and experimental axial magnetic flux densities

field distribution obtained from CST simulation is imported into the beam-wave interaction model, and the Ohmic loss due to surface roughness of the metal is also considered.

Figure 5.28 illustrates the physical principle of the RCRO. It can be seen that the electron beam transports along the $+z$ direction, excites electromagnetic waves in the MSWS, interacts with them, amplifies them, and finally couples them out from the coaxial output coupler near the electron gun.

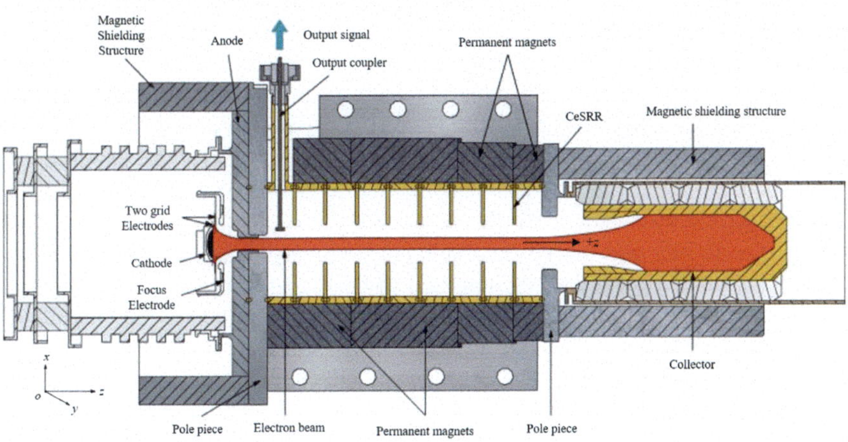

Fig. 5.28 Schematic diagram of the RCRO

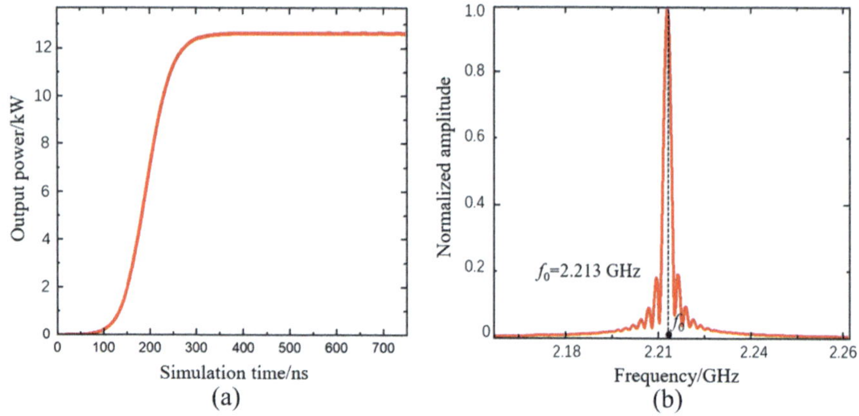

Fig. 5.29 Simulated results of beam-wave interaction. **a** Output power. **b** Spectrum of the output signal

CST simulated results indicate that the output power is 12.3 kW and the electron efficiency is approximately 23.7% at 2.213 GHz when the beam voltage is − 26 kV, the beam current is 2 A, and the magnetic flux density on the central axis of the MSWS is approximately 962 G, as shown in Fig. 5.29. This confirms that the device indeed generates the RCR.

5.3.2 *Hot Test Experiment*

Based on the beam-wave interaction analysis mentioned above, an RCRO is designed. It includes the MSWS, coaxial output coupler, grid-controlled electron gun, magnetic focusing configuration, and collector. These components were then manufactured and assembled. Finally, the complete RCRO sample was formed, as shown in Fig. 5.30. The sample was evacuated and the required vacuum degree for the hot test experiment, approximately 1.7×10^{-6} to 2.5×10^{-6} Pa, was achieved.

A hot test platform was set up, as depicted in Fig. 5.31a. Hot test experiments were conducted using many instruments, including a high-voltage power supply, signal analyzer, power meter, 50 dB attenuator, 20 dB attenuator, and directional coupler. The RCRO was placed on a suitable stationary table and was grounded, the electron gun lead wires were connected to the high-voltage power supply, and the titanium pump was connected to the power supply. After the coaxial output port was converted to N-type connector using a TNC converter, it was connected to the directional coupler through the coaxial line, and the output signal was connected to the signal analyzer and power meter through the directional coupler, as shown in Fig. 5.31b.

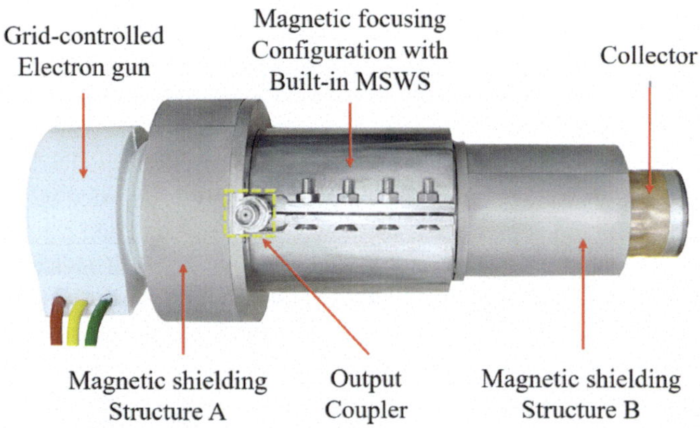

Fig. 5.30 The sample of RCRO

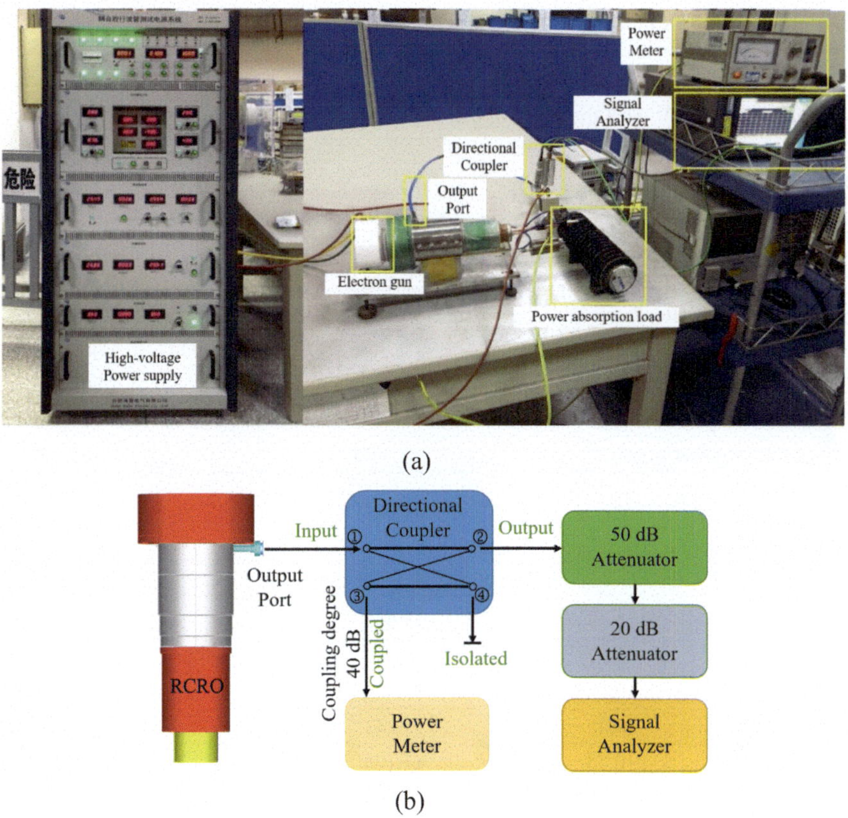

Fig. 5.31 **a** Hot test platform. **b** Diagram of the hot test

Here is a brief overview of the hot test process. Firstly, the high-voltage power supply is launched, and the instruments are checked for normal display. The power supply duty cycle is set to be one-thousandth, and the cathode heater voltage and current are adjusted to the predetermined values. The cathode is preheated for five minutes. The parameters for the hot test experiment are listed in Table 5.6. After preheating, the high-voltage output of the power supply is adjusted to 24 kV, and the high voltage button is pressed. If the power supply system and external testing instruments are stably connected, and the failure lamp is off, it indicates that the 24 kV high-voltage is stable. Subsequently, the pulse button is pressed to apply high voltage to the RCRO. At this point, the RCRO starts to operate, and microwave signals can be detected at the output port. Parameters such as the high voltage of the power supply, cathode heater voltage, cathode heater current, control grid electrode operating voltage, cathode pulse current, output power, and operating frequency are recorded. After the high voltage of the power supply and control grid electrode operating voltage are adjusted, all experimental data are recorded again. The test results for different beam currents when the cathode voltage is − 26 kV are listed in Table 5.7.

The test results are shown in Fig. 5.32. When the beam voltage varies between −24 and − 26.5 kV, and the control grid electrode voltage is adjusted to maintain a

Table 5.6 Test parameter settings

Item	Symbol	Test value	Unit	Remark
Heater voltage of cathode	U_h	9.5~10.8	V	Adjustable
Heater current of cathode	I_h	2.68~2.9	A	Adjustable
Warm-up time of cathode	t_{kyr}	5	min	Adjustable
Cathode voltage	U_c	− 24 to − 26.5	kV	Adjustable
Electrode voltage for the control grid	U_{gp}	10~430	V	Adjustable
Duty cycle of power supply	D	0.1%		Adjustable

Table 5.7 Test results for cathode voltage of − 26 kV

Cathode current/A	Output power/kW	Electron efficiency/%	Frequency/GHz
1.8	6.38	13.626	2.22288
1.9	9.32	18.886	2.21550
2.0	10.16	19.538	2.22079
2.2	10.43	18.243	2.22079
2.4	10.73	17.196	2.22068
2.6	10.89	16.110	2.22055
2.8	11.60	15.934	2.22012
3.0	12.47	15.987	2.22000
3.1	12.67	15.720	2.21998

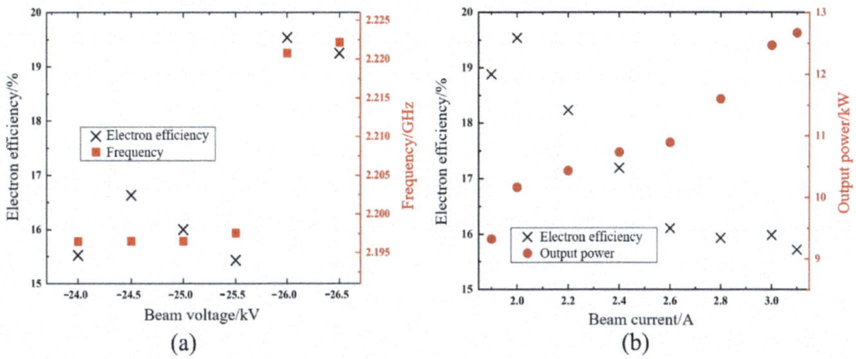

Fig. 5.32 Hot test experimental results. **a** Test results when the beam current is 2A. **b** Test results when the beam voltage is − 26 kV

constant beam current of 2 A, the output power ranges from 7.45 to 10.2 kW. At a beam voltage of − 26 kV and a beam current of 2 A, the output power is 10.16 kW at an operating frequency of 2.221 GHz, with the highest electron efficiency of approximately 19.54%. The 3 dB operating bandwidth of the RCRO is approximately 27 MHz.

At the beam voltage of − 26 kV, the operating frequency of the RCRO is 2.221 GHz, slightly lower than the frequency predicted by the synchronization condition, which is 2.291 GHz, as shown in Fig. 5.33. According to the synchronization condition, the phase velocity of the electromagnetic wave should be slightly less than the velocity of the electron beam, resulting in an actual operating frequency lower than the frequency corresponding to the intersection of the beam line and the dispersion curve.

Additionally, calculations confirm that the signal analyzer used in the hot test experiment can accurately detect the input signal and precisely analyze its frequency. The frequency readout accuracy of the signal analyzer is given by ± (displayed frequency × frequency reference accuracy + 0.1% × span + 5% × resolution bandwidth + 2 Hz + 0.5 × horizontal resolution). A displayed frequency of 2.20548 GHz is taken as an example. In this case, the frequency reference accuracy is $\pm 9 \times 10^{-8}$, the span is 200 MHz, the resolution bandwidth is 1.8 MHz, and the horizontal resolution is 200 kHz, resulting in an accuracy of ± 0.39 MHz. This indicates that the signal analyzer can provide accurate frequency data. Moreover, based on the measured power data, the output power of the RCRO ranges from 66.2 to 71.53 dBm. After the signal passes through a 50 dB attenuator and a 20 dB attenuator before entering the signal analyzer, the signal power ranges from − 3.8 to 1.53 dBm. From the signal analyzer's user manual, the input signal can be accurately detected by the signal analyzer.

The cross-sectional dimension of the MSWS in the RCRO is only approximately 0.33λ (here λ is the wavelength in free space) and the longitudinal dimension is only about 0.78λ, while the SWS of traditional BWOs has cross-sectional dimensions

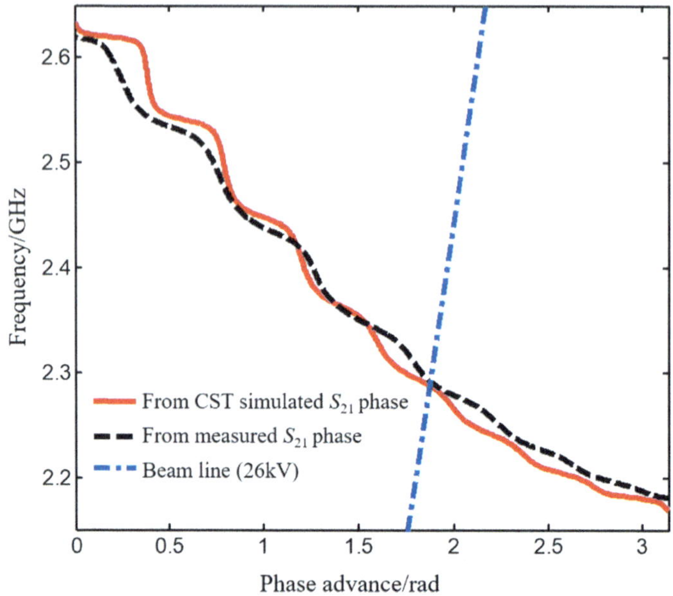

Fig. 5.33 Dispersion curves and beam line (26 kV)

of approximately λ to 1.7λ and longitudinal dimension ranging from 0.6λ to 1.6λ [45–47]. Traditional BWOs typically have an electron efficiency of less than 15% [45–47], while the RCRO can achieve an electron efficiency of 19.54%. From the experimental results mentioned above, it is evident that the RCRO has the obvious advantages of miniaturization and high efficiency, stemming from the creative combination of metamaterials and SWSs. Metamaterials can offer some advantages such as subwavelength and strong resonance. Subwavelength reduces the cross-section's size of the MSWS, leading to its compactness. Strong resonance increases the axial electric field intensity in the MSWS, strengthening the beam-wave interaction and achieving higher electron efficiency. Of course, techniques like the phase velocity jump [48] and multistage depressed collectors [49] can further enhance the device efficiency of the RCRO.

5.4 Reversed Cherenkov Radiation Amplifier

The hot test experiment of the RCRO confirms the feasibility of RCR in metamaterials for generating high-power microwave signals. Based on the theoretical analysis and experimental research of the RCRO, this section will explore an RCRA with high gain, high efficiency, and miniaturization characteristics, using existing circular double-bridge CeSRRs and an electron-optical system [50, 51].

The RCRA is constructed based on the RCRO. Here, the RCRA consists of six components: an electron gun, the first segment MSWS, a cut-off region, the second segment MSWS, a magnetic focusing configuration, and a collector. The whole device model is shown in Fig. 5.34. Both the first and second segment MSWSs consist of 16 all-metal double-bridge CeSRRs loaded periodically into a hollow circular waveguide, respectively. A coaxial electric probe is an important element of input and output couplers for efficiently coupling electromagnetic waves. The RCRA is designed with three waveguide ports: the input port, output port 1, and output port 2. The period of the first segment MSWS is P_1, with a total of $N_1 = 15$ periods. The width of the output cavity near output port 1 is D_{r1}, and the width of the input cavity near the input port is D_{r2}. The period of the second segment MSWS is P_2, with a total of $N_2 = 15$ periods. The width of the output cavity near output port 2 is D_{r3}, and the width of the cavity near the collector is D_{r4}. The inner and outer radii of the coaxial line are denoted as r and R, respectively. The height and width of the coaxial electric probe are h and L_1, respectively, and the supporting medium between the inner and outer conductors of the coaxial line is made of PTFE. The relevant structural parameters are provided in Table 5.8.

The electron beam generated by the electron gun passes through the first segment MSWS, the cut-off region, and then enters the second segment MSWS. After

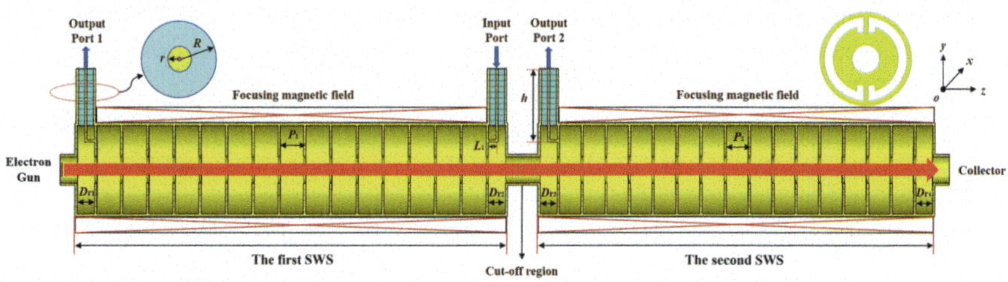

Fig. 5.34 Schematic diagram of the RCRA

Table 5.8 Structural parameters of the RCRA

Parameter	Value/mm
P_1	12
P_2	12
D_{r1}	8
D_{r2}	8
D_{r3}	8
D_{r4}	8
R	3.5
r	0.5
h	33
L_1	4.5

inputting the signal through the input port, it interacts with the electron beam in the first segment MSWS. In this beam-wave interaction, the high-frequency field modulates the velocity of the electron beam, achieving preliminary velocity modulation and density modulation. The modulated electron beam exhibits good bunching characteristics. The electron beam then enters the second segment MSWS through the cut-off region, which allows only the electron beam to pass through while the high-frequency field is cut off. This effectively suppresses the self-oscillation. When the electron beam with good bunching characteristics enters the second segment MSWS, it will generate RCR, re-exciting a high-frequency electromagnetic field. The electron beam continues to interact with the high-frequency field, exchanging energy, and thus amplifying the high-frequency electromagnetic signal. The operating principle of the electron beam within the second segment MSWS is similar to that of an RCRO [52–54].

Subsequently, the beam-wave interaction is simulated for the RCRA. Under the conditions of input power of 5.2 W, axial magnetic flux density of 0.2 T, and electron beam voltage and current of − 38 kV and 0.45 A, respectively. The simulated results show that the output power at output port 1 is 244 W, and the saturated output power at output port 2 is 2.89 kW, with a saturated gain of 27.45 dB at 2.286 GHz, and a 3 dB operating bandwidth of approximately 21 MHz (2.277–2.298 GHz), as shown in Fig. 5.35. The total electron efficiency is approximately 18.33%. Therefore, the RCRA simultaneously achieves high-power and low-power microwave signals, both at the same operating frequency, making it suitable for applications requiring high-power and low-power microwave signals at the same frequency.

To further improve the output power and gain of the RCRA, the phase velocity jump technique is applied to optimize the period of the second segment MSWS, resulting in an optimized P_2 of 10.9 mm. Subsequently, the beam-wave interaction analysis is performed under the input power of 7.8 W, axial magnetic flux density of 0.2 T, and beam voltage and current of − 38 kV and 0.45 A, respectively. The

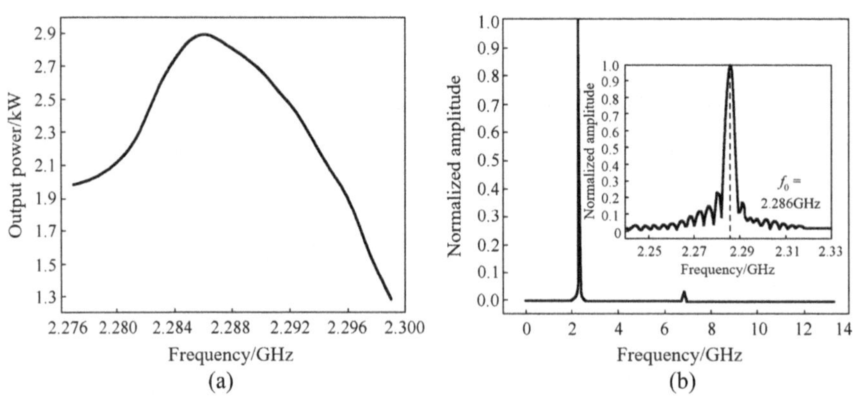

Fig. 5.35 Simulated results of beam-wave interaction. **a** The output power of output port 2 varies with frequency. **b** The signal spectrum for output port 2

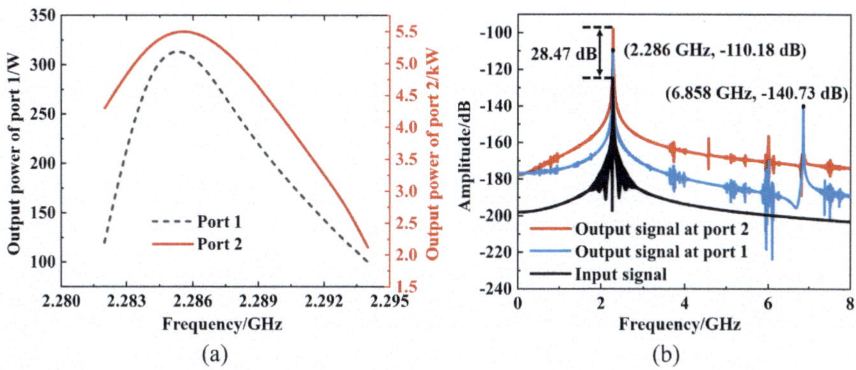

Fig. 5.36 Simulated results of beam-wave interaction. **a** The output power versus frequency. **b** The signal spectrum

simulated results show that the output power at output port 1 is 307 W, while the saturated output power at output port 2 is 5.48 kW, with a saturated gain of 28.47 dB at 2.286 GHz, and a 3 dB operating bandwidth of approximately 11 MHz (2.282–2.293 GHz), as shown in Fig. 5.36. The total electron efficiency is approximately 33.84%. In addition, the maximum amplitude of the higher harmonics (6.858 GHz) at port 1 is 30 dB lower than the amplitude of the central operating frequency (2.286 GHz), which demonstrates the effectiveness of the RCRA in suppressing the intrinsic self-oscillation [55].

From the simulated results of the RCRA with $P_2 = 12$ mm and the RCRA with $P_2 = 10.9$ mm, it can be found that the latter, optimized with the phase velocity jump technique, exhibits a notable improvement in saturated gain and electron efficiency compared to the former.

The radius of the RCRA's MSWS is only 20 mm, approximately two-thirds of the radius of the SWS employed in traditional S-band coupled-cavity TWTs. Its longitudinal length is 410 mm, comparable to that of traditional counterparts. Consequently, it significantly reduces in size relative to traditional coupled-cavity TWTs [56]. Furthermore, the optimized RCRA achieves a saturated gain of 28.47 dB and an electron efficiency of 33.84%, surpassing traditional S-band coupled-cavity TWTs (typically around 14~24%) and helix TWTs (approximately 10 ~ 20%) [56, 57]. Therefore, this RCRA not only shares the characteristics of miniaturization and high efficiency with the RCRO but also offers high gain.

The RCR within metamaterials can be employed not only for the generation but also for the amplification of electromagnetic waves. Compared to traditional counterparts, it exhibits distinct advantages in gain, electron efficiency, and miniaturization [58]. The RCRO and RCRA presented here have important applications in narrowband microwave signal generation and amplification. In fact, metamaterials are not limited to the microwave frequency range, which can be extended to millimeter wave, terahertz, and even optical frequency ranges for promising research.

References

1. Wang WX. Vacuum Electron Devices. Beijing: National Defense Industry Press. 2012 (In Chinese).
2. Veselago VG. The electrodynamics of substances with simultaneously negative values of ε and μ. Physics-Uspekhi. 1968; 10:4–509.
3. Duan ZY, Guo C, Guo X, et al. Double negative-metamaterial based terahertz radiation excited by a sheet beam bunch. Physics of Plasmas. 2013; 20:9–093301.
4. Duan ZY, Wu BI, Lu J, et al. Reversed Cherenkov radiation in unbounded anisotropic double-negative metamaterials. Journal of Physics D: Applied Physics. 2009; 42:18–185102.
5. Chen HS, Chen M. Flipping photons backward: Reversed Cherenkov radiation. Materials Today. 2011; 14:1-2-34.
6. Lyu ZF, Luo HY, Wang X, et al. Compact reversed Cherenkov radiation oscillator with high efficiency. Applied Physics Letters. 2022; 120:5.053501.
7. Duan ZY, Gong YB, Wang WX, et al. Accurate tape analysis of the attenuator-coated helical slow-wave structure. IEEE Transactions on Electron Devices. 2006; 53:4–903.
8. Duan ZY, Gong YB, Lü MY, et al. Optimization design of helix pitch for efficiency enhancement in the helix travelling wave tubes. Chinese Physics Letters. 2008; 25:3–934.
9. Duan ZY, Gong YB, Wei YY, et al. Theoretical research on the performance of a practical helix travelling wave tube. Chinese Physics B. 2008; 17:7–2484.
10. Chernin D, Antonsen TM Jr, Levush B, et al. A three-dimensional multifrequency large signal model for helix traveling wave tubes. IEEE Transactions on Electron Devices. 2001; 48:1–3.
11. Chong CK, Menninger WL. Latest advancements in high-power millimeter-wave helix TWTs. IEEE Transactions on Plasma Science. 2010; 38:6–1227.
12. Ji DX, Zhu L, Huang WC, et al. Development of a high efficiency coupled-cavity traveling wave tube. IEEE International Vacuum Electronics Conference. Busan. 2019:1–2.
13. Pershing DE, Nguyen KT, Abe DK, et al. Demonstration of a wideband 10-kW Ka-band sheet beam TWT amplifier. IEEE Transactions on Electron Devices. 2014; 61:6–1637.
14. Kowalski EJ, Shapiro MA, Temkin RJ. An overmoded W-band coupled-cavity TWT. IEEE Transactions on Electron Devices. 2015; 62:5.1609.
15. Gilmour AS Jr. Klystrons, Traveling Wave Tubes, Magnetrons, Crossed-Field Amplifiers, and Gyrotrons. Boston: Artech House. 2011.
16. Mendel JT. Helix and coupled-cavity traveling-wave tubes. Proceedings of the IEEE. 1973; 61:3–280.
17. Duan ZY, Shapiro MA, Schamiloglu E, et al. Metamaterial-inspired vacuum electron devices and accelerators. IEEE Transactions on Electron Devices. 2019; 66:1–207.
18. Shapiro MA, Trendafilov S, Urzhumov Y, et al. Active negative-index metamaterial powered by an electron beam. Physical Review B. 2012; 86:8–085132.
19. Hummelt JS, Lu XY, Xu HR, et al. Coherent Cherenkov-cyclotron radiation excited by an electron beam in a metamaterial waveguide. Physical Review Letters. 2016; 117:23–237701.
20. Lu XY, Stephens JC, Mastovsky I, et al. High power long pulse microwave generation from a metamaterial structure with reverse symmetry. Physics of Plasmas. 2018; 25:2–023102.
21. Lu XY, Shapiro MA, Mastovsky I, et al. Generation of high-power, reversed-Cherenkov wakefield radiation in a metamaterial structure. Physical Review Letters. 2019; 122:1–014801.
22. Picard J, Mastovsky I, Shapiro MA, et al. Generation of 565 MW of X-band power using a metamaterial power extractor for structure-based wakefield acceleration. Physical Review Accelerators and Beams. 2022; 25:5.051301.
23. Yurt SC, Fuks MI, Prasad S, et al. Design of a metamaterial slow wave structure for an O-type high power microwave generator. Physics of Plasmas. 2016; 23:12–123115.
24. Prasad S, Yurt SC, Shipman KA, et al. A compact high-power microwave metamaterial slow-wave structure: From computational design to hot test validation. Computing and Electromagnetics International Workshop. Barcelona, Spain; 2017:61–62.

25. de Alleluia AB, Abdelshafy AF, Ragulis P, et al. Experimental testing of a 3-D-printed metamaterial slow wave structure for high-power microwave generation. IEEE Transactions on Plasma Science. 2020; 48:12–4356.
26. Tang XF, Duan ZY, Ma XW, et al. Dual band metamaterial Cherenkov oscillator with a waveguide coupler. IEEE Transactions on Electron Devices. 2017; 64:5.2376.
27. Tang XF, Li XQ, Wang QF, et al. Miniature metamaterial backward wave oscillator with a coaxial coupler. IEEE Transactions on Electron Devices. 2022; 69:3–1389.
28. Wang YS, Duan ZY, Tang XF, et al. All-metal metamaterial slow-wave structure for high-power sources with high efficiency. Applied Physics Letters. 2015; 107:15.153502.
29. Duan ZY, Wang YS, Huang X, et al. Miniaturized all-metal slow-wave structure. US 9425020 B2. 2016.
30. Duan ZY, Wang YS, Huang X, et al. A miniaturized all-metal slow-wave device. ZL 201410280414.4. 2016 (In Chinese).
31. Wang YS, Duan ZY, Wang F, et al. S-band high-efficiency metamaterial microwave sources. IEEE Transactions on Electron Devices. 2016; 63:9–3747.
32. Ghione G, Naldi CU. Coplanar waveguides for MMIC applications: Effect of upper shielding, conductor backing, finite-extent ground planes, and line-to-line coupling. IEEE Transactions on Microwave Theory and Techniques. 1987; 35:3–260.
33. Smith DR, Vier DC, Kroll N, et al. Direct calculation of permeability and permittivity for a left-handed metamaterial. Applied Physics Letters. 2000; 77:14–2246.
34. Szabó Z, Park GH, Hedge R, et al. A unique extraction of metamaterial parameters based on Kramers–Kronig relationship. IEEE Transactions on Microwave Theory and Techniques. 2010; 58:10–2646.
35. Marqués R, Martel J, Mesa F, et al. Left-handed-media simulation and transmission of EM waves in subwavelength split-ring-resonator-loaded metallic waveguides. Physical Review Letters. 2002; 89:18–183901.
36. Esteban J, Camacho-Peñalosa C, Page JE, et al. Simulation of negative permittivity and negative permeability by means of evanescent waveguide modes—theory and experiment. IEEE Transactions on Microwave Theory and Techniques. 2005; 53:4–1506.
37. Rosa EB. The Self and Mutual Inductances of Linear Conductors. Washington: U.S. Bulletin of the National Bureau of Standards. 1907.
38. Fujisawa K. General treatment of klystron resonant cavities. IRE Transactions on Microwave Theory and Techniques. 1958; 6:4–344.
39. Dishal M. Design of dissipative band-pass filters producing desired exact amplitude-frequency characteristics. Proceedings of the IRE. 1949; 37:9–1050.
40. Curnow HJ. A general equivalent circuit for coupled-cavity slow-wave structures. IEEE Transactions on Microwave Theory and Techniques. 1965; 13:5.671.
41. Kirley MP, Booske JH. Terahertz conductivity of copper surfaces. IEEE Transactions on Terahertz Science and Technology. 2015; 5:6–1012.
42. Hammerstad EO, Jensen O. Accurate models for microstrip computer-aided design. IEEE MTT-S International Microwave Symposium Digest. Washington, DC, USA. 1980.
43. Groiss S, Bardi I, Biro O, et al. Parameters of lossy cavity resonators calculated by the finite element method. IEEE Transactions on Magnetics. 1996; 32:3–894.
44. Electronic Tube Design Manual Editorial Committee. O-Type Backward Wave Tube Design Manual. Beijing: National Defense Industry Press. 1985:25.30 (In Chinese).
45. Johnson HR. Backward-wave oscillators. Proceedings of the IRE. 1955; 43:6–684.
46. Sullivan JW. A wide-band voltage-tunable oscillator. Proceedings of the IRE. 1954; 42:11–1658.
47. Tien PK. Bifilar helix for backward-wave oscillators. Proceedings of the IRE. 1954; 42:7–1137.
48. Jung SS, Soukhov AV, Jia BF, et al. Positive phase-velocity tapering of broadband helix traveling-wave tubes for efficiency enhancement. Applied Physics Letters. 2002; 80:16–3000.
49. Latha AM, Ghosh SK. Design and development of a novel, compact, and light-weight multistage depressed collector for space TWTs. IEEE Transactions on Electron Devices. 2016; 63:1–481.

50. Heffner H. Analysis of the backward-wave traveling-wave tube. Proceedings of the IRE. 1954; 42:6–930.
51. Liu SG. Introduction to Microwave Electronics. Beijing: National Defense Industry Press. 1985:226–235 (In Chinese).
52. Currie MR, Whinnery JR. The cascade backward-wave amplifier: A high-gain voltage-tuned filter for microwaves. Proceedings of the IRE. 1955; 43:11–1617.
53. Paoloni C, Carlo AD, Brunetti F, et al. Design and fabrication of a 1 THz backward wave amplifier. Terahertz Science and Technology. 2011; 4:4–149.
54. Paoloni C, Carlo AD, Bouamrane F, et al. Design and realization aspects of 1-THz cascade backward wave amplifier based on double corrugated waveguide. IEEE Transactions on Electron Devices. 2013; 60:3–1236.
55. Wang CC, Li XY, Lyu ZF, et al. Reversed Cherenkov radiation amplifier with compact structure and high efficiency. Physics of Plasmas. 2023; 30:9–093301.
56. Li WJ, Xu Z, Lin YZ, et al. Cold-test experimental research of an S-band broadband high power CCTWT. Chinese Physics C. 2008; 32:S1–178.
57. Abrams RH, Levush B, Mondelli AA, et al. Vacuum electronics for the 21st century. IEEE Microwave Magazine. 2001; 2:3–61.
58. Wang YS. Research on novel radiation sources based on metamaterials. Chengdu: University of Electronic Science and Technology of China. 2017 (In Chinese).

Chapter 6
Vacuum Electron Devices Based on Transition Radiation

V. L. Ginzburg and I. M. Frank theoretically predicted the existence of TR [1]. TR is a type of electromagnetic radiation generated when charged particles pass through a discontinuous interface between two different media, as shown in Fig. 6.1. This electromagnetic radiation is caused by the collective effect of the matter surrounding the trajectories of the charged particles to reshape the electromagnetic fields excited by the charged particles. In the first medium far from the boundary, the charged particles have certain fields that characterize the motion of the charged particles and the properties of the first medium. When the charged particles enter the second medium, the fields also have the properties that characterize the motion of the charged particles and the properties of the second medium. If the electromagnetic properties of the two media are different, the initial and final fields will be different even if the charged particles are moving at the same velocity. The fields must change as the charged particles approach and pass through the interface, and some of the field is emitted in the form of TR. TR can be used to realize a new type of radiation source operating at X-rays or even shorter wavelengths [2–4]. Therefore, it is widely used to detect the relativistic energy and the magnetic moments of microscopic charged particles [5, 6]. However, conventional TR utilizes energetical charged particles to bombard a target to generate spatially distributed electromagnetic radiation with a certain angle, and its radiation density will be limited by vacuum boundaries, finite mass foils, and laminated metal-plastic targets [3]. The energy of the charged particles decreases rapidly after bombarding the target, and thus the intensity of TR decreases dramatically.

In a broad sense, the generation of TR is not only excited by the passage of charged particles through the interface of two different media but also can be generated based on transition time effects [7], such as the perturbations generated by charged particles passing through a resonant cavity [8, 9]. As shown in Fig. 6.2, the standing wave at the transit gap of the resonant cavity is utilized to perturb the charged particles, thus realizing the energy exchange, which is a typical application of TR in VEDs [10]. In these VEDs, the charged particles do not bombard the medium but rather

Z. Duan, *Metamaterial-Based Electromagnetic Radiations and Applications*,
https://doi.org/10.1007/978-981-97-8108-9_6

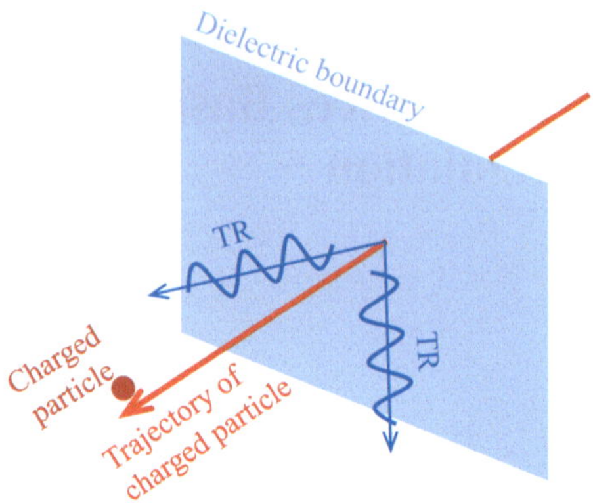

Fig. 6.1 Schematic diagram of narrowly-defined TR

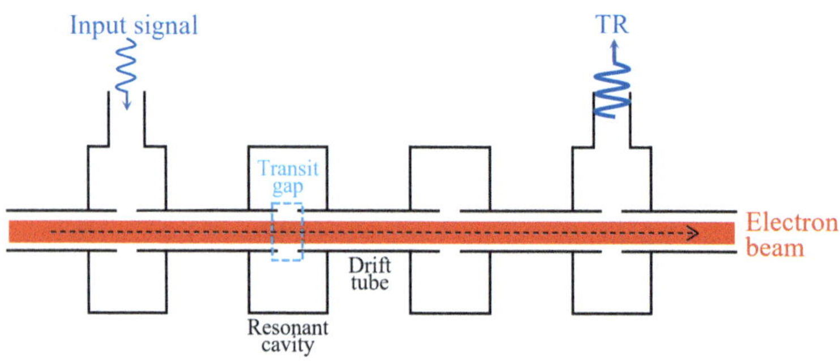

Fig. 6.2 Schematic diagram of broadly-defined TR in a klystron

interact with the electromagnetic field at the transit gap of the resonant cavity to generate a high-power coherent microwave output [11]. In addition, in the context of exploring applications for stronger coherent radiation (for klystrons, they are used in tokamak devices for the International Thermonuclear Experimental Reactor (ITER) program) and higher acceleration gradients (for accelerators, they are used in large-scale scientific facilities such as the Spallation Neutron Source (SNS)), emerging functional materials [12] and metamaterials [13–15] are being used to develop different types of high-power microwave radiation sources.

TR is an important physical mechanism for electromagnetic radiation using an electron beam. Klystron, extended interaction oscillator (EIO), and extended interaction amplifier (EIA) have been invented based on this mechanism. R. H. Varian and S. F. Varian are recognized as the inventors of the two-cavity klystron in 1937 [16, 17]. K. Fujisawa proposed the concept of EIO in 1954 [18], followed by M. Chodorow and T. Wessel-Berg who proposed the EIA [19, 20]. They shorted the ends of a conventional SWS, thus constructing a novel extended interaction resonant cavity (EIRC), which can improve the characteristic impedance of the resonant cavity. The performance of the klystron can be further improved by using the

EIRC instead of the single-gap resonant cavity [20]. Theoretically, the characteristic impedance of an EIRC with *N* transit gaps is about *N* times that of a single-gap resonant cavity. Increasing the characteristic impedance of the resonant cavity can improve the gain-bandwidth product and electron efficiency of the klystron [21].

Klystrons are widely used in radar, communication, medical equipment, industrial heating, and large-scale scientific devices due to the advantages of high power, high efficiency, high gain, and low phase noise [22–24]. Klystrons are widely used as high-energy microwave sources in controlled thermonuclear fusion devices, synchrotron radiation light sources, scattered-slit neutron sources, and other large-scale scientific facilities, which are not only numerous, but also most of them operate at low frequencies, such as the P, L, S, and C bands [25, 26]. This puts higher demands on klystrons in output power, electron efficiency, gain, and miniaturization. Therefore, it is imperative to reduce device size and costs of the high-performance microwave radiation sources [27–31]. In recent years, metamaterials, as a new type of subwavelength structures, have attracted much attention from researchers in the microwave and optical fields due to their special physical properties [32–40]. The creative combination of metamaterial unit cells with klystrons or extended interaction devices [41, 42] is not only capable of exploring new coherence-enhanced TR from the physical mechanism, but also is expected to realize the miniaturization and high efficiency of klystrons and extended interaction devices. It is of great scientific significance and engineering value.

6.1 Resonant Cavity Loaded with Metamaterial Unit Cells

6.1.1 Electromagnetic Parameters of Resonant Cavity

According to the number of transit gaps in the resonant cavity, the resonant cavity can be divided into single-gap resonant cavity and multi-gap EIRC. Due to its simple structure and stable operation, the single-gap resonant cavity is more suitable for high-power applications requiring high beam voltage and high beam current excitation. For the multi-gap EIRC, when the electron beam is injected into the resonant cavity, the electrons are subjected to more velocity modulation at the transit gap, which promotes better bunching of the electrons, thus improving the electron efficiency of the extended interaction device. At the same time, the enhanced velocity modulation in the resonant cavity can accelerate the electron bunching, which is conducive to shortening the longitudinal length of the extended interaction device. However, the multi-gap structure leads to more types of modes in the resonant cavity, which is easy to cause mode competition at high power, so it is suitable for low-power cases. The electromagnetic parameters of the resonant cavity mainly include the resonant frequency f, the quality factor Q, the coupling coefficient M, the characteristic impedance R/Q, and the normalized electron beam loading conductance G_e/G_0.

(1) Resonant frequency

Resonant frequency f is the frequency at which a mode field in the resonant cavity reaches its strongest point or the stored energy reaches its maximum, which characterizes the frequency selectivity of the resonant cavity. The resonant frequency f_{mnp} of a rectangular waveguide resonant cavity can be expressed as [43]:

$$f_{mnp} = \frac{\omega_{mnp}}{2\pi} = \frac{k_{mnp}}{2\pi\sqrt{\mu\varepsilon}} = \frac{1}{2\sqrt{\mu\varepsilon}}\sqrt{\left(\frac{m}{a}\right)^2 + \left(\frac{n}{b}\right)^2 + \left(\frac{p}{l}\right)^2}. \tag{6.1}$$

The resonant frequency of the circular waveguide resonant cavity can be expressed separately according to the mode [43]:

$$f_{TE_{mnp}} = \frac{1}{2\sqrt{\mu\varepsilon}}\sqrt{\left(\frac{\mu_{mn}'}{\pi R}\right)^2 + \left(\frac{p}{l}\right)^2}, \tag{6.2}$$

$$f_{TM_{mnp}} = \frac{1}{2\sqrt{\mu\varepsilon}}\sqrt{\left(\frac{\mu_{mn}}{\pi R}\right)^2 + \left(\frac{p}{l}\right)^2}, \tag{6.3}$$

where for the rectangular waveguide resonant cavity, m, n, and p are the number of semi-standing wave wavelengths distributed on the long side, wide side, and the longitudinal direction of the rectangular waveguide resonant cavity. ω_{mnp} and k_{mnp} denote the angular frequency and the amplitude of wave number of the resonant cavity in a specific mode, respectively. a, b, and l indicate the broad side and narrow side of the rectangular waveguide cross-section and the longitudinal length, respectively. For the circular waveguide resonant cavity, m, n, and p are the period number of standing waves distributed along the entire circumference, the number of semi-standing wave wavelengths distributed over the radius, and the number of semi-standing wave wavelengths distributed over the longitudinal length. R is the radius of the circular waveguide resonant cavity. μ_{mn} and μ'_{mn} are the nth root of mth-order Bessel function of the first kind and the derivative and its derivative, respectively. l is the longitudinal length of the resonant cavity. ε and μ denote the permittivity and the permeability of the dielectric that fills the resonant cavity, respectively.

(2) Quality factors

Quality factors characterize the energy storage capacity or frequency selectivity of a microwave resonant cavity, which can be divided into the intrinsic quality factor Q_0, the loaded quality factor Q_L, and the external quality factor Q_e according to the different definitions [43]. Q_0 is defined as the ratio of the average total energy storage in the resonant cavity to the loss of the resonant cavity in one high-frequency period when the resonant cavity is worked in a stable resonant state, reflecting the ability of the resonant cavity to store energy:

$$Q_0 = 2\pi\frac{W}{W_s} = \omega\frac{W}{P_s}, \tag{6.4}$$

$$P_s = W_s f, \tag{6.5}$$

where ω is the angular frequency of the resonant cavity, W is the average total energy storage of the resonant cavity, W_s is the energy loss in one high-frequency period of the resonant cavity, and P_s is the average power loss of the resonant cavity in one high-frequency period.

To describe the coupling characteristics of the resonant cavity and the external circuit, the loading quality factor Q_L and the external quality factor Q_e are defined, respectively. When the resonant cavity is coupled with the external circuit, the microwave energy is lost both inside the resonant cavity and in the external circuit. At this time, the Q_L of the resonant cavity is

$$Q_L = \omega \frac{W}{P_s + P_e}, \tag{6.6}$$

where P_s represents the average power loss of the resonant cavity itself during one high-frequency period, and P_e is the power loss on the external circuit load coupled to the resonant cavity. (6.6) can be rewritten as:

$$\frac{1}{Q_L} = \frac{P_s}{\omega W} + \frac{P_e}{\omega W} = \frac{1}{Q_0} + \frac{1}{Q_e}. \tag{6.7}$$

Therefore, Q_e is

$$Q_e = \omega \frac{W}{P_e}, \tag{6.8}$$

where P_e is usually specified as the microwave power loss by the matched load. So, Q_e is the external quality factor under the matched load conditions.

(3) Effective characteristic impedance

For the average power loss P_s in (6.5), this parameter can also be expressed as:

$$P_s = \frac{1}{2} G U_m^2, \tag{6.9}$$

where G is the equivalent conductance of the resonant cavity, and U_m is the voltage amplitude at the both ports of the equivalent circuit. G not only depends on the voltage U_m, but also on the resonant cavity power loss P_s. P_s is not only related to the resonant cavity shape and material but also to the processing process (the processing situation will affect the surface resistance of the inner wall of the resonant cavity and the size of the surface area). So, it is difficult to determine P_s and G. The other important parameter of the resonant cavity, Q_0, also considers the loss in the resonant cavity. To characterize the equivalent resistance R (i.e., the generated electric field intensity) that may be generated inside the resonant cavity more conveniently, a new parameter, the characteristic impedance R/Q, is defined with the following form [44]:

$$R/Q = \frac{U_{\mathrm{m}}^2}{2\omega W} = \frac{\left[\int_{-\infty}^{+\infty} E_z(z)dz\right]^2}{2\omega W}, \tag{6.10}$$

where *R*/*Q* depends only on the shape of the resonant cavity, geometry, and field distributions in the resonant cavity, and has nothing to do with the loss in the resonant cavity, so *R*/*Q* can also be called shape factor or geometry factor. It means the high-frequency electric field intensity can be established on the transit gap when f and the energy storage of the resonant cavity are fixed.

M is defined as the ratio of the modulation voltage felt by the electron beam through the transit gap to the modulation voltage actually loaded on the transit gap, and is expressed as follows:

$$M(\beta_{\mathrm{e}}) = \frac{\int_{-\infty}^{+\infty} E_z(z)e^{\mathrm{j}\beta_{\mathrm{e}}z}dz}{\int_{-\infty}^{+\infty} E_z(z)dz}, \tag{6.11}$$

where β_{e} is the propagation coefficient of the electron beam. The physical meaning of M determines that its value is constantly less than 1. For resonant cavities loaded with a single-bridge CeSRR, its *R*/*Q* and M correlate negatively within a certain range when the gap transit angle θ varies. θ is expressed as follows:

$$\theta = \beta_{\mathrm{e}} l_1 = \frac{\omega}{v_0} l_1 = \frac{2\pi f}{c_0\sqrt{1-\frac{1}{\gamma^2}}} l_1, \tag{6.12}$$

where l_1 is the length of the transit gap, v_0 is the Electron beam speed, c_0 is the velocity of light in vacuum, and $\gamma = 1 + U/511000$ is the relativistic factor, where U is the electron beam voltage. When the *R*/*Q* of the resonant cavity is high, M may be at a lower level. When the transit gap can establish a large longitudinal electric field intensity, the electric field acting on the electron beam may be not large. Therefore, it is not comprehensive to use *R*/*Q* alone to describe the electromagnetic characteristics of the resonant cavity. Here, the effective characteristic impedance M^2R/Q is used to characterize the actual longitudinal electric field acting on the electron beam when it passes through the transit gap, and this parameter is positively correlated with the modulation amplitude obtained when the electron beam passes through the transit gap. At the same time, there exists an optimal θ such that the M^2R/Q of the resonant cavity is maximized, and it can be predicted that a klystron with such resonant cavities can achieve a large electron efficiency.

Increasing the number of transit gaps in the EIRC can also increase its effective characteristic impedance M^2R/Q, but a higher M^2R/Q will subject the electron beam to a stronger modulation, resulting in the loss of kinetic energy of the electrons in the output cavity exceeding their initial kinetic energy, which leads to the electrons returning. In this case, the electrons will flow back from the output cavity to the input cavity or even to the electron gun, thus devastating the transport of the electron beam and the beam-wave interaction. Therefore, an optimal number of transit gaps is helpful for maximizing the electron efficiency and minimizing the electrons returning.

(4) Normalized electron beam loading conductance

M and *R*/*Q* are used to measure the degree of interaction of the high-frequency field with the electron beam, while whether the electron beam absorbs the energy of the high-frequency field or transfers DC energy to the high-frequency field needs to be characterized by the normalized electron beam loading conductance, G_e/G_0. It is expressed as:

$$G_e/G_0 = -\frac{1}{4}\beta_e\frac{\partial|M|^2}{\partial\beta_e}, \tag{6.13}$$

It indicates the degree of energy exchange between the electron beam and the high-frequency field, and the expression is shown in (6.13) [44], where G_0 is the DC electron conductance, $G_0 = I_0 / U_0$ (U_0 and I_0 are the voltage and current of the DC electron beam, respectively).

If $G_e/G_0 > 0$, the electron beam absorbs energy from the high-frequency field. If $G_e/G_0 < 0$, the electron beam gives energy to the high-frequency field. Therefore, for EIOs, when the electron beam voltage is given, it is possible to make the operating mode $G_e/G_0 < 0$ and the non-operating mode $G_e/G_0 > 0$. For EIAs, where the electron beam absorbs energy from the high-frequency field in the input cavity, the operating mode and its adjacent competing modes need to satisfy $G_e/G_0 > 0$. The output cavity acts as a structure for the modulated electron beam to transfer energy, and then the operating mode of the output cavity needs to satisfy $G_e/G_0 < 0$.

6.1.2 Electromagnetic Properties of Resonant Cavity Loaded with Single-Bridge CeSRRs

To study the effect of CeSRRs on the fundamental mode (operating mode) of a circular waveguide resonant cavity, a hollow circular waveguide resonant cavity is constructed, and the electric field distributions of its fundamental mode are simulated in the CST Eigenmode Solver, as shown in Fig. 6.3. It can be seen that the electric field of this mode has only a longitudinal component, which is concentrated near the central axis of the resonant cavity, and there is no directional change along the longitudinal direction for the TM_{010} mode. Due to the study of the 714 MHz metamaterial-inspired klystron, *f* of the TM_{010} mode is adjusted to 714 MHz for comparison.

When the drift tubes are loaded on both sides of the short-circuited surface of the hollow circular waveguide resonant cavity to construct the electron beam channel and the transit gap, the electric field will be concentrated at the transit gap. The electric

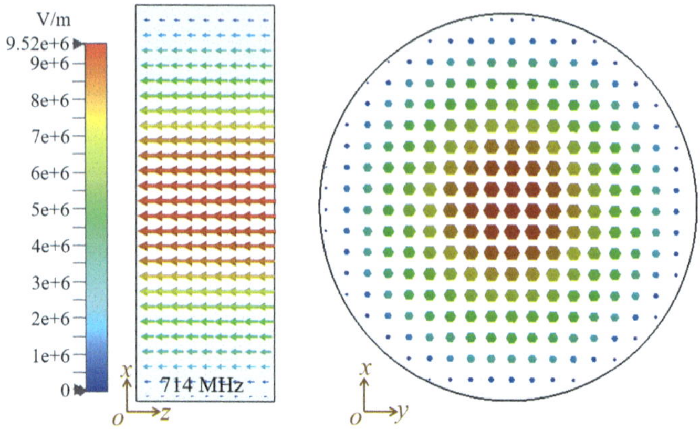

Fig. 6.3 For the fundamental mode, the electric field distributions of hollow circular waveguide resonant cavity in *xoz* plane (left) and *xoy* plane (right)

field will be cut off inside the drift tubes, which is also the reentrant resonant cavity structure used in conventional klystrons, as shown in Fig. 6.4. This resonant cavity facilitates the electron beam to be strongly velocity modulated by the longitudinal electric field at the transit gap. In addition, it can be seen from the electric field distributions of the fundamental mode that the loading of the drift tubes causes the electric field in its vicinity to be distributed perpendicular to the drift tubes, i.e., a radial component of the electric field appears, thus the fundamental mode is quasi-TM_{010} mode. At the same time, the loading of the drift tubes also increases the equivalent capacitance inside the resonant cavity, so that f of the fundamental mode decreases. The final simulated result of f is 621 MHz.

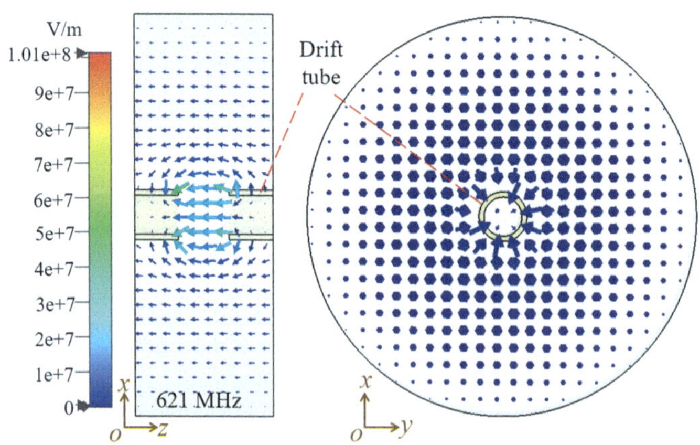

Fig. 6.4 The electric field distributions of the fundamental mode of the resonant cavity with drift tubes in *xoz* plane (left) and *xoy* plane (right)

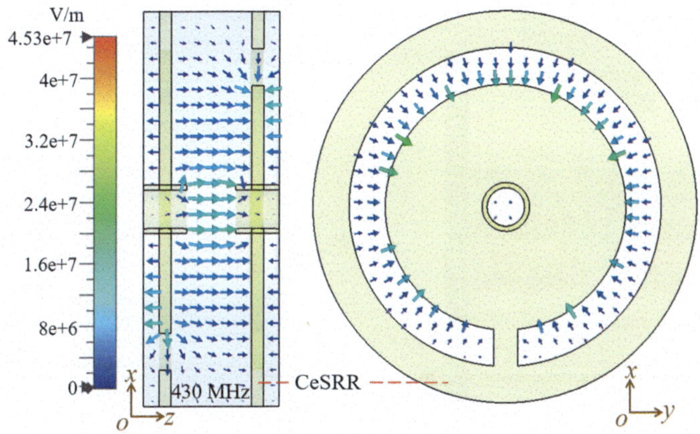

Fig. 6.5 The electric field distributions of the fundamental mode of the resonant cavity with drift tubes and CeSRRs in *xoz* plane (left) and *xoy* plane (right), respectively

Based on the resonant cavity structure shown in Fig. 6.4, a single-bridge CeSRR is loaded at the longitudinal center of each drift tube, and its electric field distributions of the fundamental mode are simulated by the CST Eigenmode Solver, as shown in Fig. 6.5. In this model, the longitudinal electric field of the fundamental mode is still more concentrated at the transit gap. However, the local strong resonance characteristic of the CeSRR affects the electric field distributions in the resonant cavity. In particular, the radial electric field distributed in the slot-line of the CeSRR increases the transverse electric field components in the resonant cavity, and the longitudinal electric field component in the resonant cavity is also more distributed near the CeSRR due to the strong resonance characteristics of the CeSRR. As a result, the fundamental mode is still quasi-TM_{010} mode, and its f is greatly reduced to 430 MHz due to the enhancement of the equivalent capacitance and inductance of the resonant cavity loaded with the CeSRRs. So, the metamaterial-inspired resonant cavity loaded with the single-bridge CeSRR can excite the fundamental mode at 430 MHz under the dimensions of the hollow circular waveguide resonant cavity with the fundamental frequency of 714 MHz. Thus, the miniaturization of the resonant cavity is achieved.

6.1.3 Experimental Study of Resonant Cavities Loaded with Single-Bridge CeSRRs

Based on the simulated results of the electromagnetic characteristics of the metamaterial-inspired resonant cavity described in 6.1.2, the cold test experiment of 714 MHz metamaterial-inspired input cavity is carried out. The reflection coefficient S_{11} can be tested by a vector network analyzer for the input and output cavities. First, the vector

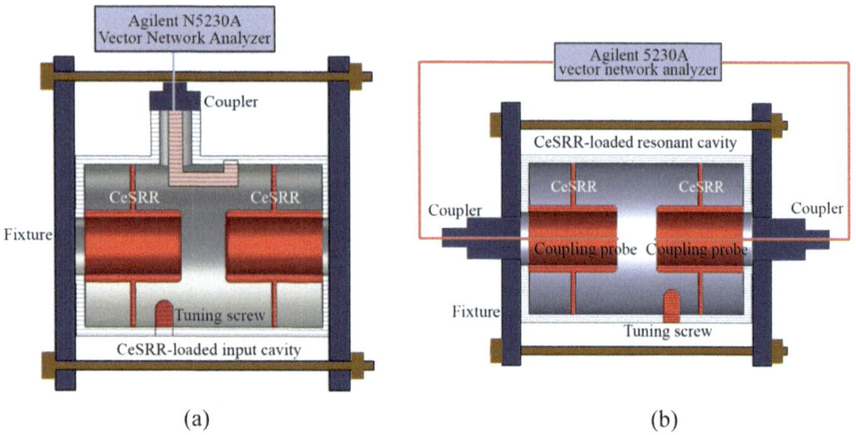

Fig. 6.6 Experimental setup of the CeSRR-loaded resonant cavities. **a** Input cavity. **b** Idler cavity

network analyzer needs to be calibrated before testing. Second, a standard coaxial line is connected to the standard N-type connector of the metamaterial-inspired input cavity, and the coupling probe is used to excite the electric field at the transit gap through the coaxial input coupler. The S_{11} curve displayed on the vector network analyzer is converted into a group delay curve after the post-processing. Finally, the test data are exported from the vector network analyzer for further treatment. The measurement principle is shown in Fig. 6.6a, where the group delay τ is related to the voltage reflection coefficient Γ as [45]:

$$\tau = -\frac{d\Phi(\Gamma)}{d\omega}. \tag{6.14}$$

If the input cavity is regarded as a single-port network, $S_{11} = \Gamma$ [46], the mathematical relation between S_{11} and τ is constructed. Then the Q_e of the input cavity can be calculated:

$$Q_e = \omega\tau/4. \tag{6.15}$$

For the idler cavity loaded with two double-bridge CeSRRs, since it cannot be directly coupled to the external circuit like the input cavity, it is necessary to measure the resonant frequency f and the frequencies f_1 and f_2 of the corresponding half-power points by using both the source and receiver probes to obtain the Q_L [24], which is illustrated in the schematic diagram of Fig. 6.6b. When Q_e is much smaller than the intrinsic quality factor Q_0, Q_e can be approximated as Q_L according to the relation $1/Q_L = 1/Q_0 + 1/Q_e$, i.e., $Q_e \approx Q_L = f/(f_1 - f_2)$. In addition, Q_e can be accurately measured using the phasor method [47].

The experimental test is carried out by taking the metamaterial-inspired input cavity as an example, the components, the assembly sample, and the test platform are shown in Fig. 6.7. The components consist of a coaxial input coupler, two single-bridge CeSRRs, a middle segment of the cavity, and two side segments of the cavity. The material used is oxygen-free copper TU1 with 99.97% purity. The thickness of the resonant cavity wall is 15 mm, and the coaxial inner conductor is connected to a standard N-type connector through the input pedestal.

The simulation and test results of the single-bridge CeSRR-loaded input cavity are listed in Table 6.1. It can be seen that the simulated results are consistent with the test results. This confirms the miniaturization property of the metamaterial-inspired resonant cavity and lays the foundation for the following study of coherently enhanced TR in metamaterials.

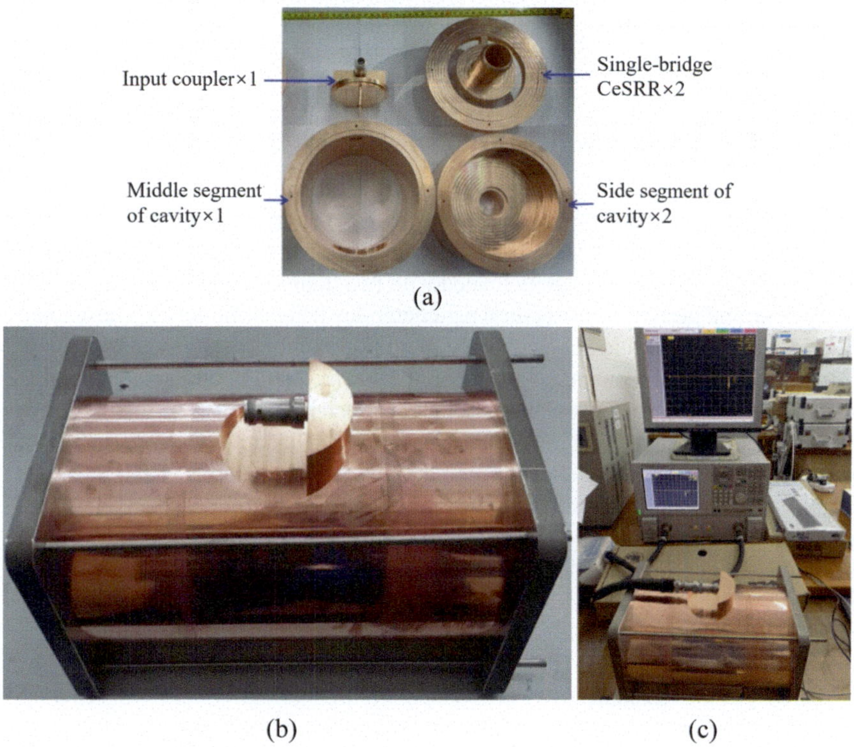

Fig. 6.7 Components, sample, and test platform of the P-band metamaterial-inspired input cavity. **a** The components. **b** The sample. **c** The test platform

Table 6.1 Simulation and test results of 714 MHz metamaterial-inspired input cavity

	f (MHz)	Q_e
Simulation	714	240.8
Test	714.24	262

6.2 Transition Radiation with High Coherence and Strong Radiation

Chapter 5 introduces the RCRO, which has the advantages of miniaturization and high electron efficiency. The helix oscillator based on the CR mechanism has broad bandwidth characteristics, while the metamaterial oscillator based on strong resonance characteristics has narrow bandwidth characteristics, which determines that it has a wide and narrow tuning band. It is well known that as a high-power amplifier, conventional klystrons are essentially narrow-bandwidth devices. The inherent advantages of metamaterials make them very suitable for klystrons, so this section investigates theoretically and experimentally the TR in metamaterial-inspired klystrons for the development of new VEDs.

6.2.1 Theoretical Analysis of Transition Radiation

To explore the advantages of TR based on metamaterial, a compact MW-level klystron with two double-bridge CeSRRs loaded in the resonant cavities is proposed [48]. A single-gap resonant cavity loaded with two double-bridge CeSRRs is shown in Fig. 6.8a, and the relevant structural parameters are listed in Table 6.2. The diameter of the constructed double-bridge CeSRR-based resonant cavity is smaller than that of the conventional resonant cavity due to the subwavelength characteristics of the double-bridge CeSRR. The diameter of the metamaterial-inspired resonant cavity is 36 mm, which is only 0.75 of the traditional reentrant resonant cavity with a diameter of 48 mm [49]. The reduction of the diameter of the metamaterial-inspired resonant cavity exemplifies the potential of miniaturization of the proposed metamaterial-inspired klystron. More importantly, its fundamental mode is mainly determined by the double-bridge CeSRRs, and the electric field is concentrated in a specific region of the resonant cavity, which is very favorable for the modulation of the electron beam [50].

Through the simulation of the CST Eigenmode Solver [51], the longitudinal electric field E_z of the resonant cavity is obtained. Figure 6.8b shows the E_z distributions of fundamental mode TM_{010} and high-order mode TM_{011}. According to (6.10) [44], the R/Q of the resonant cavity loaded with two double-bridge CeSRRs is calculated according to the E_z amplitude of the two modes, as shown in Fig. 6.8c. The R/Q of the fundamental mode is 115.7 Ω at 2.852 GHz, while the R/Q of the high-order mode is 2.1 Ω at 3.508 GHz. The lower R/Q of the high-order mode makes the high-order mode in the resonant cavity less excited by the electron beam, which indicates that the fundamental mode can interact with the electron beam more effectively than the high-order mode, thus ensuring the coherence of the TR. Q_e of the input cavity and output cavity, Q_L of the idler cavity, the diameter D and the length l of the resonant cavity, f, M, and R/Q of the fundamental mode of the resonant cavity are listed in Table 6.3. It can be seen that D is about 0.34 of the wavelength in free space, so

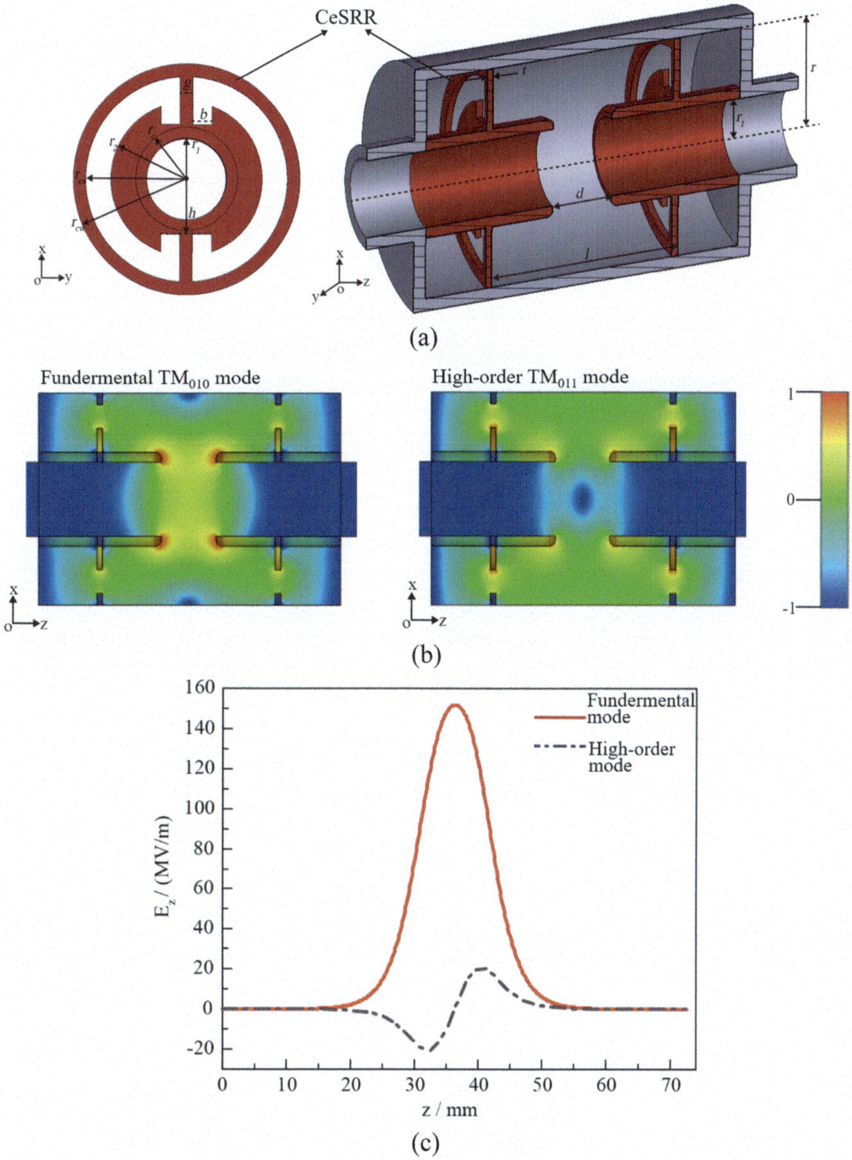

Fig. 6.8 The double-bridge CeSRR, the resonant cavity loaded with double-bridge CeSRRs, and its electromagnetic characteristics. **a** The double-bridge CeSRR and the resonant cavity loaded with double-bridge CeSRRs. **b** The E_z distributions of fundamental and high-order modes in the *xoz* plane. **c** The E_z distributions of fundamental and high-order modes at the central axis of the resonant cavity

Table 6.2 Size parameters of the double-bridge CeSRR and the resonant cavity loaded with double-bridge CeSRRs (unit: mm)

r_1	r_c	r_2	h
6.3	8	12	9
r_{ex}	r_{cw}	r	g
16	18	18	1.86
b	l	d	t
3.07	31	9.4	1

the klystron with the double-bridge CeSRR-loaded resonant cavities has obvious miniaturization characteristic.

To avoid the electrical breakdown in the output cavity loaded with two CeSRRs, the CeSRRs and the drift-tube cones are tailored. The improved CeSRR unit cell, the simulated contours of E_z, and the transverse electric field contours for the output cavity with a waveguide-based output coupler are shown in Fig. 6.9.

The metamaterial-inspired klystron can generate TR. The metamaterial-inspired klystron proposed here consists of an electron gun, resonant cavities loaded with two double-bridge CeSRRs (five in total, each connected by a drift tube), a coaxial input coupler, a waveguide output coupler, a magnetic focusing configuration, and a collector, as shown in Fig. 6.10. The physical mechanism of the TR is presented

Table 6.3 The sizes and electromagnetic parameters of the cavities

	Sizes and electromagnetic parameters						
	D (mm)	l (mm)	f_r (GHz)	M	R/Q (Ω)	Q_e	Q_L
Input cavity	36	32	2.856	0.88	107	153	–
Idler cavity 1	36	31	2.868	0.87	117	–	3221
Idler cavity 2	36	30	2.962	0.86	116	–	3310
Idler cavity 3	36	30	3.021	0.86	118	–	3314
Output cavity	31	20.7	2.856	0.79	120	11.2	–

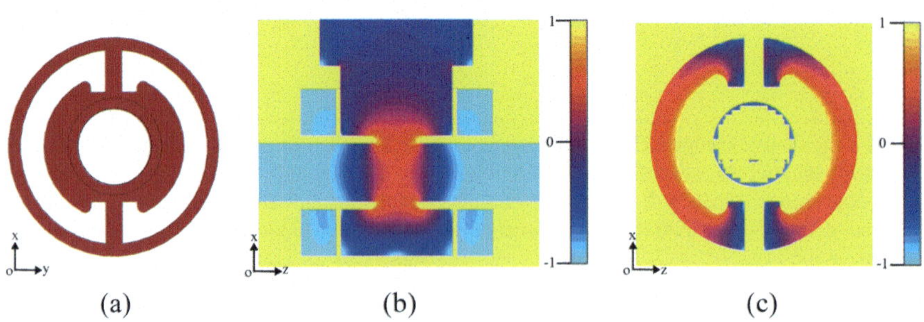

(a) (b) (c)

Fig. 6.9 A tailored CeSRR and simulated E_z contours for the output cavity with a waveguide-based output coupler. **a** The optimized CeSRR. **b** E_z contours in the output cavity. **c** Transverse electric field contours in the CeSRR

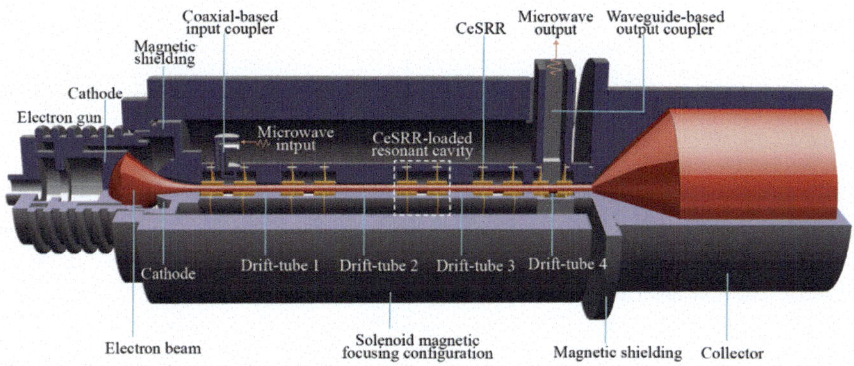

Fig. 6.10 The structure of the metamaterial-inspired klystron

here: when a uniform pencil beam emitted from the electron gun passes through the transit gap of the input cavity, the electron beam undergoes velocity modulation through the E_z interactions with the fundamental mode. The electron beam is density-modulated as it passes through drift tubes and forms electron bunching. Stronger velocity modulation occurs when the electron beam passes through transit gaps in idler cavities. The strongly bunched electrons are decelerated at the transit gap of the output cavity, which converts most of the kinetic energy of the electrons into TR, resulting in the amplification of the TR signal.

Based on the above simulation of the electromagnetic properties of the resonant cavity, the beam-wave interaction of the metamaterial-inspired klystron is simulated using the CST Particle Studio [51]. When relativistic effects are neglected, the beam voltage U_0 is given by [21]:

$$U_0 = 5.11 \times 10^5 \times \left(\frac{1}{\sqrt{1 - (\omega/\beta_e c_0)^2}} - 1 \right), \tag{6.16}$$

where $\omega = 2\pi f$. The CST simulation model of the metamaterial-inspired interaction structure is shown in Fig. 6.11. The radius and channel of the electron beam are 4.1 mm and 6.3 mm, respectively. Setting the potential of the anode as the reference potential, under the condition of − 120 kV/80 A of the electron beam, the PIC simulated results of the TR are obtained as shown in Fig. 6.12. When the power and frequency of the input signal are 24.5 W and 2.852 GHz, respectively, the saturated output power of the TR is 5.94 MW, the saturated gain is 53.85 dB, and the electron efficiency is 61.9%, as shown in Fig. 6.12a. In Fig. 6.12b, the output signal is post-processed using the normalization method with the formula:

$$\mathrm{X_n} = \frac{X}{X_{\max}}, \tag{6.17}$$

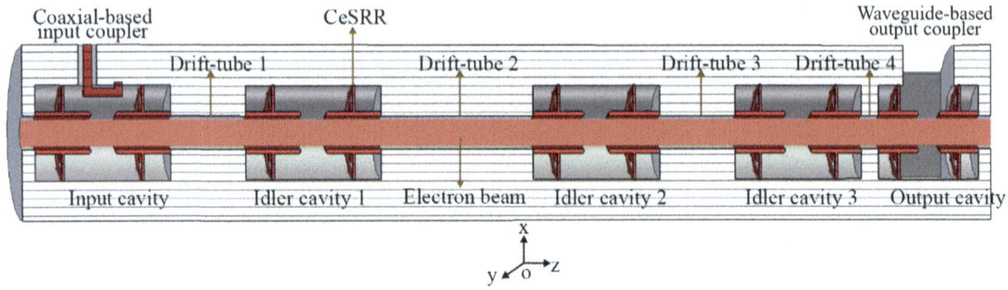

Fig. 6.11 The simulation model of the metamaterial-inspired interaction structure

where X_n is the normalized data, X is the original data, and X_{max} is the maximum value of X. The curve of output power versus frequency is shown in Fig. 6.12c. It can be seen that the 3 dB bandwidth is more than 32 MHz, which can meet the requirements of future applications such as accelerators [52]. These findings indicate that the TR is highly coherent and strongly electromagnetic. The gain and electron efficiency are varied by adjusting the input power with other parameters unchanged, as shown in Fig. 6.12d. It can be seen that as the input power increases, the electron efficiency increases and the gain decreases.

For the MW-level metamaterial-inspired TR source, it is necessary to analyze the electric field distributions of the whole interaction structure after the output power reaches the steady state due to its miniaturization to assess the breakdown risk. In particular, the electric field distribution of the output cavity is significantly higher than that of other resonant cavities, as shown in Fig. 6.13. By setting the electric field monitor from 98 to 100 ns, the maximum electric field is 54.8 MV/m, which is located at the position of the nose cone of the output cavity (4.833 mm, – 4.000 mm, 339.000 mm) as shown in Fig. 6.13. There is no breakdown threshold under the same conditions for comparison, so the breakdown threshold of 36 MV/m [53] for the vacuum environment at the S-band is chosen as a reference. The comparison indicates that the breakdown threshold is lower than the maximum electric field in the interaction structure, suggesting a potential risk of breakdown for the device. However, in the TR experiment, the breakdown threshold is not only related to the operating frequency of the device, but also to the vacuum degree, conductor material, surface roughness, pulse width, and other factors [54]. In this TR experiment, the selected solid-state power amplifier can be adjusted to operate with a pulse width of 3~14 μs. After several experiments, it has been found that no breakdown occurred in the metamaterial-inspired klystron when the pulse width is ~ 10 μs.

For the metamaterial-inspired TR, in order to study the generation and transport of the electron beam, an electron gun is designed to form the desired pencil beam, as shown in Fig. 6.14. A barium tungsten cathode is used for the electron gun, which operates at a temperature of (1050 ± 50) °C. The heat shield structure of the electron gun is designed and optimized by using ANSYS [55], and the temperature distribution simulated result is shown in Fig. 6.14a. When the heating power of the cathode heater is 300 W, the cathode temperature reaches 1022.9 °C, which is in line with the operating conditions of the barium tungstate cathode. The particle

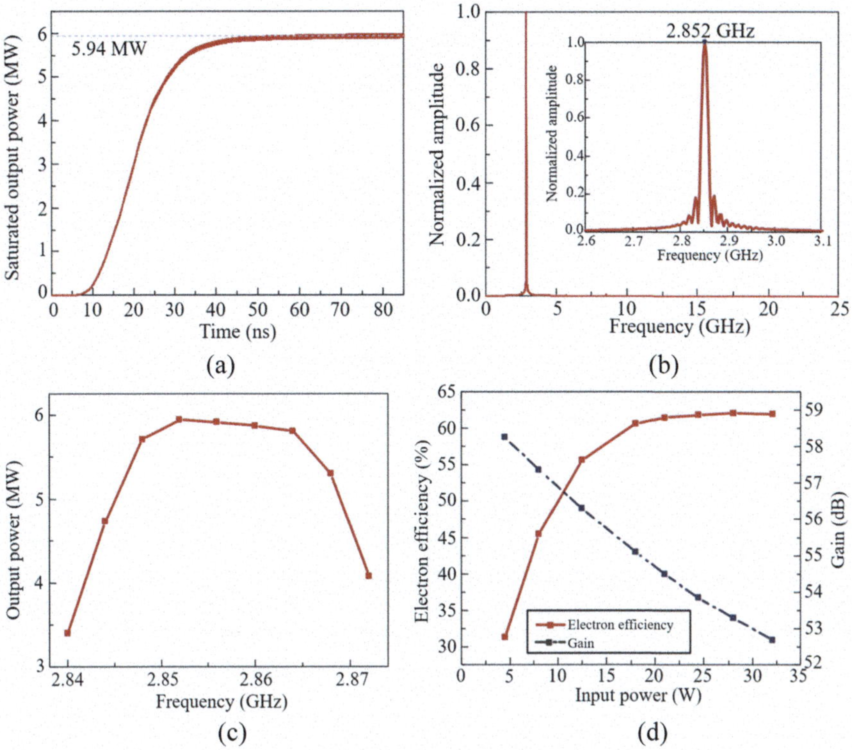

Fig. 6.12 The simulated results of the TR. **a** Saturated output power versus time. **b** Spectrum of the output signal. **c** The output power versus frequency when the input power is 24.5 W. **d** The electron efficiency and gain versus input power at 2.852 GHz

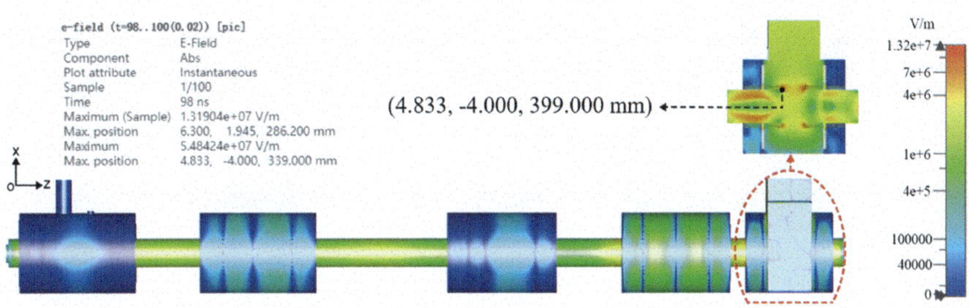

Fig. 6.13 The electric field distribution of interaction structure of the metamaterial-inspired klystron

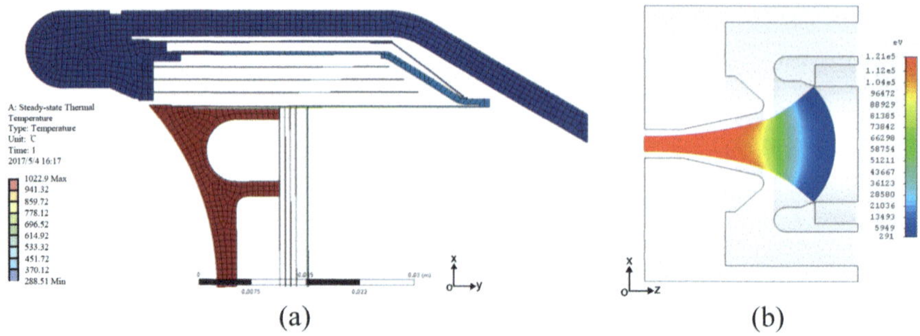

Fig. 6.14 Simulated results of the pencil electron gun. **a** Heat shield structure of the electron gun and temperature distributions simulated by ANSYS. **b** Particle trajectories of the electron beam in the electron gun simulated by CST Particle Tracking Solver

trajectories in the electron gun are simulated by the CST Particle Tracking Solver [51], as shown in Fig. 6.14b. The simulated results show that the conduction coefficient of the electron gun is about 1.9 μP when the electron beam voltage is approximately − 120 kV and the electron beam current is approximately 80 A. The beam waist radius at the entrance of the input cavity loaded with two double-bridge CeSRRs is 4.1 mm, which corresponds to a filling factor of 0.65.

In addition, a solenoidal magnetic focusing configuration is employed to ensure the transport of electron beam. The focusing theory preliminarily predicts that the longitudinal magnetic flux density B_z in the center of the focusing coil is not less than 0.1874 T [24]. After consideration of the magnetic field distributions within the transition region of the electron gun, the output cavity, and the collector, the magnetic flux density in the center of the solenoidal magnetic focusing configuration is simulated using the CST M-Static Solver [51], as shown in Fig. 6.15. The simulated results show that the effective focusing of electron beam can be realized when the peak value of B_z is 0.24 T, thus facilitating the beam-wave interaction.

6.2.2 Experimental Verification of Transition Radiation

Based on the above theoretical studies, the experiment of metamaterial-inspired TR with high coherence and strong radiation was carried out. The electron gun, collector, magnetic focusing configuration, and metamaterial-inspired interaction structure were processed, assembled, cold tested, and the TR tested. To experimentally characterize the electromagnetic characteristics of a resonant cavity loaded with two double-bridge CeSRRs, the components were fabricated, as shown in Fig. 6.16a. The output cavity is connected to a flat waveguide through a rectangular coupler, and then connected to a standard rectangular waveguide through a microwave window. A water-flowing system is designed on the H-plane to cool the output waveguide. The input cavity, idler cavities, and output cavity loaded with two double-bridge

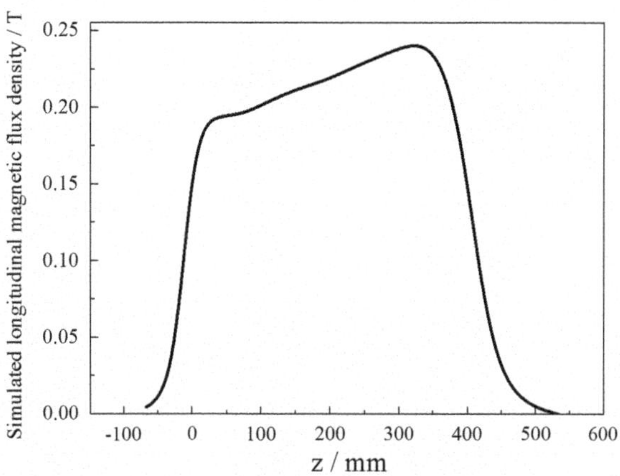

Fig. 6.15 Simulated results of magnetic flux density in the center of the solenoid magnetic focusing configuration

CeSRRs were assembled to form a metamaterial-inspired high-frequency structure, as shown in Fig. 6.16b. The test principle is detailed in 6.1.3. As can be seen from Fig. 6.16c, the measured f and corresponding τ are slightly lower than the simulated result, and the measured $Q_e = 132$ is also lower than the simulated result (153). The measurement results and the simulated results of the input cavity, idler cavity 1 (the measured results shown in Fig. 6.16d), 2, and 3, and the output cavity are summarized in Table 6.4. Through comparative analysis, it is found that the test results are in good agreement with the simulated results.

In addition, tuning screws are provided in the input cavity, idler cavity 1, and idler cavity 2 loaded with two double-bridge CeSRRs for tuning f. In order to increase the power capacity of the idler cavity 3 and the output cavity loaded with two double-bridge CeSRRs, the tuning screws are not loaded in these cavities. The surface roughness of these resonant cavities is strictly controlled during machining to minimize electromagnetic wave loss within the resonant cavities, thereby reducing heat dissipation. When manufacturing the metamaterial-inspired high-frequency structure, the material of the double-bridge CeSRR and the cavity wall is 99.97% oxygen-free copper, and the magnetic shielding material is pure iron. In the experiment, the vector network analyzer and the standard coaxial line are connected to the coaxial input coupler, and since the output power is greatly attenuated to the mW-level by the loaded attenuator, the waveguide output coupler is converted to the coaxial line output for easy measurement.

Based on the aforementioned cold test, the electron gun, solenoid magnetic focusing configuration, and collector were machined, assembled, and tested respectively. The electron gun sample is encapsulated in a glass container to ensure that the tungsten barium cathode operates properly at the required temperature for sufficient electron emission, as shown in Fig. 6.17a. The relationship between the cathode temperature and heater power is obtained through the test, as shown in Fig. 6.17b.

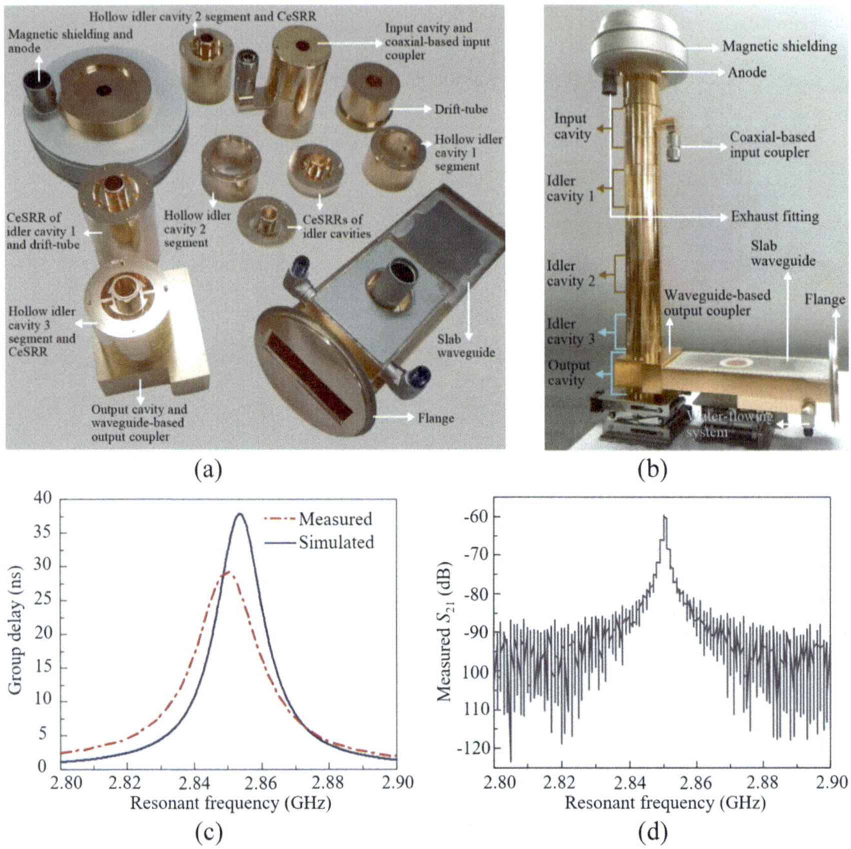

Fig. 6.16 The components and the prototype, and electromagnetic characteristics of the input cavity and idler cavity 1 loaded with double-bridge CeSRRs. **a** Fabricated components. **b** The prototype of metamaterial-inspired high-frequency structure. **c** Comparison between the simulated and measured group delays versus resonant frequency for the input cavity. **d** Measured S_{21} of the idler cavity 1

Table 6.4 Comparison of simulation value and test value of each resonant cavity

f (MHz)/Q		Input cavity	Idler cavity 1	Idler cavity 2	Idler cavity 3	Output cavity
Simulation	f (MHz)	2856	2868	2962.3	3021.2	2856.8
	Q_L/Q_{ext}	153	3954	4073	4109	11.2
Test	f (MHz)	2851	2870	2962	3025	2853
	Q_L/Q_{ext}	132	3248	3368	3421	30

From the test results, when the heater voltage and heater current are set at 6.1 V and 25 A, respectively, the cathode temperature can reach 1030 °C. At the same time, the waveforms of the beam voltage and beam current are observed, as shown in Fig. 6.17c. It can be seen that the beam voltage and beam current meet the nominal values of − 120 kV and 80 A, respectively.

The height of the solenoid magnetic focusing configuration is 458 mm, and its inner and outer diameters are 118 mm and 406 mm, respectively, as shown in Fig. 6.18a. The longitudinal magnetic flux density is measured using a Gaussian meter, and the measured result is shown in Fig. 6.18b. By adjusting the current of the coil, the B_z of the high-frequency structure region can be varied between 0.18 and 0.24 T, which is used to improve the focusing characteristics of the electron beam.

Based on the experiments conducted on resonant cavities loaded with double-bridge CeSRRs and the characteristics of electron beam generation and transmission, the experimental study of the metamaterial-inspired TR is carried out. Therefore, the test scheme and the experimental platform were constructed, as shown in Fig. 6.19. The main measurement data include input power, output power, and output spectrum.

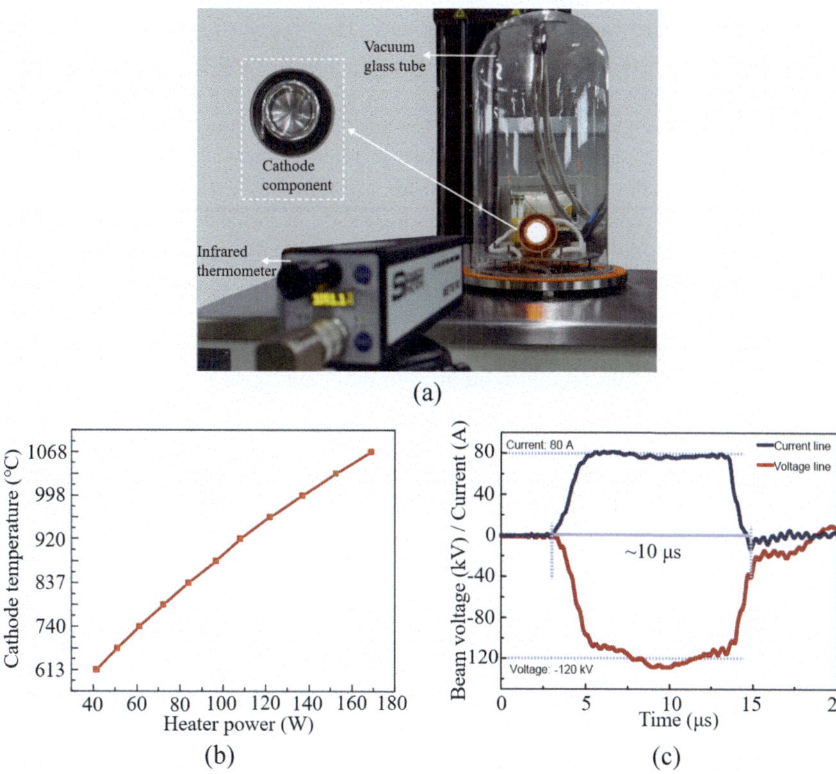

Fig. 6.17 The experimental platform and measured results of the pencil electron gun. **a** Experimental platform. **b** Cathode temperature versus heater power. **c** Measured beam voltage and beam current of the electron beam versus time

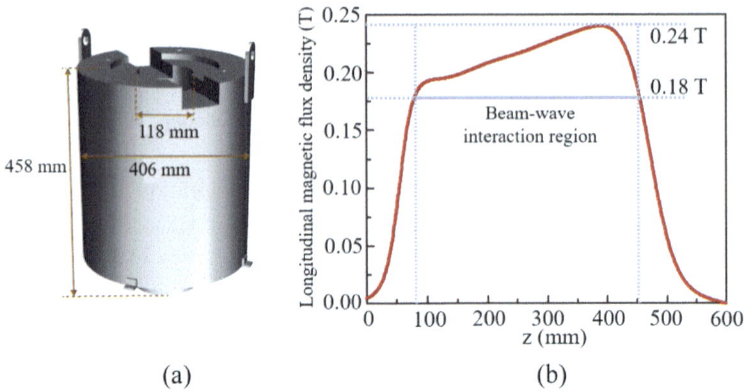

Fig. 6.18 The design model and the measured results of the solenoid magnetic focusing configuration. **a** Design model of the solenoid magnetic focusing configuration. **b** Measured longitudinal magnetic flux density along the z-axis

The experimental platform includes a power supply system, microwave input component, and microwave output component. The power supply system mainly includes the pulse switching power supply, pulse transformer, coil power supply, and cathode heater power supply. The microwave input component is composed of the signal generator, isolator, solid-state power amplifier, coaxial-based directional coupler, and power meter, which is used to provide microwave input signal and stimulate the metamaterial-inspired TR in klystron. In this experiment, the signal generator operates in pulse mode, synchronizing the input signal well with the voltage pulse of the power modulator. The input power supplied to the metamaterial-inspired klystron is measured by connecting a coaxial-based directional coupler to a power meter. Microwave output components include a waveguide-based directional coupler, two attenuators, a detector, an oscilloscope, a titanium pump, and a high-power water load. The role of the titanium pump is to maintain the high vacuum degree of the metamaterial-inspired klystron during operation and storage. The oscilloscope connected to the detector is used to measure the waveform of the output signal, while the output spectrum of the signal is obtained through the fast Fourier transform. The remaining microwave energy is absorbed by the high-power water load. In addition, to ensure the normal operation of the experiment, water-flowing systems 1 and 2 are used to cool the collector and the plate waveguide respectively.

In the TR experiment, the attenuation of the input link is 6.66 dB, and it comes from the loss of an 8-m coaxial line. Since the output power of the solid-state power amplifier can vary in the range of 30~110 W, the actual input power supplied to the metamaterial-inspired klystron varies within the range of 6.47~23.74 W. The total output link attenuation of 87.95 dB comes from two attenuators with attenuation of 19.88 dB and 19.87 dB, a waveguide output coupler with attenuation of 45.7 dB, and an 8-m coaxial line with attenuation of 2.5 dB. The gas leakage test of the assembled and welded metamaterial-inspired klystron shows that the vacuum degree is ~ 1.0×10^{-7} Pa, which meets the experimental requirements.

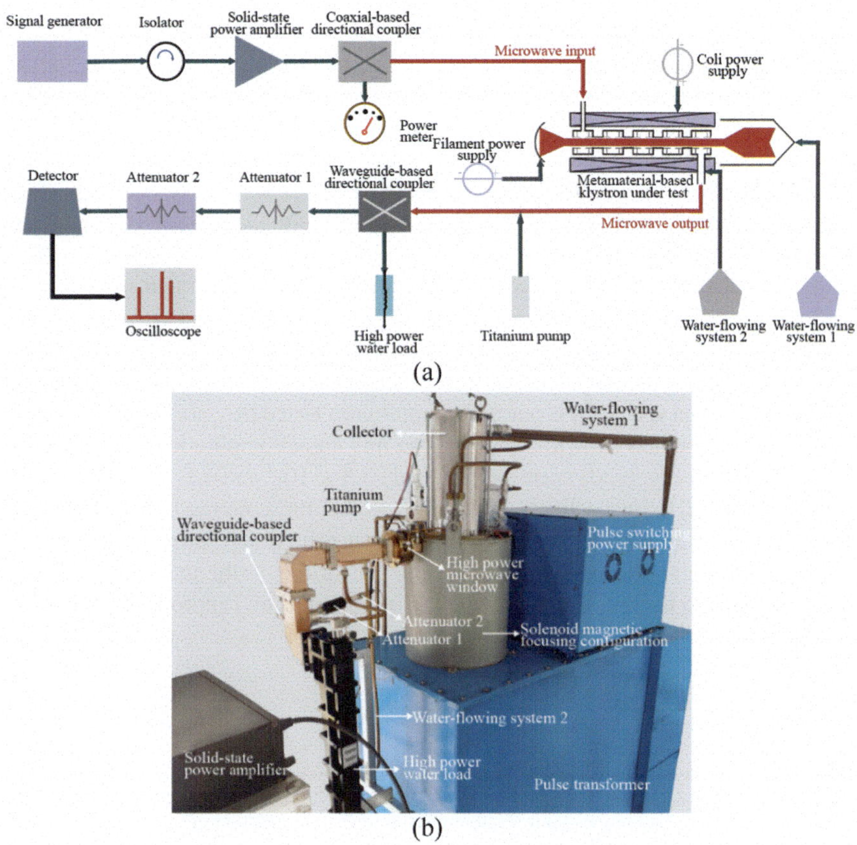

Fig. 6.19 The measurement block diagram and experimental platform of the metamaterial-inspired TR. **a** Measurement block diagram. **b** Experimental platform

The electron gun used in the TR experiment generates a pencil beam with a pulse width of ~ 10 μs, a repetition rate of 5 Hz, and a beam current of 80 A at − 120 kV. As previously mentioned, the input power provided by a solid-state power amplifier with a pulse width of 20 μs is adjustable from 6.47 to 23.74 W.

In physics, coherence is an intrinsic and extrinsic property of a wave that enables it to undergo stationary interference, a necessary condition for a wave to exhibit significant interference phenomena. If the frequency, the phase difference, and the vibration direction of two sine waves are consistent, the two waves are coherent. When interference occurs, two waves superimposed together may generate a wave with a larger amplitude than either wave (interference enhancement), or subtract from each other to generate a minimum wave (interference elimination), depending on their relative phase [56]. However, the frequency and the phase difference between non-coherent waves are different, and the interference pattern generated is unstable and cannot be clearly observed. Coherence can be classified into temporal coherence and spatial coherence. Temporal coherence describes the correlation between waves observed at different times and is related to the bandwidth of the waves. Spatial coherence describes the correlation between waves at different locations in space [57] and is related to the finite size of the wave source.

In the metamaterial-inspired TR, when the longitudinal size of the bunching structure is shorter than the wavelength of the electromagnetic wave [58], each free electron in the electron beam behaves as a coherent source with the same electromagnetic wave frequency, constant phase difference, and the same vibration direction. As a result, the coherent-enhanced electromagnetic radiation is generated upon superposition. At the same time, the amplitude of the tested TR signal depends on the frequency and power of the input signal. For the three typical operating frequencies 2.852, 2.856, and 2.858 GHz shown in Fig. 6.20a, b, and c, it can be seen that the tested output signal spectrum is pure. The amplitudes tested here are also normalized using a similar method shown in (6.17). The tiny signal amplitudes at 3.125 GHz and 4.688 GHz in Fig. 6.20a correspond to only 3.3×10^{-4} and 4.7×10^{-4} times the power at 2.852 GHz, respectively, which means that these higher order modes are negligible. It can be seen from Fig. 6.20a and b that the tested spectrum results are in good agreement with the simulated results. According to the theory in Ref. [3], for incoherent TR, the spectrum is frequency-independent and approximately flat. In summary, what is observed here is coherent TR.

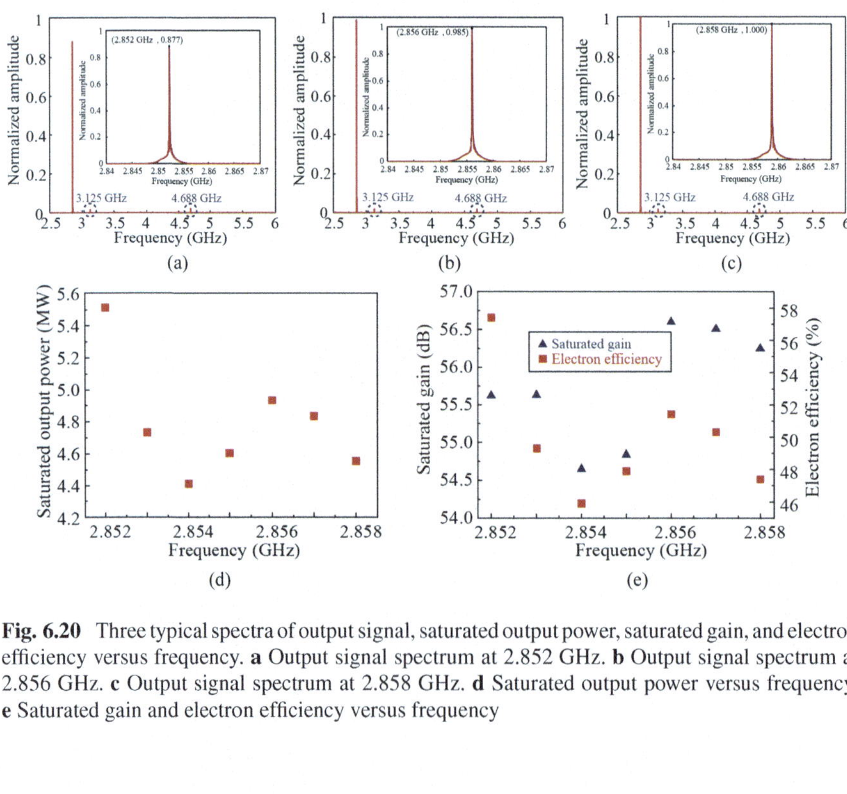

Fig. 6.20 Three typical spectra of output signal, saturated output power, saturated gain, and electron efficiency versus frequency. **a** Output signal spectrum at 2.852 GHz. **b** Output signal spectrum at 2.856 GHz. **c** Output signal spectrum at 2.858 GHz. **d** Saturated output power versus frequency. **e** Saturated gain and electron efficiency versus frequency

In addition, the saturated gain and electron efficiency of the metamaterial-inspired klystron can be obtained by examining the relationship between the tested saturated output power and the frequency of the input signal, as shown in Fig. 6.20d and e respectively. As can be seen from Fig. 6.20d and e, the tested saturated output power is greater than 4.41 MW from 2.852 to 2.858 GHz, the saturated gain is more than 54.7 dB, and the electron efficiency is more than 45.9%. Especially at 2.852 GHz, the tested saturated output power is 5.51 MW, the saturated gain is 55.6 dB, and the electron efficiency is 57.4%, which closely approximates the corresponding analog values of 5.94 MW, 53.85 dB, and 61.9%. As predicted by theory, the intensity of coherent TR is several orders of magnitude higher than that of incoherent TR, because the intensity of coherent radiation is quadratic to the total charge, rather than linear as incoherent radiation [59]. Therefore, the theory reasonably explains the experimental results, that is, the coherent TR with MW-level radiation intensity is observed here.

The metamaterial-inspired klystron holds significant promise for applications in SNS, Tokamak device used in ITER, positive and negative ion collider, cancer treatment equipment, medical equipment for disinfection, food preservation, industrial irradiation, material and device modification, and customs container detection device [60].

6.3 P-Band Metamaterial-Inspired Klystron

In the field of particle accelerators, klystrons are widely used as microwave drive sources for accelerators because of their higher efficiency compared to solid-state amplifiers, more stable operation and lower phase noise than induction output tubes. In order to meet the needs of electron–positron colliders, industrial irradiation, and medical linear accelerators, it is urgent to develop various types klystron with miniaturization and high-power. For proton linac and accelerator storage rings in SNS [26], a large number of P-band (0.23~1 GHz) high-power klystrons are required to accelerate the particles to the required high energy level [61, 62]. For example, the P-band long pulse klystron is the important device to drive the proton linac in the SNS, the pulse power is in the range of hundreds of kW to several MW and operates with pulse width ranging from several ms to tens of ms. The P-band CW klystron is employed in the accelerator storage ring and its output power ranges from hundreds of kW to MW. However, the P-band klystron has a large volume and heavy weight [61]. Considering the European Scattering Neutron Source (ESNS) as an example, its core is composed of a proton linear accelerator that is driven by a variety of high-power microwave devices, including klystron, multi-beam klystron, and multi-beam induction output tube [25]. Among these, the medium-β section of the proton linac is equipped with 36 P-band MW-level klystrons [63]. The size of each klystron is approximately $3.9 \times 1.5 \times 1\ m^3$ and weighs approximately 2.5 tons [61], this results in a length of 76.7 m for the medium-β section [63], and the total weight of the klystron is approximately 90 tons.

At present, one of the ways to solve this problem is to develop the multi-beam klystron, which can reduce the volume of the interaction structure and the weight of the klystron [64]. From the standpoint of engineering, this section offers an alternative solution to the aforementioned challenges. It results in the construction of a metamaterial-inspired interaction structure [65, 66] based on single-bridge CeSRRs [50], leveraging the TR mechanism described in Sect. 6.2 of the metamaterial-inspired klystron. This approach aims to achieve an MW power output for the metamaterial-inspired klystron at 714 MHz. The volume of the P-band high-power klystron is much larger than that of the S-band, so, loading the single-bridge CeSRRs into the P-band klystron can achieve more significant miniaturization characteristics than the S-band, and it has more important engineering value. The study results demonstrate that the radius and length of the interaction structure of the P-band metamaterial-inspired klystron are 1/2 and 2/3 of the conventional interaction structure [67], respectively. For the performance and size of the klystron in the medium-β section of ESS [63], the length of the metamaterial-inspired interaction structure can be reduced by approximately 560 mm. Because the working klystron is perpendicular to the ground, the bottom area of the metamaterial-inspired interaction structure will be reduced by three-quarters, which can greatly save space. In addition, the weight of the metamaterial-inspired interaction structure will be 5/6 less than that of the conventional interaction structure. In summary, P-band metamaterial-inspired klystron can effectively reduce the manufacturing cost and footprint of proton linac, and save operating costs [68, 69].

6.3.1 High-Frequency Characteristics of Metamaterial-Inspired Klystron

A single-gap metamaterial-inspired resonant cavity is constructed based on the single-bridge CeSRRs, which can achieve more significant miniaturization characteristics for the resonant cavity [50]. Considering that the output power of the metamaterial-inspired klystron reaches the MW level, the grooves of the single-bridge CeSRR in Ref. [50] are removed, and the drift tube nose cone in the resonant cavity is chamfered to reduce the potential risk of breakdown. This improved CeSRR is referred to as the single-bridge non-groove CeSRR. The structural dimensions of the single-bridge CeSRR without grooves are $r_1 = 64.69$ mm, $r_2 = 15.8$ mm, $r_3 = 40.63$ mm, $r_4 = 52.63$ mm, and $h = 5.3$ mm, as shown in Fig. 6.21a. The two single-bridge CeSRRs are rotated 180° relative to each other and loaded in a circular resonant cavity to maintain uniform distribution of the fundamental mode E_z in the electron beam channel region, thus facilitating the interaction between the electron beam and electromagnetic waves, as shown in Fig. 6.21b. In addition, the parameters in the resonant cavity are set to $t = 10$ mm, $l_1 = 41.2$ mm, $l_2 = 125$ mm, $l_3 = 208.8$ mm, $l_4 = 15$ mm, $l_5 = 30$ mm.

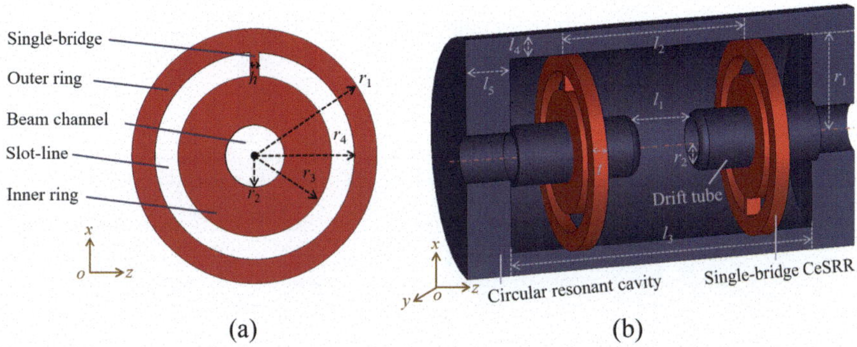

Fig. 6.21 The single-bridge CeSRR without grooves and the metamaterial-inspired resonant cavity loaded with two single-bridge CeSRRs without grooves. **a** The structure with its size parameters of a single-bridge CeSRR without grooves. **b** The metamaterial-inspired resonant cavity loaded with two single-bridge CeSRRs without grooves placed 180° rotated from each other

The CST Eigenmode Solver is used to calculate the quasi-TM_{010} mode of the single-bridge CeSRRs-loaded resonant cavity, corresponding to f of 714 MHz. Using the amplitude of the fundamental mode E_z depicted in Fig. 6.22a, the variation of M^2R/Q with the transition angle θ is calculated by using (6.10) and (6.11), as shown in Fig. 6.22b. The resulting R/Q is 117 Ω, and M^2R/Q corresponds to 92.6 Ω for θ of 3.4 rad, where M is 0.89. Based on the resonant cavity structure loaded with single-bridge CeSRRs, an input cavity (Fig. 6.23a), an idler cavity 1, an idler cavity 2, and an output cavity (Fig. 6.23b) loaded with two single-bridge CeSRRs without grooves are constructed. The electromagnetic characteristics of each resonant cavity are listed in Table 6.5. The experimental testing of the resonant cavities is detailed in Sect. 6.1.3.

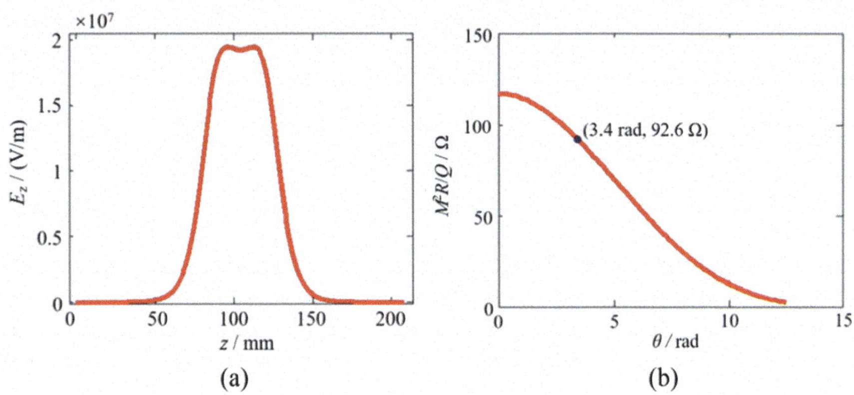

Fig. 6.22 The electromagnetic characteristics of the metamaterial-inspired resonant cavity loaded with two single-bridge CeSRRs without grooves. **a** The E_z amplitude versus z. **b** M^2R/Q versus θ

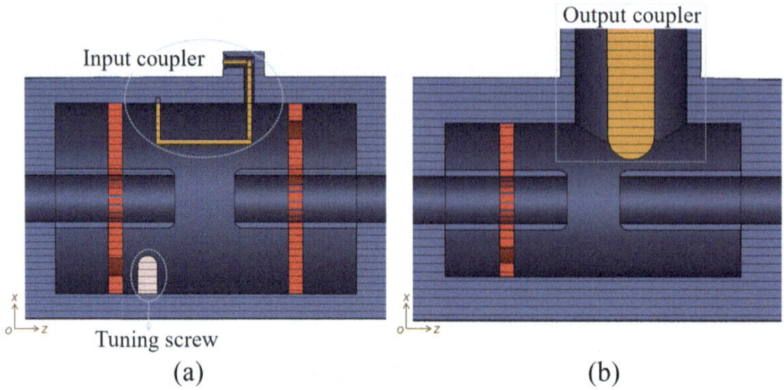

Fig. 6.23 The input and output cavities loaded with single-bridge CeSRRs without grooves. **a** Input cavity. **b** Output cavity

Table 6.5 Simulated results of electromagnetic characteristics of metamaterial-inspired resonant cavities

Parameter	Input cavity	Idler cavity 1	Idler cavity 2	Output cavity
f	714	716	729	714
R/Q	115.2	117	116	104
M	0.8872	0.8874	0.8803	0.9035
Q_L or Q_e	244.8	3151	3025	26

6.3.2 Beam-Wave Interaction of Metamaterial-Inspired Klystron

Based on the simulation and experimental investigation of the electromagnetic characteristics of resonant cavities loaded with single-bridge CeSRRs, a four-cavity metamaterial-inspired klystron operating at 714 MHz is proposed, as shown in Fig. 6.24. It comprises an electron gun, a metamaterial-inspired interaction structure, a magnetic focusing configuration, and a collector. The resonant cavity diameter and longitudinal length of this structure are approximately 0.32 λ and 2.9 λ (λ is the wavelength in free space), respectively. When the beam voltage is -100 kV, the beam current is 40 A, and the axial magnetic flux density is 720 G, the klystron based on the single-bridge CeSRRs is simulated. Figure 6.25 shows that the output power reaches 2.28 MW at 714 MHz, corresponding to an electron efficiency of 57%, a saturated gain of 46.6 dB, and a 1 dB bandwidth of 4.7 MHz. The output signal shown in Fig. 6.25a is of MW magnitude and the spectrum is pure near 714 MHz. According to the realization method of S-band metamaterial-inspired klystron in Sect. 6.2 and considering various engineering application requirements, the complete assembly of a P-band metamaterial-inspired klystron can be envisaged for future development.

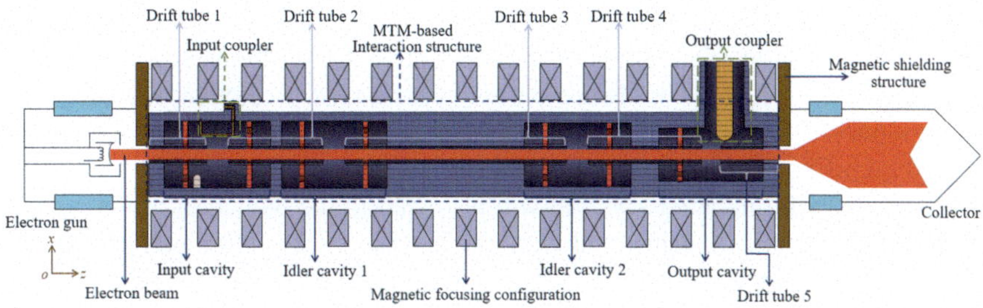

Fig. 6.24 The simulation model of the four-cavity interaction structure of the metamaterial-inspired klystron based on single-bridge CeSRRs without grooves

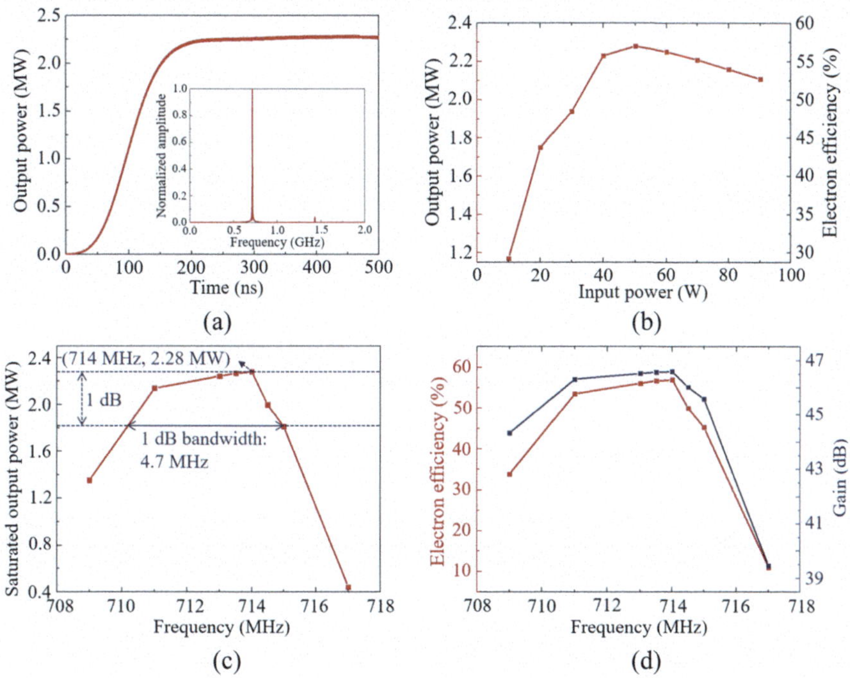

Fig. 6.25 The PIC simulated results of the four-cavity metamaterial-inspired klystron based on the single-bridge CeSRRs without grooves. **a** The output power versus time (inset: the frequency spectrum of the output signal). **b** The output power and electron efficiency versus the input power. **c** The saturated output power versus the frequency of the input signal. **d** The electron efficiency and gain versus the frequency of the input signal

The metamaterial-inspired klystron proposed in this chapter exhibits a smaller size and higher gain than the conventional klystron based on the reentrant resonant cavities of 700 MHz/1 MW [67]. The specific comparison results are listed in Table 6.6. In terms of size, the longitudinal length of the metamaterial-inspired high-frequency structure is approximately 2/3 that of the conventional high-frequency structure, while the beam voltage of the metamaterial-inspired klystron is close to that of the conventional klystron. The radius of the conventional high-frequency structure is not provided in Ref. [67], but based on the approximate prediction, the radius of the conventional high-frequency structure is about twice that of the metamaterial-inspired high-frequency structure. It should be pointed out that the miniaturization of the metamaterial-inspired klystron will also lead to the reduction of the size of the solenoid magnetic focusing configuration. In terms of the output performance of the device, the saturated gain of the metamaterial-inspired klystron is slightly higher than that of the conventional klystron, but the electron efficiency of the metamaterial-inspired klystron is lower. The high electron efficiency of conventional klystron in Ref. [67] is mainly due to: (1) Low conductivity coefficient; (2) The introduction of second harmonic bunching technology; (3) More idler cavities for more in-depth velocity and density modulation of the electron beam.

In summary, integrating metamaterial miniaturization technology with TR facilitates the development of compact and lightweight klystron, addressing challenges such as high manufacturing complexity, large footprint, and transportation difficulties of medical accelerators. In addition, the idea of constructing the metamaterial-inspired klystron in this frequency band can also be used for reference in large-scale scientific facilities. Taking the ESS as an example, the medium-β section of the proton linac comprises 36 klystrons operating at 704 MHz, each with an output power in the MW level [61, 67]. These klystrons are approximately $3.9 \times 1.5 \times 1\ m^3$ in size and weigh approximately 2.5 tons [61]. Consequently, the length of the medium-β section is approximately 76.7 m and the total weight of the klystrons is approximately 90 tons. Since the 714 MHz klystron studied in this chapter is close to the klystron described above in terms of operating frequency, the longitudinal length of

Table 6.6 Comparison of metamaterial-inspired klystron and conventional klystron

Parameter	Klystron loaded with CeSRRs	Klystron based on reentrant cavities [67]
f (MHz)	714	700
Number of cavities	4	6 (with a second harmonic cavity)
Longitudinal length of high-frequency structure (mm)	1236	~ 1800
Transverse radius of high-frequency structure (mm)	64.7	~ 135 (estimated value)
Beam voltage (kV)/beam current (A)	– 100/40	95/16.6
Perveance (μP)	1.26	0.57
Output power (MW)	2.28 (Pulse)	1.1 (CW)
Electron efficiency (%)	57	69.8
Gain (dB)	46.6	45.6
1 dB bandwidth (MHz)	4.7	~ 3.3

the metamaterial-inspired high-frequency structure can be reduced by approximately 560 mm. Because the klystron works vertically, the bottom area of the metamaterial-inspired high-frequency structure will be reduced by 3/4, which can greatly reduce the footprint. In addition, the weight of the metamaterial-inspired high-frequency structure will be reduced by 5/6 compared to conventional high-frequency structure. Therefore, the miniaturization characteristics of the metamaterial-inspired klystron also have important engineering value in large-scale scientific facilities.

6.3.3 Thermal Analysis of Metamaterial-Inspired Klystron

Reducing the size of the high-frequency structure in a high-power device may pose a temperature rise from a thermal stability perspective. Compared with other components in the high-frequency structure, the current of the fundamental mode reaches the maximum value in the metamaterial-inspired output cavity. Due to the surface resistance, the induced current generates higher surface heat loss in the output cavity. Therefore, thermal analysis of the metamaterial-inspired output cavity is significantly important. In this section, ANSYS is used to simulate the steady-state heat distributions of the 714 MHz output cavity, both with and without single-bridge CeSRR.

In the thermal analysis simulation process, key parameters include the material characteristics of the output cavity, the heat transfer coefficient α of the water-flowing system, and the average output power. The thermal conductivity of oxygen-free copper is set to 340 W/m·K. The expression for α is as follows:

$$Re = \frac{du\rho}{\lambda_t}, \tag{6.18}$$

$$\alpha = 0.023m\frac{\lambda_t}{d}Re^{0.8}P_r^n, \tag{6.19}$$

where d is the diameter of the water-flowing channel, flow speed of water u is set to 0.4 m/s, ρ is the density of water, λ_t is the thermal conductivity, m is the fixed coefficient under special conditions, Prandtl number P_r is set to 6.22, corresponding to the input water temperature is 22 °C, and n is 4 when the water is heated [70]. A simple water-flowing channel is constructed at the drift tube and coaxial inner conductor which are prone to temperature rise. According to (6.18) and (6.19), the α of these two locations is 2000 W/m^2 K and 1000 W/m^2 K, respectively. At the same time, the pulse output power of the metamaterial-inspired klystron is 2.28 MW, and the average output power is 4.56 kW when the pulse width is 20 μs and the repetition rate is 100 Hz.

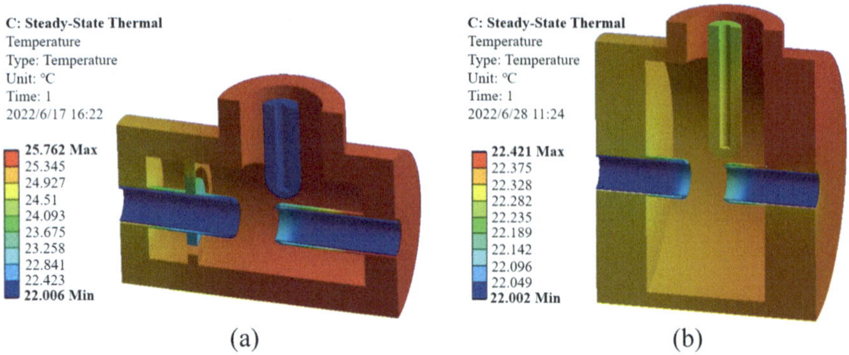

Fig. 6.26 Surface temperature distributions of two output cavities. **a** With a single-bridge CeSRR. **b** Without single-bridge CeSRR

In the ANSYS Workbench, the multi-physical field simulation is carried out for the output cavity loaded with and without single-bridge CeSRR. First, the CST simulation model of the output cavity is imported into the geometry module. Secondly, the electric field distributions in the output cavity and the surface heat loss distributions are simulated in the ANSYS Workbench. Finally, the temperature distributions on the output cavity surface are simulated in a steady-state thermal environment. Figure 6.26a shows the surface temperature distributions of the CeSRR-loaded output cavity. The good heat dissipation performance of copper, water-flowing systems, and the low average output power of the cavity, make the maximum temperature on the surface of the output cavity loaded with single-bridge CeSRR approximately 25.8 °C, much lower than the melting temperature of the copper. At the same time, the temperature rise of approximately 3.8 °C can only increase f of the output cavity loaded with single-bridge CeSRR by 0.05 MHz, which can meet the operation requirements of the metamaterial-inspired klystron.

For comparison, a conventional reentrant output cavity with the same f, length between short-circuited surfaces, length of transit gap, radius of electron beam channel, and radius of inner and outer conductors of coaxial coupler as those in the metamaterial-inspired output cavity in Fig. 6.26a is constructed, and thermal analysis is carried out, as shown in Fig. 6.26b. It can be seen that when the average output power is 4.56 kW, the maximum surface temperature of the conventional output cavity is 22.4 °C, and the temperature rise is 0.4 °C. Due to the significantly reduced radius of the metamaterial-inspired output cavity, its temperature rise is higher compared to that of the conventional output cavity. However, the miniaturization advantage is also more significant.

6.4 Metamaterial-Inspired Extended Interaction Oscillator

The EIO is a kind of VED based on the mechanism of TR. In order to achieve miniaturization and high electron efficiency of the device, the CeSRR is employed and creatively loaded into the EIRC. The double-bridge CeSRR generates strong resonance characteristics due to its internal equivalent capacitance and inductance, leading to localized electric field enhancement [42]. This enhanced electric field is beneficial for the interaction between the electron beam and electromagnetic waves. Therefore, the TR intensity excited in the metamaterial-inspired structure will be greater than that excited in the conventional structure [71, 72]. To study the coherent enhanced TR excited in the metamaterial-inspired extended interaction device, it is necessary to investigate the most important component of the extended interaction device the metamaterial-inspired EIRC. It intercepts a segment of the metamaterial-inspired SWS and closes its ends with metal walls, serving as the location for energy exchange between the electron beam and electromagnetic waves. Its electromagnetic characteristics directly determine the output power, electron efficiency, bandwidth, and other properties of the TR in the extended interaction device.

6.4.1 High-Frequency Characteristics of Metamaterial-Inspired Extended Interaction Oscillator

For example, in the S-band, a triple-gap EIRC based on an all-metal CeSRR is proposed, as shown in Fig. 6.27. Its structural parameters are set as follows: $p =$ 21 mm, $r_1 = 4$ mm, $r_2 = 11$ mm, $r_{ex} = 14.5$ mm, $r_{cw} = 16.5$ mm, $h = 8.3$ mm, $b =$ 3 mm, $g = 2$ mm, $t = 1$ mm.

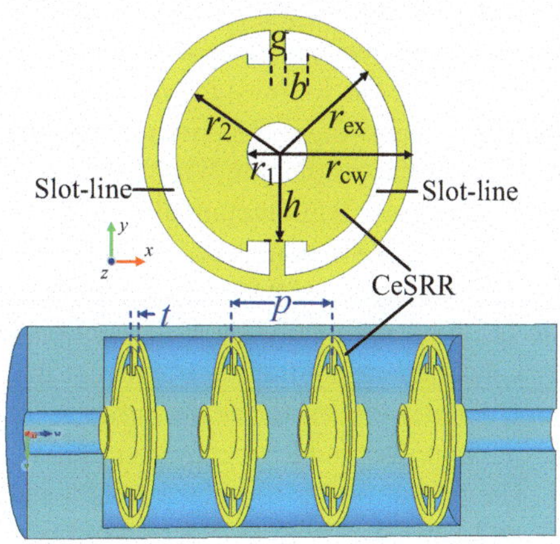

Fig. 6.27 The double-bridge CeSRR unit cell and the MEIRC loaded with CeSRRs

The simulation study indicates that the extended interaction structure composed of the EIRC based on the double-bridge CeSRR can generate TR in a very short length, converting the kinetic energy of the electron beam into microwave energy. In order to demonstrate the advantages of the metamaterial-inspired EIRC, a comparative study between a conventional double-gap EIRC and a metamaterial-inspired double-gap EIRC is conducted. The essential differences are as follows: (1) The transverse dimension of the metamaterial-inspired EIRC is approximately $\lambda/4$, while those of the conventional EIRC are approximately $\lambda/2$. This indicates that the metamaterial-inspired EIRC has a significant miniaturization; (2) The double-bridge CeSRR exhibits strong resonance characteristics, allowing a circular waveguide loaded with the double-bridge CeSRR array to operate in a frequency range below the cutoff frequency of the corresponding empty waveguide TM mode; (3) Due to the miniaturization and strong resonance characteristics of the double-bridge CeSRR, the metamaterial-inspired EIRC has a higher electromagnetic energy density and thus a higher effective characteristic impedance. Therefore, the TR generated by the constructed metamaterial-inspired EIO will have high output power and high electron efficiency [73].

6.4.2 Beam-Wave Interaction of Metamaterial-Inspired Extended Interaction Oscillator

For an EIO consisted of a single EIRC, when the electron beam passes through the EIRC loaded with CeSRRs, it can be regarded as the electron beam being perturbed by the standing-wave field at the transit gaps. Within the EIRC, the electron beam undergoes velocity modulation first, followed by density modulation. This process leverages the TR mechanism, facilitating energy exchange between the electron beam and the electromagnetic field.

Taking a 5-gap metamaterial-inspired EIO with a waveguide output coupler as an example, its optimized simulation model is shown in Fig. 6.28 [74, 75]. The optimized structural parameters are as follows: the distance from the double-bridge CeSRR to the short-circuited end is $L_0 = 21.5$ mm, the optimized period is $L = 26.5$ mm, and the dimensions of the coupler are $Wca = 54$ mm and $Wcb = 24.5$ mm, while the output port uses a standard BJ32 rectangular waveguide.

The beam-wave interaction of the metamaterial-inspired EIO is simulated using the CST PIC Solver [51]. The research results indicate that the metamaterial-inspired EIRC can generate TR in an exceptionally short interaction length. This process effectively converts the kinetic energy of the electron beam into microwave energy. When the electron beam voltage is 130 kV, the beam current is 80 A, and the magnetic flux density is 0.2 T, the output power and frequency spectrum of the TR are obtained, as shown in Fig. 6.29. The output power reaches a steady state after 80 ns and is approximately 4.9 MW at 2.866 GHz, with an electron efficiency of approximately 46%, and the spectrum is pure. This indicates that the TR excited in the metamaterial-inspired EIO is coherent and enhanced. The transverse diameter of the metamaterial-inspired EIO is only 37 mm (~ 0.35 λ), and the longitudinal length is approximately

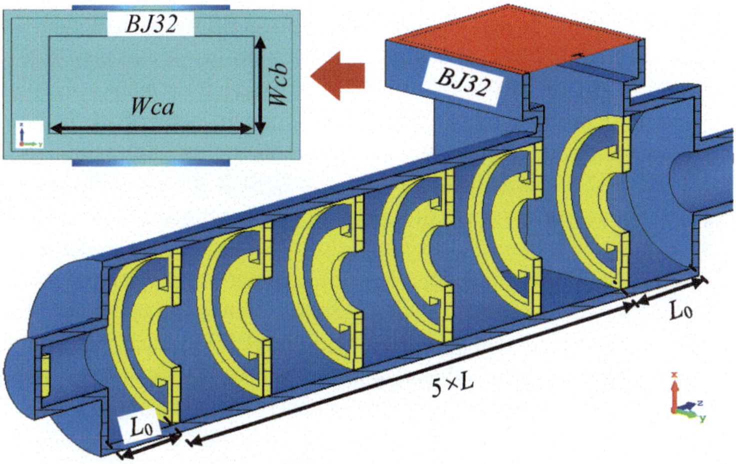

Fig. 6.28 Simulation model of a 5-gap metamaterial-inspired EIO with waveguide output coupler

180 mm (~ 1.71 λ). Therefore, the metamaterial-inspired EIO exhibits significant miniaturization characteristics.

Furthermore, an improved 5-gap metamaterial-inspired EIO is studied, as shown in Fig. 6.30 [76]. Here, a single-bridge CeSRR [50] is employed, known for its enhanced subwavelength characteristics compared to the double-bridge CeSRR shown in Fig. 6.30a. This is due to the longer slot-line of the single-bridge CeSRR, which exhibits a negative correlation with the frequency of the CeSRR. It can be inferred that the miniaturization characteristics of the EIRC based on the single-bridge CeSRR will be more significant. Additionally, this oscillator employs magnetic coupling for output. Under the conditions of the beam voltage and beam current of 40 kV and 4 A, respectively, and the magnetic flux density of 0.3 T, the

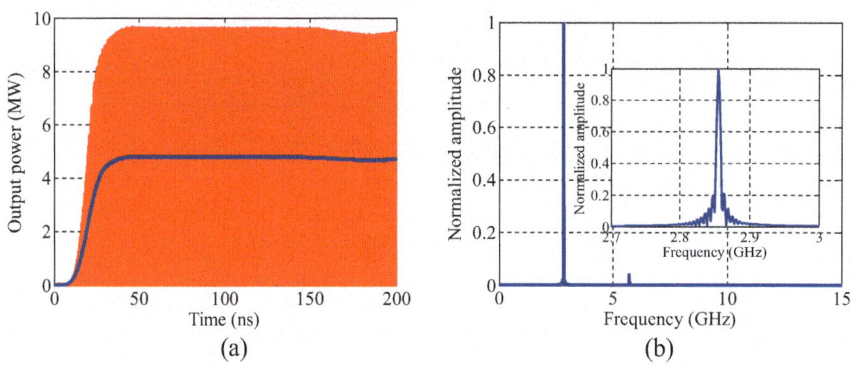

Fig. 6.29 The output power and output signal spectrum of TR. **a** The output power of TR. **b** The output signal spectrum of TR

TR output power of this 5-gap metamaterial-inspired EIO is 50 kW, with an electron efficiency of 31.2%. The output signal has no spurious signals near 2.362 GHz, indicating that the TR excited in this metamaterial device is coherent, as shown in Fig. 6.31. The entire metamaterial-inspired EIRC has a diameter of 30 mm (~ 0.24 λ) and a longitudinal length of 90 mm (~ 0.71 λ), demonstrating further miniaturization compared to the EIO based on the double-bridge CeSRR shown in Fig. 6.28.

From the above research results, it is evident that in terms of the mechanism for generating output power in microwave radiation sources, the metamaterial-inspired EIO is based on the TR mechanism, while the conventional backward-wave oscillator relies on the CR mechanism. Due to the differences in operating principles and interaction structures, the metamaterial-inspired EIO presents potential advantages in structural dimensions and device performance. Particularly, as a standing-wave device compared to the traveling-wave device of the backward-wave oscillator, the metamaterial-inspired EIO can achieve electron beam velocity modulation and density modulation in a shorter interaction length (only 0.57 λ), thereby achieving coherent enhanced TR while reducing the size of the device.

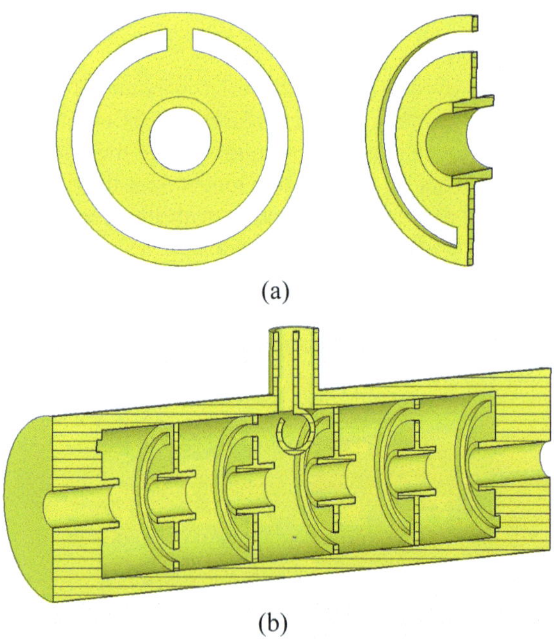

Fig. 6.30 Simulated models of TR of the metamaterial-inspired EIO loaded with single-bridge CeSRRs. **a** Plane and longitudinal cross-section of the single-bridge CeSRR. **b** Interaction structure of the metamaterial-inspired EIO loaded with single-bridge CeSRRs

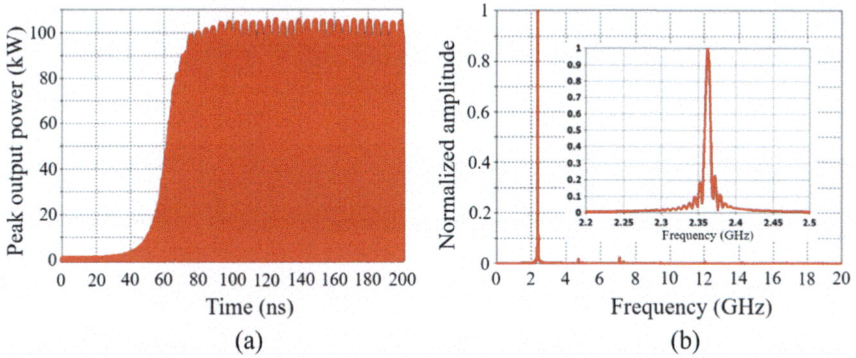

Fig. 6.31 The output power and the spectrum of the output signal of TR. **a** The output power versus time. **b** The spectrum of the output signal

6.5 Metamaterial-Inspired Extended Interaction Amplifier

Section 6.4 explores the significant advantages of coherent enhanced TR excited by metamaterials in achieving miniaturization of high-frequency structures and improving electron efficiency. This section focuses on optimizing the design of the input and output couplers for the metamaterial-inspired EIRC, exploring TR induced by metamaterials in the EIA, and analyzing the coherence and enhancement of the TR [42, 77–80]. The research shows that the metamaterial-inspired EIA exhibits miniaturization characteristics.

6.5.1 High-Frequency Characteristics of Metamaterial-Inspired Extended Interaction Amplifier

In Sect. 6.4.1, based on the research on the metamaterial-inspired EIRC, the optimization of the metamaterial-inspired EIA is carried out. A double-gap input cavity and a triple-gap output cavity are proposed, as shown in Fig. 6.32. The input and output couplers are based on coaxial lines with a metal rod, facilitating the cavities' assembly and test. The structural parameters of the input and output cavities are set as follows: $p_1 = 20$ mm, $s_1 = 13$ mm, $d = 7$ mm, $h_1 = 11.5$ mm, $h_2 = 2$ mm, $d_{r1} = 4$ mm, $r_{co} = 2.7$ mm, $r_{ci} = 0.65$ mm, $p_2 = 19$ mm, $s_2 = 12.5$ mm, $d_s = 6.5$ mm, $h_3 = 11$ mm, $h_4 = 3$ mm, and $d_{r2} = 4$ mm. The resonant frequency of the fundamental mode of the input and output cavities are both 2.45 GHz, which is simulated by the CST Eigenmode Solver [51]. The external quality factors of the double-gap input cavity and the triple-gap output cavity are 284 and 322, respectively.

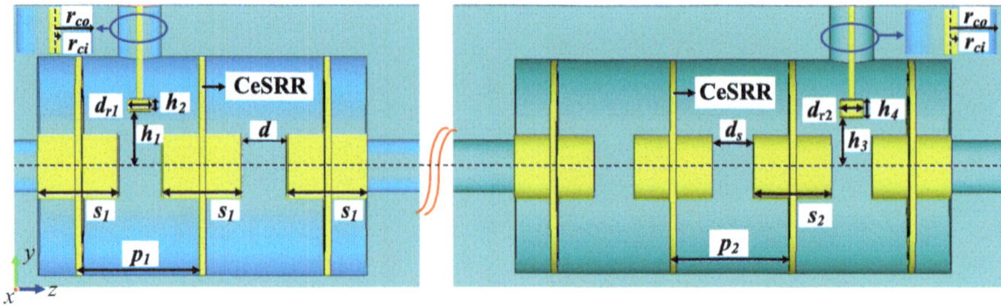

Fig. 6.32 The metamaterial-inspired double-gap input cavity and triple-gap output cavity

The operating frequency band of the metamaterial-inspired EIRC is determined by the E_z calculation. Figure 6.33a shows the E_z distributions of the metamaterial-inspired triple-gap EIRC for the π mode (fundamental mode), π/2 mode, and 2π mode, with resonant frequencies of 2.45 GHz, 2.7 GHz, and 4.1 GHz, respectively. The input cavity and the idler cavity need to operate in the amplification range of the EIRC. Based on the variation of G_e/G_0 with the transit angle θ in Fig. 6.33b and the amplifier operating mode as described in Sect. 6.4.1, it can be observed that θ is in the range of 3–3.3 rad. While increasing the number of gaps in the input cavity can enhance M^2R/Q, it may also cause mode competition due to the narrowing of the frequency space between adjacent modes, thereby affecting the stability of the input cavity or even the entire device. Therefore, considering the electromagnetic characteristics and operating conditions of the metamaterial-inspired EIRC, it is optimized to design the input cavity as a double-gap metamaterial-inspired EIRC, while the idler and output cavities are both triple-gap metamaterial-inspired EIRCs.

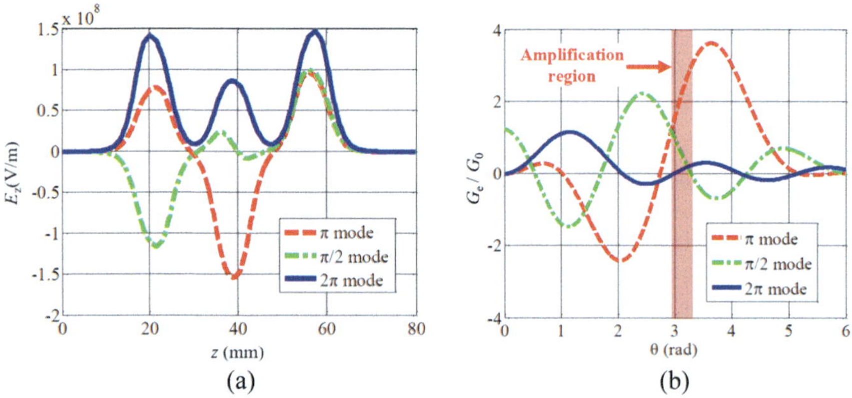

Fig. 6.33 E_z amplitudes and G_e/G_0 of the MEIRC. **a** E_z of π mode, π/2 mode and 2π mode. **b** The variation of G_e/G_0 corresponds to different modes versus θ

6.5.2 Experimental Investigation of Metamaterial-Inspired Extended Interaction Amplifier

Based on the research of the metamaterial-inspired EIRC at S-band, the experimental study of the double-cavity metamaterial-inspired EIA shown in Fig. 6.32 is conducted. A test platform is constructed through component processing, cleaning, and assembly, as shown in Fig. 6.34. The length of the drift tube in the input cavity is 13 mm, the length of the drift tube in the output cavity is 12.5 mm, and the thickness of the cavity body is 4 mm. The height of the rod probe of the coaxial coupler in the input and output cavity is 2 mm and 3 mm, respectively, and the diameters of both are 3 mm, which are connected to the standard SMA coaxial line via flanges.

Notably, coupling probes of different lengths are processed to adjust the resonant frequency of the EIRC. Furthermore, an assembled model of the double-cavity metamaterial-inspired extended interaction structure with input and output cavities is obtained, as shown in Fig. 6.35, where the drift tube length between the input and output cavities is 50 mm, with a total length of the interaction structure approximately 200 mm.

(1) Resonant frequency and external quality factor

The reflection coefficients (S_{11} and S_{22}) of the double-cavity metamaterial-inspired extended interaction structure were measured by using an Agilent 5230A vector network analyzer. The comparison between the test and simulated results of the double-gap metamaterial-inspired EIRC (input cavity) and the triple-gap metamaterial-inspired EIRC (output cavity) is shown in Fig. 6.36. The external quality factors Q_e of the output cavity and input cavity were directly measured by using the 3 dB bandwidth method. Q_e is calculated as f_0 / f_{3dB}, where f_0 is the resonant frequency and f_{3dB} is the 3 dB bandwidth, the measured Q_e of the output cavity and input cavity are 287 and 281, respectively. Additionally, Q_e is also calculated using the group delay method by using the CST Time Domain Solver [51], as shown in (6.15).

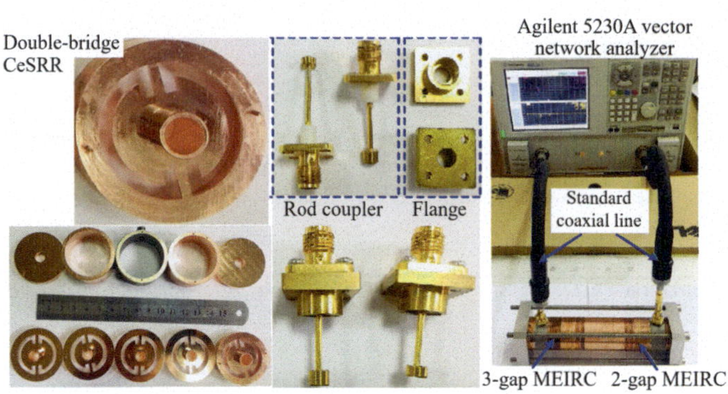

Fig. 6.34 Components and test platform for double-cavity metamaterial-inspired EIA

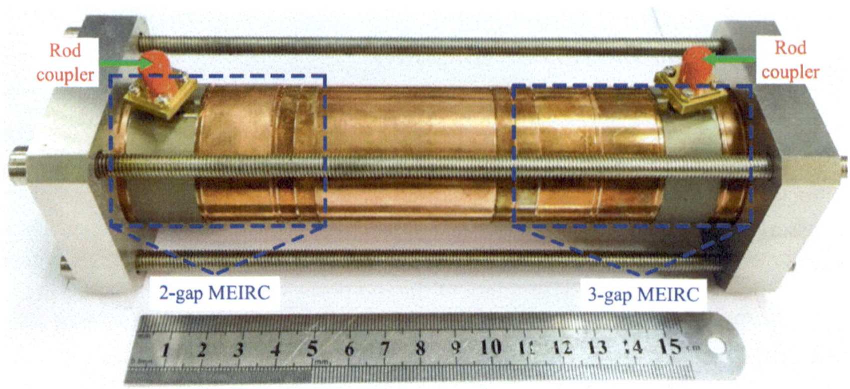

Fig. 6.35 The photograph of the double-cavity metamaterial-inspired extended interaction structure with an input and an output cavity

The simulated Q_e of the output cavity and input cavity are 322 and 284, respectively, which are highly consistent with the measured results, as shown in Table 6.7. Furthermore, the simulated fundamental resonant frequencies of the input cavity and output cavity are 2.452 GHz and 2.453 GHz, respectively. Through comparative analysis, it is found that the simulated results are in good agreement with the experimental measurements.

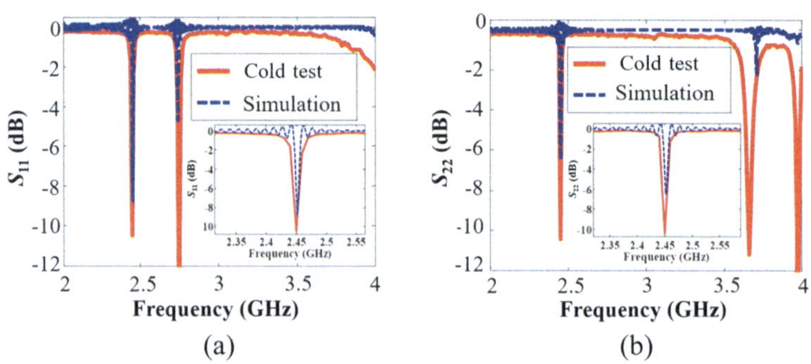

Fig. 6.36 Measured and simulated results of double-cavity metamaterial-inspired extended interaction structure. **a** Reflection coefficients of measured and simulated results for input cavity. **b** Reflection coefficients of measured and simulated results for output cavity

Table 6.7 Simulation and test results of S-band double-cavity metamaterial-inspired EIA

Parameter	Input cavity		Output cavity	
	Simulation	Test	Simulation	Test
f (GHz)	2.452	2.45	2.453	2.45
Q_e	284	281	322	287

(2) Characteristic impedance

In order to verify whether the *R*/*Q* of the designed resonant cavity meets the design index, it is necessary to carry out experimental measurements of *R*/*Q*. According to classical perturbation theory, if a small perturbation object is introduced into the resonant cavity, the perturbation will be generated to the energy storage of the resonant cavity, resulting in a resonant frequency offset of the resonant cavity, so that the *R*/*Q* of the resonant cavity can be measured indirectly. In this study, the dielectric cylinder perturbation method is used to test the *R*/*Q* of the resonant cavity. For an extended interaction resonant cavity with *n* gaps, assuming the electric field distributions is the same for each gap, its *R*/*Q* can be expressed as:

$$\frac{R}{Q} = \frac{nd}{\pi(\varepsilon_r - 1)\varepsilon_0 S} \frac{\Delta f}{f_0^2}, \tag{6.20}$$

where d is the length of the transit gap, ε_r is the relative dielectric constant of the medium, ε_0 is the dielectric constant of the vacuum, S is the cross-section area of the dielectric cylinder, f_0 is the resonant frequency of the empty resonant cavity, and Δf is the resonant frequency offset caused by the dielectric cylinder.

The resonant frequency and the designed *R*/*Q* of the double-gap metamaterial-inspired input cavity are 2.454 GHz and 185.2 Ω, respectively. The *R*/*Q* simulation of the dielectric cylinder perturbation method is carried out, and the PTFE (ε_r = 2.2) is selected as the material of the dielectric cylinder. The loading method of the dielectric cylinder is shown in Fig. 6.37.

According to the drawing design and component processing and assembly, the components of the metamaterial-inspired double-gap input cavity and the dielectric cylinder are shown in Fig. 6.38a. The S_{11} of the input cavity is measured by using the Agilent E8363B vector network analyzer. The test platform is shown in Fig. 6.38b)

Dielectric cylinders of different thicknesses are inserted into the drift tube of EIRCs, and the resonant frequencies are tested by a vector network analyzer. The *R*/*Q* is calculated according to (6.20), the test results are summarized in Table 6.8, and the average values are taken as the experimental result. Therefore, the *R*/*Q* of the metamaterial-inspired double-gap input cavity is 182.5 Ω, which is only a 1.5% error from the design value.

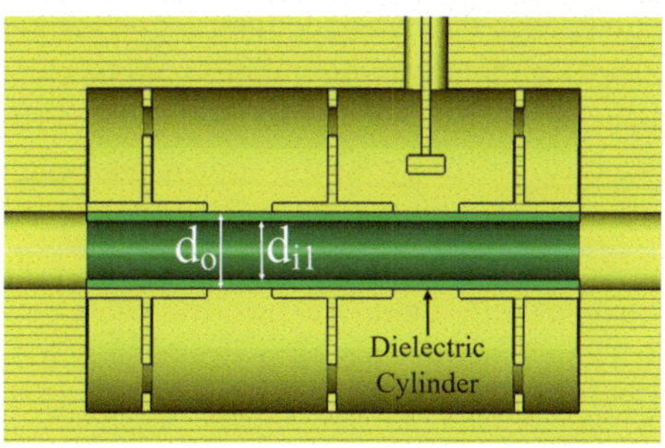

Fig. 6.37 The metamaterial-inspired double-gap input cavity loaded with dielectric cylinder

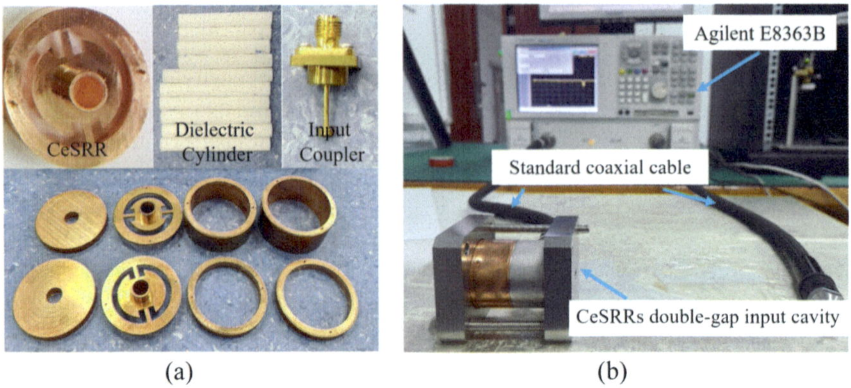

Fig. 6.38 *R*/*Q* testing of metamaterial-inspired double-gap input cavity. **a** Components. **b** Test platform

Table 6.8 *R*/*Q* test results of resonant cavity loaded with different thickness cylinders

d_{i1} (mm)	Frequency offset (MHz)	Test values of *R*/*Q* (Ω)	Design values of *R*/*Q* (Ω)	Error of *R*/*Q* (%)
7.0	28.9	190.5	185.2	2.9
6.8	32.7	182.1		1.2
6.6	37.4	180.8		2.4
6.4	41.1	176.4		4.8

6.5.3 Beam-Wave Interaction of Metamaterial-Inspired Extended Interaction Amplifier

The sectional view of a three-cavity metamaterial-inspired EIA is shown in Fig. 6.39. The input signal couples into the input cavity of metamaterial-inspired extended interaction to create the specific electromagnetic field modes. The electron beam undergoes periodic velocity modulation of the electromagnetic field at the transit gap. Subsequently, the modulated electron beam enters the drift tube, where it undergoes density modulation due to inertial effects. The modulated electron beam induces a stronger electromagnetic field in the extended interaction idler cavity, and continues to perform stronger velocity and density modulation on the electron beam under the action of the local enhanced electric field caused by the CeSRRs. The acceleration and deceleration of the electron beam at the transit gap gradually form electron bunches in the drift tube. These bunches eventually enter the extended interaction output cavity, where most electrons are in the deceleration region, while a few are accelerated by the electromagnetic field. The electron beam converts kinetic energy into RF energy through TR and couples it to the external circuit by the output coupler,

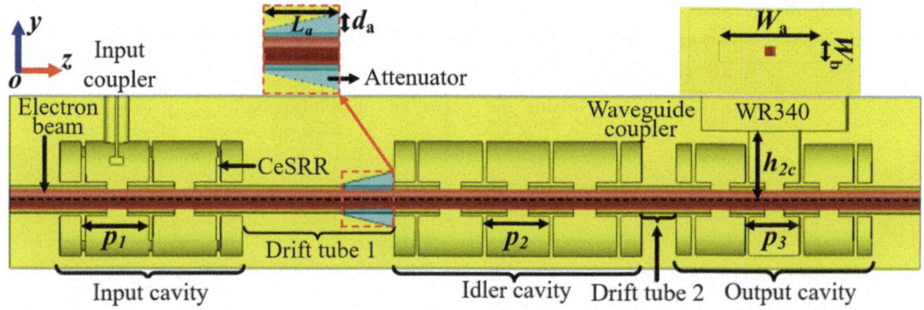

Fig. 6.39 Sectional view of a three-cavity metamaterial-inspired EIA in the *yoz* plane

thereby completing the input, amplification, and output of the entire electromagnetic wave.

Based on the above parameters, the CST PIC Solver is used to simulate the TR. Set the electron beam voltage and current to 30 kV and 3 A, the B_z to 1000 G, the power and frequency of the input port signal of the metamaterial-inspired EIA are 1.2 W and 2.453 GHz, the electron beam radius is 3 mm, and the electron beam channel radius is 4 mm. The simulated results are shown in Fig. 6.40a and b, the output signal of the TR reaches a steady state after 80 ns, the saturated output power is 56 kW, and the corresponding center frequency of the output signal is 2.453 GHz. The frequency spectrum is pure near 2.453 GHz, which indicates that the TR excited by the metamaterial in the EIA is coherent. At the same time, the 3 dB bandwidth is 24 MHz, as shown in Fig. 6.40d, which indicates that the metamaterial-inspired EIA has narrow-band characteristics. By adjusting the input power, the electron efficiency and gain of the TR in the metamaterial-inspired EIA can be obtained, as shown in Fig. 6.40c. When the input power is 1.1 W, the electron efficiency reaches a maximum of 62%, and the corresponding saturated gain is 47 dB. The saturated gain of conventional amplifiers is ~ 45 dB, and the electron efficiency is 30–45%, which can reach 54% with the method of improving the electron efficiency [80]. Therefore, the electron efficiency and saturated gain of TR of the metamaterial-inspired amplifier are higher than the conventional amplifiers in the same frequency band, that is, the TR with high coherence and strong radiation is shown.

The electric field amplitude of the metamaterial-inspired high-frequency structure is monitored when the saturated output power reaches a steady state with a monitoring time of 190–195 ns. From Fig. 6.41, it can be seen that the maximum electric field amplitude of the metamaterial-inspired high-frequency structure is 12.99 MV/m, located at the transit gap of the output cavity. The electric field amplitude in the double-bridge CeSRR slot-line reaches up to 11.24 MV/m, located at the cusp of the slot-line. Both of which are less than the vacuum breakdown threshold of 36 MV/m for short-pulse power output [53]. However, for long-pulse or CW output power, the maximum electric field in the output cavity is within the breakdown threshold range of 10–15 MV/m [81]. Therefore, the simulated results indicate a breakdown risk for the metamaterial-inspired EIA in the case of long-pulse or CW output. The

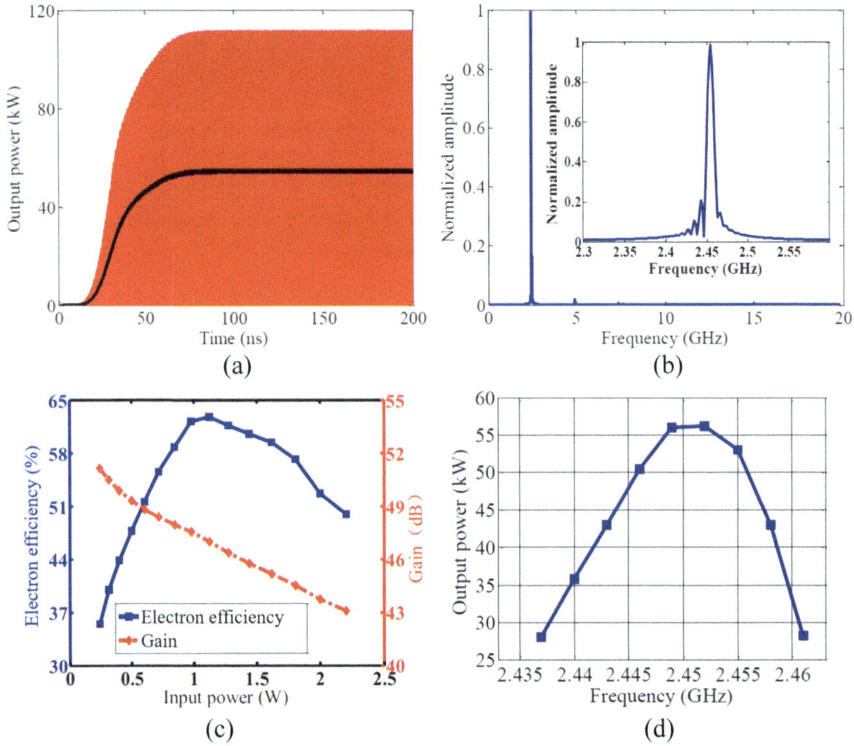

Fig. 6.40 The simulated TR results of metamaterial-inspired EIA. **a** The output power versus time. **b** Normalized spectrum of the output signal. **c** Electron efficiency and gain versus input power. **d** Output power versus different frequencies of input signals

breakdown risk exists at the transit gaps of the output cavity, including the nose cone of the drift tube and the cusp of the slot-line of the double-bridge CeSRR unit cell. The monitoring results show that the maximum electric field is located at the nose cone of the double-bridge CeSRR in the output cavity (0.5, 4.5, and 273.7 mm). Therefore, for further research on the coherent enhanced TR of the metamaterial-inspired EIA, it is necessary to increase the slot-line width of the double-bridge CeSRR unit cell, round the cusp of the slot-line, and improve the surface finish. Moreover, it is necessary to increase the thickness of the nose cone of the output cavity.

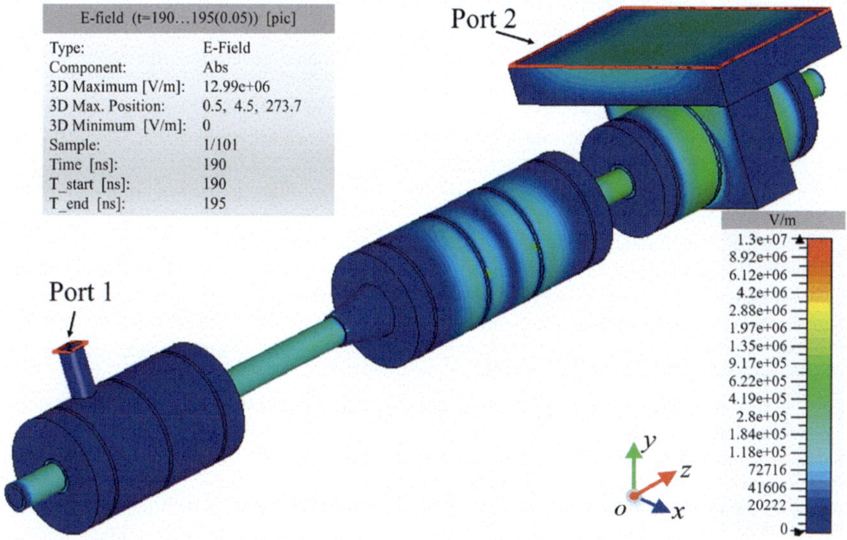

Fig. 6.41 Electric field distributions of the metamaterial-inspired high-frequency structure

References

1. Ginzburg VL, Frank IM. Radiation of a uniformly moving electron due to its transition from one medium into another. Journal of Physics. 1945; 9:353.
2. Alikhanian AI, Kankanian SA, Oganessian AG, et al. Detection of X-ray transition radiation of 31-GeV electrons. Physical Review Letters. 1973; 30:3–109.
3. Liao GQ, Li YT, Zhang YH, et al. Demonstration of coherent terahertz transition radiation from relativistic laser-solid interactions. Physical Review Letters. 2016; 116:20–205003.
4. Shibata Y, Ishi K, Takahashi T, et al. Observation of coherent transition radiation at millimeter and submillimeter wavelength. Physical Review A. 1992; 45:12–8340.
5. Dolgoshein B. Transition radiation detectors. Nuclear Instruments and Methods in Physics Research Section A: Accelerators, Spectrometers, Detectors and Associated Equipment. 1993; 326:3–434.
6. Ivanov IP, Karlovets DV. Detecting transition radiation from a magnetic moment. Physical Review Letters. 2013; 110:26–264801.
7. Marcum J. Interchange of energy between an electron beam and an oscillating electric field. Journal of Applied Physics. 1946; 17:1–4.
8. Coleman PD, Lerttamrab M, Gao J. Generation of single-frequency coherent transition radiation by a prebunched electron beam traversing a vacuum beam tunnel in a periodic medium. Physical Review E. 2002; 66:6–066502.
9. Gold SH, Nusinovich GS. Review of high-power microwave source research. Review of Scientific Instruments. 1997; 68:11–3945.
10. Parker RK, Abrams RH, Danly BG, et al. Vacuum electronics. IEEE Transactions on Microwave Theory and Techniques. 2002; 50:3–835.
11. Zhang JD, Ge XJ, Zhang J, et al. Research progresses on Cherenkov and transit-time high-power microwave sources at NUDT. Matter and Radiation at Extremes. 2016; 1:3–163.
12. Teo KBK, Minoux E, Hudanski L, et al. Carbon nanotubes as cold cathodes. Nature. 2005; 437:7061–810007.

13. Hummelt JS, Lu X, Xu H, et al. Coherent Cherenkov-cyclotron radiation excited by an electron beam in a metamaterial waveguide. Physical Review Letters. 2016; 117:23–237701.
14. Yu LH, Babzien M, Ben-Zvi I, et al. High-gain harmonic-generation free-electron laser. Science. 2000; 289:5481–932.
15. Lu XY, Shapiro MA, Mastovsky I, et al. Generation of high-power, reversed-Cherenkov wakefield radiation in a metamaterial structure. Physical Review Letters. 2019; 122:1–014801.
16. Varian RH. Electrical translating system and method. US 2242275A. 1941.
17. Varian RH, Varian SF. A high frequency oscillator and amplifier. Journal of Applied Physics. 1939; 10:5–321.
18. Fujisawa K. The Laddertron-A new millimeter wave power oscillator. IEEE Transactions on Electron Devices. 1964; 11:8–381.
19. Wessel-Berg T. A general theory of klystrons with arbitrary, extended interaction fields. Report: Stanford Microwave Laboratory, CA. 1957.
20. Chodorow M, Wessel-Berg T. A high-efficiency klystron with distributed interaction. IRE Transaction Electron Devices. 1960; 8:1–44.
21. Preist DH, Leidigh WJ. Experiments with high-power CW klystrons using extended interaction catchers. IEEE Transactions on Electron Devices. 1963; 10:3–201.
22. Vacuum Electronics Design Manual Editorial Board. High-Power Klystron Design Manual. Beijing: Defense Industry Press. 1979 (In Chinese).
23. Xie JL, Zhao YX. Bunching Theory of Klystrons. Beijing: Science Press. 1966 (In Chinese).
24. Ding YG. Design, Manufacture and Application of High-Power Klystron Tubes. Beijing: Defense Industry Press. 2010 (In Chinese).
25. Ding YG, Zhang ZQ. High-frequency systems and high-power microwave devices for Spallation Neutron Source. Vacuum Electronics. 2017; 06:12 (In Chinese).
26. Ding YG. The status and progress of high-power klystron tubes. Vacuum Electronics. 2020; 01:1 (In Chinese).
27. Normile D. Accelerator boom hones China's engineering expertise. Science. 2018; 359: 6375–507.
28. Zhu XL, Chen M, Weng SM, et al. Extremely brilliant GeV γ-rays from a two-stage laser-plasma accelerator. Science Advances. 2020; 6:22-eaaz7240.
29. Sapra NV, Yang KY, Vercruysse D, et al. On-chip integrated laser-driven particle accelerator. Science. 2020; 367: 6437–79
30. Wang WT, Feng K, Ke LT, et al. Free-electron lasing at 27 nanometres based on a laser wakefield accelerator. Science. 2021; 595:7868–516.
31. Nanni EA. Cascaded particle accelerators reach new energy. Nature Photonics. 2021; 15:6–405.
32. Shelby RA, Smith DR, Schultz S. Experimental verification of a negative index of refraction. Science. 2001; 292:5514–77.
33. Seddon N, Bearpark T. Observation of the inverse Doppler effect. Science. 2003; 302:5650–1537.
34. Smith DR, Padilla WJ, Vier DC, et al. Composite medium with simultaneously negative permeability and permittivity. Physical Review Letters. 2000; 84:18–4184.
35. Lin X, Easo S, Shen YC, et al. Controlling Cherenkov angles with resonance transition radiation. Nature Physics. 2018; 14:8–816.
36. Duan ZY, Tang XF, Wang ZL, et al. Observation of the reversed Cherenkov radiation. Nature Communications. 2017; 8:14901.
37. Adamo G, Ou JY, So JK, et al. Electron-beam-driven collective-mode metamaterial light source. Physical Review Letters. 2012; 109:217401.
38. Duan ZY, Shapiro MA, Schamiloglu E, et al. Metamaterial-inspired vacuum electron devices and accelerators. IEEE Transactions on Electron Devices. 2019; 66:1–207.
39. Yao J, Liu ZW, Liu YM, et al. Optical negative refraction in bulk metamaterials of nanowires. Science. 2008; 321:5891–930.
40. Choudhury PK. Metamaterials Technology and Applications. Boca Raton: CRC Press (Taylor & Francis Group, UK). 2021

41. Wang X, Tang XF, Li SF, et al. Recent advances in metamaterial klystrons. EPJ Applied Metamaterials. 2021; 8:9–1.
42. Wang X, Li SF, Zhang XM, et al. Novel S-band metamaterial extended interaction klystron. IEEE Electron Device Letters. 2020; 41:10–1580.
43. Pozar DM. Microwave Engineering. 4th ed. Hoboken: John Wiley & Sons Inc. 2011.
44. Caryotakis G. High power klystrons: Theory and practice at the Stanford Linear Accelerator Center, 2004; SLAC-PUB-10620.
45. Zhang D, Cao J, Miao YZ, et al. The group delay method solves the appearance Q_(ext) of the output cavity of the klystron. Journal of Vacuum Science and Technology. 2007; 05:391 (In Chinese).
46. Chen BY. Radio Frequency Communication Circuits. Beijing: Science Press. 2002 (In Chinese).
47. Zhou QY. Microwave Measurements. Beijing: Defense Industry Press. 1964 (In Chinese).
48. Wang X, Zhang XM, Zou JJ, et al. Experimental demonstration of compact S-band MW-level metamaterial-inspired klystron. IEEE Electron Device Letters. 2023; 44:1–152.
49. Zhang ZQ, Luo JR, Zhang ZC. Analysis and suppression of high-order mode oscillation in an S-band klystron. IEEE Transactions on Plasma Science. 2015; 43:2–515.
50. Wang X, Duan ZY, Zhan XR, et al. Characterization of metamaterial slow-wave structure loaded with complementary electric split-ring resonators. IEEE Transactions on Microwave Theory and Techniques. 2019; 67:6–2238.
51. CST Studio Suite. Dassault Systems, Vèlizy-Villacoublay, France. 2016.
52. Communications & Power Industries, VKS-8262 S-band pulsed klystrons. 2020.
53. Vlieks AE, Allen MA, Callin RS, et al. Breakdown phenomena in high-power klystrons. IEEE Transactions on Electrical Insulation. 1989; 24:6–1023.
54. Hassanein A, Insepov Z, Norem J, et al. Effects of surface damage on rf cavity operation. Physical Review Special Topics-Accelerators and Beams. 2006; 9:6–062001.
55. ANSYS, ANSYS Inc. Pennsylvania, United States. 2017.
56. Born M, Wolf E. Principles of Optics, 7th ed. Cambridge University Press. 1999.
57. Hecht E. Optics, 3rd ed. Addison Wesley Longman. 1998.
58. Leemans WP, Geddes CGR, Faure J, et al. Observation of terahertz emission from a laser-plasma accelerated electron bunch crossing a plasma-vacuum boundary. Physical Review Letters. 2003; 91:7–074802.
59. Schroeder CB, Esarey E, Tilborg J, et al. Theory of coherent transition radiation generated at a plasma-vacuum interface. Physical Review E. 2004; 69:1–16501.
60. Ding YG, Liu PK, Zhang ZC, et al. The technical status and research progress of high-power klystron tubes. Vacuum Electronics. 2010; 06:1 (In Chinese).
61. Aymar G, Eisen E, Stockwell B, et al. Development and production of a 704 MHz, 1.5 MW peak power klystron. IEEE International Vacuum Electronics Conference. London, UK. 2017.
62. Beunas A, Marchesin R, Rampnoux E, et al. 2.8 MWp 352 MHz long pulse klystron for proton linac. IEEE International Vacuum Electronics Conference. Monterey, CA, USA. 2012.
63. Garoby R, Danared H, Alonso I, et al. The European Spallation Source design. Physica Scripta. 2018; 93:1–014001.
64. Ding YG, Zhang ZQ, Shen B, et al. The design and calculation of P-band 1.2 MW multi-beam klystron. IEEE International Vacuum Electronics Conference. London, UK. 2017.
65. Zhang XM, Wang S, Zou J, et al. Metamaterial-based interaction structure for MW-level klystron at 714 MHz. IEEE Transactions on Electron Devices. 2022; 69:11–6336.
66. Duan ZY, Zhang XM, Wang X, et al. A miniaturized high-power klystron based on a single-bridge CeSRR unit. CN 113838727A. 2021.
67. Chung B, Hong J, Jeon J, et al. A CW K-klystron of 700 MHz and 1 MW for PEFP. Journal of the Korean Physical Society. 2008; 52:9–761.
68. Zhang XM, Wang SZ, Zou JJ, et al. Miniaturized P-band MW-level klystron with CeSRRs for SNS linac application. IEEE Transactions on Electron Devices. 2023; 70:9–4878.
69. Wang X. Research on metamaterial extended interaction devices. Chengdu: University of Electronic Science and Technology of China. 2021 (In Chinese).

70. Lu ZJ, Fukuda S, Zhou ZS. Design and development of radio frequency output window for Circular Electron-Positron Collider klystron. Chinese Physics B. 2018; 27:11–118402.
71. Alekhina TY, Tyukhtin AV, Grigoreva AA. Cherenkov-transition radiation in a waveguide partly filled with a resonance dispersion medium. Physical Review Special Topics-Accelators and Beams. 2015; 18:9–091302.
72. Galyamin SN, Tyukhtin AV, Kanareykin A, et al. Reversed Cherenkov-transition radiation by a charge crossing a left-handed medium boundary. Physical Review Letters. 2009; 103:19–194802.
73. Wang X, Zhang XM, Luo HY, et al. Compact and high-efficiency metamaterial extended interaction oscillator. IEEE International Vacuum Electronics Conference. Monterey, CA, USA. 2020.
74. Duan ZY, Wang X, Luo HY, et al. Metamaterial-based vacuum electronic devices with miniaturization. 2020 URSI Regional Conference on Radio Science. Varanasi, India. 2020.
75. Wang YS, Duan ZY, Tang XF, et al. All-metal metamaterial slow-wave structure for high-power sources with high efficiency. Applied Physics Letters. 2015; 107:15–153502.
76. Wang X, Duan ZY, Yang G, et al. An S-band miniaturized metamaterial extended interaction oscillator. CN 109256309B. 2021.
77. Duan ZY, Wang X, Zhan XR, et al. Left-handed material extension interaction klystron. CN 110233091B. 2021.
78. Guha R, Wang X, Tang XF, et al. Metamaterial assisted microwave tubes: A review. Journal of Electromagnetic Waves and Applications. 2022; 36:9–1189.
79. Duan ZY, Wang X, Zhan XR, et al. Left-handed material extended interaction klystron. US 10418219B2. 2019.
80. Luo JR, Feng JJ, Gong YB. A review of microwave vacuum devices in China: Theory and device development including high-power klystrons, spaceborne TWTs, and gyro-TWTs. IEEE Microwave Magazine. 2021; 22:4–18.
81. Ding YG. Theoretical and Computational Simulation of High-Power Klystron Tubes. Beijing: Defense Industry Press. 2008 (In Chinese).

Epilogue

Since the experimental verification of metamaterials was done in 2000, the research in this field has undergone revolutionary development. To date, it has expanded from the initial microwaves to optics, acoustics, mechanics, thermodynamics, and others. USA, Japan, European Union, Russia, South Korea, Singapore, China, and so on, have provided long-term funding for metamaterials research. In USA, the Defense Advanced Research Projects Agency (DARPA) defined metamaterials as Thrust Areas in 2012 [1], and the Air Force Office of Scientific Research (AFOSR) listed metamaterials among the Ten Focus Areas in 2012 [2]. In China, the National Natural Science Foundation of China (NSFC) has provided funding supports. International academic organizations such as IEEE, Optica, APS, and EPS, along with academic conferences such as the Metamaterials Congress and the International Conference on Metamaterials, Photonic Crystals, and Plasmonics, have promoted international exchange and collaboration in the field of metamaterials. In China, the Metamaterials branch of the Chinese Materials Research Society was established in 2018, organizing the national conferences on metamaterials regularly. The related companies have been founded including Metalenz and Alpha Cen in USA, TeraView in UK, Mitsubishi Group in Japan, and Kuangchi Science in China. Currently, my group is closely collaborating with the Guoli Vacuum (GLVAC) Industrial Technology Research Institute of High Power Devices Co., Ltd. to develop metamaterial-inspired P-band klystrons, which are intended for China Spallation Neutron Source.

Without exaggeration, the emergence of metamaterials has significantly promoted the development of the related disciplines including physics, chemistry, engineering, and materials science. The novel characteristics of metamaterials such as negative refraction, reversed Doppler effect, and RCR, have been experimentally confirmed. Metamaterial applications are flourishing, drawing considerable attention from researchers worldwide (as discussed in Chap. 1). In my viewpoint, metamaterials have become one of the most popular emerging interdisciplinary research directions.

Certainly, the field of metamaterials has been accompanied by skepticism and criticism, as briefly outlined in the development history of metamaterials in Chap. 1. Here, I would like to suggest readers carefully read B. A. Munk's monograph [3] to gain

Z. Duan, *Metamaterial-Based Electromagnetic Radiations and Applications*,
https://doi.org/10.1007/978-981-97-8108-9

an alternative perspective on metamaterials. In fact, skepticisms and criticisms are accompanied by the development of any new research fields. For instance, although the negative refraction in the microwave frequency was experimentally verified in 2001 [4], Munk remained skeptical about the existence of negative refraction. [3]. In 2023, A. J. Sternbach's group reported the negative refraction in the optical band [5]. The publication of this work indicates the significance of long-term, comprehensive research, and many experimental validations of the negative refraction. In international and national conferences, I often encountered highly sensitive questions: What exactly are metamaterials? Are they not just periodic structures? Why use the term Metamaterial, which will lead to misunderstanding? What are their successful applications? Is it possible to develop brand-new devices with better performance without metamaterials? Indeed, metamaterials are not as perfect as its name suggests. I am convinced that the ongoing research will eventually give the satisfactory answers for readers.

I have been involved in metamaterials research since 2007. During this period, I have guided more than 50 graduate students. As per my experience and lesson, I would like to share some insights into metamaterials below with readers:

(1) Basic concept

The term metamaterial was first introduced by Rodger Walser in 1999 [6]. The definition he provided is cited here: metamaterials are macroscopic composites having man-made, three-dimensional, periodic cellular architecture designed to produce an optimized combination, not available in nature, of two or more responses to specific excitation [7]. In fact, this term often confuses scientists and engineers. It clearly refers to a structure. So, why is it called a material? Which term is more accurate or concise? Moreover, it has created the illusion in the industry that the use of metamaterials will lead to the development of brand-new devices. In both the academic and industry, metamaterials defined above exhibit extraordinarily complex structures, making their fabrication extremely challenging and nearly impractical. The metamaterials introduced in this monograph (as outlined in Chaps. 4, 5, and 6) do not strictly meet the above definition of metamaterials. Instead, it can be considered as a novel subwavelength electromagnetic structure, which was proposed according to the idea of metamaterials. So, I often use these terms like metamaterial-based, metamaterial-inspired, and metamaterial-assisted.

(2) Theoretical configuration

The theoretical foundation of narrowly-defined metamaterials is from EMT. For electromagnetic metamaterials, their ultimate theory is mainly based on macroscopic electrodynamics. In other words, it is the application of electrodynamics to subwavelength electromagnetic structures. As I know, EMT is accurate when both the size and period of metamaterial unit cells are smaller than $\sim 1/10 \times \lambda$ [8–10]. However, the proposed metamaterial unit cells realized by using the resonant methods can hardly meet this condition. For example, the 2001 Science paper has unit cell dimensions of $1/4.5 \times \lambda$ [4], and the metamaterial unit cells proposed by my group have dimensions of $1/7 \times \lambda$ [11]. This implies that for the realized metamaterials, EMT is inevitably approximate (see Chap. 2 for more details). Regarding narrowly-defined electromagnetic metamaterials, an improved EMT still needs to accurately predict their electromagnetic properties. When 3D metamaterials degrade into 2D metamaterials (called metasurfaces), the effective electromagnetic parameters will lose their physical significance. Surface impedance and polarization density are usually used to characterize metasurfaces [12–14]. Currently, there is no widely accepted, systematic theory for analysing metasurface. Note that in my opinion, broadly-defined metamaterials include 1D metamaterials, 2D metasurfaces, 3D metamaterials, and even photonic crystals.

(3) Implementation methods

In this monograph, three implementation methods are first summarized by me: classical combination method and its extension, natural material composite method, and transmission line method (as detailed in Chap. 2). All of these methods are constructed based on the resonant methods. Resonant characteristics lead to strong interactions between the electromagnetic waves and the metamaterial unit cells, leading to the enhancement of the local electromagnetic field, the strong dispersion of electromagnetic waves near the resonant frequency, and large loss. The reduction of loss is a crucial challenge for the application of metamaterials. For example, J. B. Pendry once led a project called Better than Silver, aiming to find materials with lower loss than silver to further develop metamaterials, but ultimately, this goal was not achieved [15]. However, he still believes that maximizing the utilization of silver's low loss in metamaterial is worthy of further exploration. In fact, such a device operating in the microwave frequency, such as P-band and L-band, exhibits low loss which is acceptable for engineering applications [16]. However, as the frequency increases, the loss will increase. Currently, metamaterials are facing the challenges of application scenarios that demand high frequency (THz or optics), wide bandwidth, and low loss.

(4) Fabrication technique

As mentioned above, the implementation of narrowly-defined 3D metamaterials is extremely challenging. In a sense, metamaterials can achieve excellent electromagnetic properties at the cost of the structure complexity. Indeed, the complex fabrication process, difficulty in realization, and challenging characterization of metamaterials contribute to the limitations in the applications. Researchers have gradually shifted their focus from studying 3D metamaterials to 2D. They use various unit cells and planar arrangements to construct different metasurfaces, aiming to achieve specific electromagnetic properties. F. Capasso pioneered the forefront research of metasurface [17, 18], and after that the Metasurface term had been spread. Note that Metafilm or Metasurface was coined in 2003 [19–21] (see Chap. 2). In terms of implementation methods, metasurfaces share some similarities with Frequency Selective Surface (FSS). Readers may wonder: What is metasurface? How does it fundamentally differ from FSS? How should they be constructed? What would Munk, a major contributor to FSS [22], think of this problem if he were still alive? Additionally, compared with 3D metamaterials, what advantages do metasurfaces offer?

J. B. Pendry believed that the key is how to expand from metasurfaces and establish high-quality 3D metamaterials in the future [15].

(5) Application prospects

As I know, metamaterial samples often focus on specific parameters. Therefore, it is challenging to meet the requirements of practical engineering applications. Objectively, according to my research experience, metamaterials have important application values in specific scenarios, which need single-frequency or narrow-bandwidth devices. However, they may not be as wonderful as claimed in some papers and news.

As mentioned earlier, metamaterials face various criticisms, which precisely indicates the high attention they received. The idea of metamaterials is very novel, this is, subwavelength unit cells are analogous to atoms or molecules in natural materials [23]. Thus, new functional structures are constructed and novel physical properties are achieved in the labs. Looking ahead to the future development of metamaterials, the following points are worthy of consideration:

1. New implementation methods: Since metamaterials realized by the resonant methods inevitably leads to strong dispersion, narrow bandwidth, and large loss, is it possible to propose non-resonant methods? According to the definition of metamaterials, they are not inherently associated with resonance. In other words, implementing metamaterials using resonant methods are just one way. Therefore, researchers should further explore other implementation methods for metamaterials [24–26].
2. Intelligent design: AI is currently one of the fastest-growing technologies in the field of electronic information. It has permeated every aspect of the national economy, and brought tremendous changes in the development of human society. Various new applications based on AI continue to emerge. Introducing AI technology into the forward or reverse design process of metamaterials [27], coupling with global optimization algorithms, developing intelligent design software for metamaterials, establishing industry standards, and achieving on-demand customization can help to advance the fast design and industrialization of metamaterials.
3. 3D printing technology: From the perspective of design, metamaterials have greatly liberated human creativity from the constraints of natural material properties. With the increasing complexity of metamaterial structures and the growing variety of metamaterials, advanced 3D printing technology has become a powerful tool for realizing multifunctional integrated metamaterials. Currently, there have been some reports of using 3D printing technology to fabricate metamaterials in various fields [28, 29]. However, current commercial 3D printing technology suffers from issues such as slow printing speed, low precision, and low interlayer strength. Therefore, it is far from reaching the level of industrial applications [30]. I believe that with the rapid development of 3D printing technology, it holds the potential to be used for precision fabrication of metamaterials, especially all-metal metamaterials.

The metamaterials mentioned in this monograph are all time-invariant metamaterials. J. B. Pendry suggests that exploring the time dimension of metamaterials will be one of the important directions in the future [15]. The time dimension of metamaterials not only refers to the changes in the physical structures but also refers to certain characteristics that change over time. Currently, there are experimental reports on achieving super-resolution imaging using time-variant metamaterials in the microwave frequency [31].

As research on metamaterials is continually deepening, it marks an advancement in humanity's ability to understand, modify, and utilize the existing natural materials. It is hoped that researchers engaged in metamaterials will maintain a scientific attitude, a spirit of inquiry, and continue to drive the development of metamaterials.

References

1. DARPA DSO Thrust Areas. 2012.
2. Parra R. AFOSR Overview. 2012.
3. Munk BA. Metamaterials: Critique and Alternatives. Hoboken: John Wiley & Sons. 2009.
4. Shelby RA, Smith DR, Schultz S. Experimental verification of a negative index of refraction. Science. 2001; 292:5514–177.
5. Sternbach AJ, Moore SL, Rikhter A, et al. Negative refraction in hyperbolic hetero-bicrystals. Science. 2023; 379:6632–555.
6. Ziolkowski RW. Metamaterials: The early years in the USA. EPJ Applied Metamaterials. 2014; 1:5–1.
7. Walser RM. Electromagnetic metamaterials. Proceeding SPIE Complex Mediums II: beyond Linear Isotropic Dielectrics. San Diego. 2001; 4467:1–15.
8. Smith DR, Padilla WJ, Vier DC, et al. Composite medium with simultaneously negative permeability and permittivity. Physical Review Letters. 2000; 84:18–4184.
9. Smith DR. Analytic expressions for the constitutive parameters of magneto-electric metamaterials. Physical Review E. 2010; 81:3–6605.
10. Cui TJ. Electromagnetic metamaterials—from effective media to field programmable systems. Scientia Sinica Information. 2020; 50:10-1427 (In Chinese).
11. Duan ZY, Tang XF, Wang ZL, et al. Observation of the reversed Cherenkov radiation. Nature Communications. 2017; 8:14901–1.
12. Fong BH, Colburn JS, Ottusch JJ, et al. Scalar and tensor holographic artificial impedance surfaces. IEEE Transactions on Antennas and Propagation. 2010; 58:10–3212.
13. Kadlec C, Skoromets V, Kadlec F, et al. Electric-field tuning of a planar terahertz metamaterial based on strained $SrTiO_3$ layers. Journal of Physics D: Applied Physics. 2018; 51:5–054001.
14. Staude I, Chen HT, Miroshnichenko A, et al. Metasurfaces for photonic devices. Journal of Applied Physics. 2022; 132:19–401.

15. Guo CZ, Luo Y. Light People: Prof. Sir John Pendry, father of metamaterials, spoke about the future of meta. Light: Science & Applications. 2023; 12:45–1.
16. Wang X, Zhang XM, Zou JJ, et al. Experimental demonstration of compact S-band MW-level metamaterial-inspired klystron. IEEE Electron Device Letters. 2023; 44:1–152.
17. Yu NF, Genevet P, Kats MA, et al. Light propagation with phase discontinuities: Generalized laws of reflection and refraction. Science. 2011; 334:6054–333.
18. Chang G. Surfing the metasurface: a conversation with Din Ping Tsai. Advanced Photonics. 2023; 5:6–060502.
19. Kuester EF, Mohamed MA, Piket-May M, et al. Averaged transition conditions for electromagnetic fields at a metafilm. IEEE Transactions on Antennas and Propagation. 2003; 51:10–2641.
20. Sievenpiper DF, Schaffner JH, Song HJ, et al. Two-dimensional beam steering using an electrically tunable impedance surface. IEEE Transactions on Antennas and Propagation. 2003; 51:10–2713.
21. Holloway CL, Dienstfrey A, Kuester EF, et al. A discussion on the interpretation and characterization of metafilms/metasurfaces: The two-dimensional equivalent of metamaterials. Metamaterials. 2009; 3:2–100.
22. Munk BA. Frequency Selective Surfaces: Theory and Design. Hoboken: John Wiley & Sons. 2000.
23. Pendry JB. Negative refraction. Contemporary Physics. 2004; 45:3–191.
24. Zheng B, Zhang RR, Zhou M, et al. Broadband subwavelength imaging using non-resonant metamaterials. Applied Physics Letters. 2014; 104:7–3502.
25. Magnus F, Wood B, Moore J, et al. A d.c. magnetic metamaterial. Nature Materials. 2008; 7–295.
26. Sun JB, Shalaev MI, Litchinitser NM. Experimental demonstration of a non-resonant hyperlens in the visible spectral range. Nature Communications. 2015; 6:7201–1.
27. Chen J, Hu SS, Zhu SN, et al. Metamaterials: From fundamental physics to intelligent design. Interdisciplinary Materials. 2022; 2:1–5
28. Alleluia de AB, Abdelshafy AF, Ragulis P, et al. Experimental testing of a 3-D-printed metamaterial slow wave structure for high-power microwave generation. IEEE Transactions on Plasma Science. 2020; 48:12–4356.
29. Xie YB, Ye SR, Reyes C, et al. Microwave metamaterials made by fused deposition 3D printing of a highly conductive copper-based filament. Applied Physics Letters. 2017; 110:18–1903.
30. Peng HX, Zhou J, Cui TJ. Metamaterials. China Railway Publishing House. 2020 (In Chinese).
31. Guan FX, Guo XD, Zeng KB, et al. Overcoming losses in superlenses with synthetic waves of complex frequency. Science. 2023; 381:6659–766.

Index

A
Absorber, 21, 25, 34, 42, 59
Active device, 35
Amplitude, 3, 9, 17, 40, 45, 50, 52, 53, 57, 59, 79–81, 87–90, 94, 111, 120, 124, 128, 129, 140, 150, 160, 161, 179, 187, 188, 194, 205, 206, 209, 220, 225
Anomalous radiation pressure, 4, 5, 16, 20
Artificial atom, 5, 38
Artificial electromagnetic material, 2
Artificial Intelligence (AI), 58, 60, 234
Artificial material, 5

B
Babinet principle, 50, 58, 107
Backward-Wave Oscillator (BWO), 22, 218
Beam waist position, 165
Bi-anisotropic metamaterial, 43, 55
Boundary conditions, 2, 43, 52, 54, 82, 92, 99, 110, 165, 168
Bruggeman method, 42

C
Cavity perturbation method, 45
Cherenkov Radiation (CR), 19, 65
Circuit theory, 58, 146
Classical combination method, 34–36, 149, 233
Classical perturbation theory, 223
Clausius-Mossotti relationship, 57
Coaxial discontinuity method, 45
Coaxial input-output coupler, 155, 157, 159, 170
Coherent radiation, 14, 100, 184, 207
Cold test experiment, 146, 158, 160–162, 191
Complementary electric Split Ring Resonator (CeSRR), 50–52, 55, 107–110, 112, 115, 117, 125, 129, 130, 146–152, 155, 159, 176, 188, 189, 191, 192, 194–196, 200, 201, 203, 208–212, 214–218, 224, 226
Complementary Metal Oxide Semiconductor (CMOS), 21
Composite right/left-handed transmission line, 39–41
Coupling impedance, 25
Crossed-strip array, 37
Curnow formula, 153

D
D'Alembertian equation, 75, 76
Density of state, 45
Dielectric medium, 8
Dielectric metasurface, 57, 59
Dielectric substrate, 15, 34, 36, 107
Diffraction, 7, 17, 21
Digital coding technique, 35
Dipole antennas, 40
Dispersion characteristic, 18, 22, 52, 54, 110, 111, 141, 150, 155
Dispersion equation, 54, 153, 154
Dispersive media, 4, 13
Displacement current, 1
Double-Negative Material (DNM), 3, 22, 66–70, 72–75, 77–79, 81, 82, 85–88, 90, 91, 93, 94, 97, 98, 101, 102, 106
Drude model, 37, 67, 91
Drude response, 51

Z. Duan, *Metamaterial-Based Electromagnetic Radiations and Applications*,
https://doi.org/10.1007/978-981-97-8108-9

Dual composite right/left-handed transmission line, 39
Dynamic controlling principle, 1

E

Effective diffusion constants, 33
Effective electromagnetic property, 6
Effective Medium Theory (EMT), 5, 33, 35, 36, 42–44, 50, 57, 58, 108, 122, 123, 232, 233
Effective modulus, 33
Effective permeability, 5, 6, 11, 12, 14, 15, 33, 34, 37, 39, 42, 43, 45, 46, 51, 52, 55, 66, 88, 105, 107–109, 122, 149, 150, 154
Effective permittivity, 5, 6, 9–12, 14, 15, 33, 34, 36–39, 42–46, 50–52, 55, 58, 66, 105, 107–109, 122, 149, 150, 154
Effective plasma frequency, 10, 11, 51, 71–73, 91
Effective refractive index, 13–15, 17, 33, 35, 43, 46, 48, 49, 105
Effective stiffness, 33
Effective thermal conductivity, 33
Electrical breakdown, 196
Electrical loss, 9
Electrically reconfigurable metasurface, 57
Electric boundary conditions, 148, 154, 170
Electric conductor, 14, 58, 91
Electric response, 15, 34, 55, 108, 109
Electromagnetic field averaging method, 45
Electromagnetic tunneling effect, 14
Electromagnetic wave propagation retrieval method, 45
Electron beam channel, 36, 37, 125, 138, 165, 189, 208, 214, 225
Electron-optical system, 163, 170, 176
Electrons returning, 188, 189
Electrostatic controlling principle, 1
Energy flux density, 4, 77
Epsilon-Near-Zero Metamaterial (ENZM), 13, 14
Equivalent circuit theory method, 146, 148
Evanescent wave, 2, 9, 16, 17, 70, 93
Extended Interaction Amplifier (EIA), 184, 189, 219, 221, 222, 224–226
Extended Interaction Oscillator (EIO), 184, 189, 215–218
Extended Interaction Resonant Cavity (EIRC), 184, 185, 188, 215–221, 223

F

Facile wet impregnation method, 39, 40
Ferrimagnet, 3
Ferromagnetic resonance, 3
Figure Of Merit (FOM), 13
Filter, 39, 59
Free electron gas model, 9
Free-space method, 45

G

Gain-bandwidth product, 185
Generalized averaged transition condition, 58
Generalized Snell's law, 57, 58
Grid-controlled electron gun, 163–166, 168, 172
Group delay method, 221
Group velocity, 4, 7, 23, 34, 39, 54, 66, 77, 106, 110, 120

H

High-frequency characteristics, 110, 146, 150, 154, 155, 208, 215, 219
High-frequency structure, 21, 22, 24, 25, 37, 201–203, 211–213, 219, 225
High-Impedance Surface (HIS), 56
High refractive index material, 9, 14, 15, 57
Holography, 59, 60
Huygens' metasurfaces, 59
Huygens' principle, 58

I

Incident wave, 3, 10, 16, 43, 51, 58, 59
Induction output tubes, 207
Information metamaterial, 35, 36
Integrated Circuit (IC), 2
Interaction impedance, 111–113, 150, 153–155
Interference fringe, 7
Inversion algorithm, 45
Invisibility cloak, 6
Isotropic medium, 8

K

Klystron, 1, 152, 184, 185, 188–190, 194, 196–199, 204, 207, 208, 210–214, 231
Kramers-kronig relation, 9, 49, 108

L
Laminar flow, 165
Left-Handed Materials (LHM), 3, 4
Left-handed transmission line, 39
Logarithmic mixture formula, 42
Lorentz local field, 42
Lorentz response, 51
Lower phase noise, 207

M
Magnetic boundary conditions, 148
Magnetic conductor, 58
Magnetic flux density, 115–117, 119, 132–135, 166, 168, 169, 171, 172, 178, 200, 201, 203, 204, 210, 216, 217
Magnetic focusing configuration, 115, 125, 131–133, 163, 166–170, 172, 177, 196, 200, 201, 203, 204, 210, 212
Magnetic loss, 9, 68, 71, 75
Magnetic medium, 8
Magnetic response, 5, 12, 14, 15, 34, 38, 39, 51, 108
Magnetic shielding structures, 167
Magnetization, 8, 55
Magnetron, 1
Master and slave boundary conditions, 154
Mathematical continuity, 43
Maximum magnetic energy product, 168
Maxwell-Garnett method, 42
Maxwell's equations, 1, 2, 42, 46, 53, 75, 98
Mean free path, 45
Mechanically reconfigurable metasurface, 57
Mechanical wave, 4
Medium, 2–4, 8, 9, 12, 13, 19–22, 33, 36, 37, 42–47, 69, 70, 72, 77, 78, 80, 94, 95, 97, 100, 101, 124, 177, 183, 207, 208, 212, 223
Metaline, 39, 41
Metamaterial, 2, 4–9, 11–17, 21–27, 33–46, 48, 50–52, 55–58, 66, 74, 90, 100, 106–110, 114, 122–124, 127, 129, 131, 141, 142, 145, 146, 149, 176, 179, 184, 185, 189, 191–194, 196–226, 231–235
Metamaterial prism, 17
Metamaterial Slow-Wave Structure (MSWS), 52–55, 107–117, 125, 127, 128, 130, 132, 134, 138, 141, 146, 147, 150, 153–160, 163, 168, 169, 171, 172, 176–179
Metasurface, 21, 27, 55–60, 82, 233, 234
Micro-Electro-Mechanical Systems (MEMS), 25
Microwave, 1, 2, 4, 5, 7, 8, 13, 14, 21, 24, 25, 27, 34, 41, 59, 60, 66, 93, 112, 122, 127, 130, 145, 158, 163, 174, 176, 178, 179, 184–187, 200, 204, 207, 216, 218, 231–233, 235
Microwave early-warning radar, 1
Microwave oscillations, 1
Mie resonance theory, 38
Mu-and-epsilon-Near-Zero Metamaterial (MENZM), 13, 14
Multi-cavity magnetron, 1
Multistage depressed collectors, 176
Mu-Near-Zero Metamaterial (MNZM), 13

N
Natural material composite method, 34, 38, 40, 233
Natural materials, 2, 4–6, 14, 26, 33, 34, 38, 45, 234, 235
Near-zero refractive index material, 13
Negative index materials, 12
Negative permeability material, 3, 9, 11, 37
Negative permittivity material, 2, 9, 37
Negative refraction, 4–8, 16, 17, 231, 232
Negative refractive index, 7, 8, 13, 105, 106
Negative refractive index materials later, 3
Non-uniformity, 7
Normalized phase velocity, 154, 155
Nose cone, 198, 208, 226
Numerical computation, 2

O
Ohmic loss, 50, 155, 171
Operational frequency, 1
Optically reconfigurable metasurface, 57
Orthogonal direction, 44, 59

P
Pancharatnam-Berry (P-B) metasurface, 59
Parallel plate capacitor formula, 151, 152
Passive device, 21, 39
Periodic boundary conditions, 154
Permeability, 2–5, 8, 9, 13, 14, 36, 42–45, 50, 51, 53, 55, 58, 66–68, 71, 82, 89, 91, 98, 99, 122, 147, 186
Permittivity, 2–5, 8–10, 13, 14, 36, 37, 42, 45, 50, 53, 66, 67, 71, 81, 82, 89, 91, 92, 98, 99, 122, 124, 127, 147, 186

Phase, 4, 7–9, 13, 16, 17, 19, 21, 34, 36, 39, 40, 45, 50, 51, 54, 57–59, 65, 66, 69–72, 87, 88, 106, 107, 110, 111, 120, 123, 130, 150, 153, 160, 175, 185, 205, 206
Phase velocity, 4, 13, 19, 34, 39, 65, 66, 69, 72, 87, 88, 106, 107, 110, 120, 123, 154, 155, 175
Phase velocity jump, 176, 178, 179
Phasor method, 192
Photonic crystals, 5, 231, 233
Planar lenses, 59
Plasmonic metasurface, 57
Point-contact germanium transistor, 2
Polarization, 8, 10, 50–52, 55, 57, 59, 108, 233
Polytetrafluoroethylene (PTFE), 16, 124, 127, 177, 223
Power reconfigurable metasurface, 57
Power supply duty cycle, 167, 174
Propagating and evanescent wave, 17
Propagation characteristics, 5, 7, 37, 55

Q
Quasimode theory, 45
Quasi-TM mode, 37, 55, 108, 122, 149

R
Radiation pressure, 20
Recoil permeability, 168
Reentrant resonant cavity, 190, 194, 211
Reflective-type metasurface, 59
Refracted wave, 16
Refractive index, 9, 12–16, 49, 57, 59, 69, 71, 72, 88, 100
Remanence, 168
Reversed Cherenkov Radiation (RCR), 4, 19, 65, 66, 74, 90, 105, 119, 125, 145
Reversed Cherenkov Radiation Amplifier (RCRA), 145, 176–179
Reversed Cherenkov Radiation Oscillator (RCRO), 145, 163, 166, 169–172, 174–176, 178, 179, 194
Reversed Doppler effect, 4, 5, 16, 18, 231
Right-Handed Materials (RHM), 2–4, 7, 8, 16, 18, 19, 80, 81, 88
Right-handed transmission line, 39, 40
Right-hand rule, 2, 4
Rotation mode, 59

S
Second harmonic bunching technology, 212
Semiconductor, 2, 57
Short-circuited transmission line, 147, 148
Slow-Wave Structure (SWS), 25, 52, 107, 110–112, 114, 146, 175, 176, 179, 184
Snell's law, 8, 57, 100
Sommerfeld radiation, 4
Space charge effect, 98, 100, 163, 165
Space charge forces, 164
S-parameters retrieval method, 43–45, 48, 50, 51, 55, 149
Spark gap oscillations experiment, 1
Spatial limiting current, 25
Split-Ring Resonator (SRR), 5, 6, 8, 11, 12, 14, 21, 22, 24, 26, 34, 36–38, 55, 66, 75, 78, 82, 87, 88, 93, 107
Standing wave, 183, 186
Subwavelength unit cell, 4, 6, 9, 26, 27, 34, 45, 56, 234
Surface impedance, 58, 233
Surface Plasmon Polariton (SPP), 41
Surface roughness, 119, 156–158, 171, 198, 201

T
Terahertz, 15, 23, 25, 27, 34, 41, 56, 59, 60, 78, 81, 88, 145, 179
Thermally reconfigurable metasurface, 57
Thermionic electron emission mechanism, 164
Transit gap, 183–185, 188–192, 197, 214, 216, 223–226
Transition Radiation (TR), 21, 183–185, 194, 196–200, 203–208, 215–219, 225, 226
Transition time effects, 183
Transmission characteristics, 24, 25, 45, 51, 106, 127–130, 148, 155–159
Transmission line method, 34, 39, 233
Transmission line theory, 34, 39, 148, 151
Transmission phase shift, 130, 131, 160, 161, 163
Trapezoidal rule, 49
Traveling-Wave Tubes (TWT), 22, 25, 65, 145, 179

U
Unit cell, 2, 14, 15, 22, 24, 25, 27, 34, 35, 37, 38, 42–44, 50, 51, 56, 58–60, 75,

105, 107, 146, 147, 185, 196, 215, 226, 232, 233

V

Vacuum degree, 127, 166, 172, 198, 204
Vacuum diode, 1
Vacuum Electron Device (VED), 1, 2, 21, 22, 26, 36–38, 66, 107, 113, 115, 163, 164, 170, 183, 215
Vacuum pentode, 1
Vacuum tetrode, 1
Vacuum triode, 1
Valence electrons, 9
Visible light, 14, 27, 35, 59
Vortex beam, 59

W

Wavefront, 7, 8, 19, 59, 65
Waveguide method, 45
Wave vector, 2, 4, 44, 66, 69, 72, 76, 85
Wire grid medium, 5
Wireless radio detection, 1
Wireless transmitters, 1

SPRINGER NATURE

GPSR Compliance

The European Union's (EU) General Product Safety Regulation (GPSR) is a set of rules that requires consumer products to be safe and our obligations to ensure this.

If you have any concerns about our products, you can contact us on ProductSafety@springernature.com

In case Publisher is established outside the EU, the EU authorized representative is:

Springer Nature Customer Service Center GmbH
Europaplatz 3
69115 Heidelberg, Germany

The manufacturer's authorised representative in the EU is Springer Nature Customer Service Centre GmbH, Europaplatz 3, 69115 Heidelberg, Germany. If you have any concerns regarding our products, please contact ProductSafety@springernature.com

Printed and bound by CPI Group (UK) Ltd, Croydon, CR0 4YY
30/09/2026
02225304-0005